Apollo 16

The NASA Mission Reports
Volume One

Compiled from the archives & edited
by Robert Godwin

Special thanks to:
John Young
Steve Garber
Benny Cheney

All rights reserved under article two of the Berne Copyright Convention (1971).
We acknowledge the financial support of the Government of Canada through the
Book Publishing Industry Development Program for our publishing activities.
Published by Apogee Books an imprint of Collector's Guide Publishing Inc., Box 62034, Burlington, Ontario, Canada, L7R 4K2
Printed and bound in Canada
Apollo 16 - The NASA Mission Reports Volume One
by Robert Godwin
ISBN 1-896522-58-0
©2002 Apogee Books
All photos courtesy of NASA

Apollo 16
The NASA Mission Reports
(from the archives of the National Aeronautics and Space Administration)

PRESS KIT

TABLES AND ILLUSTRATIONS

PRE-MISSION OPERATION REPORT

POST MISSION OPERATION REPORT

TECHNICAL CREW DEBRIEFING

INTRODUCTION

The old expression in the movie business is that without drama there is no audience. The element of risk is what draws people to watch. With that axiom in mind it seems incredible that by the time Apollo 16 left for the moon in April 1972 most of the world had stopped watching. The great bulk of the American audience were completely wrapped up in what was happening in Viet Nam and showed little or no interest in what was happening in the Lunar highlands.

What *was* happening was as remarkable as any of the great voyages of exploration in human history. Just as Livingston or Shackleton had taken enormous risks to further man's knowledge of his own world, John Young and Charlie Duke were rolling the dice to find out more about our nearest neighbor.

The public seemed to have resigned themselves to the notion that nothing extraordinary was going to happen at Descartes and so they tuned out in droves. The novelty of the lunar rover had worn off, there was no high drama as epitomised by the dilemma faced by Apollo 13. Apollo 16 was a victim of its own success. Young, Duke and Mattingly flew an almost flawless mission and hardly anyone noticed.

This situation probably tells us a lot about ourselves as a society but what can we say about Apollo 16?

Apollo 16 was not about the Moon — it was about understanding the Earth. John Young and Charlie Duke were given the task of landing their lunar module in the lunar highland area of Descartes. It was approximately 7,400 feet higher elevation than Tranquillity Base. The object? To bring back samples of lunar bedrock or evidence of lunar vulcanism.

Geology is certainly not the most glamorous science. It involves clambering around in the dirt for hours and hours, digging holes, cracking rocks with a hammer and drilling core samples. For many of the general public this represents a situation beyond the threshold of boredom and they certainly can't be bothered to watch it on television. The fact that these two explorers were conducting geology on another world was lost in amongst the perfect color television pictures — Duke and Young could just as easily have been somewhere in the deserts of the United States. But is this an accurate assessment of the situation? Was there really no drama? Was there really nothing happening?

The reader is encouraged to watch the accompanying television broadcast (supplied on the CDROM). The fact is that Young and Duke were taking outrageous risks by today's standards.

If you watch the extraordinary efforts of today's space shuttle astronauts as they assemble the International Space Station you can't help but notice the almost sterile environment in which they work. The hi-tech spacesuits they use are sparkling and new and in the risk-averse environment of the new millenium no one would even consider sending a mission specialist out the air-lock if there was so much as a speck of dirt encroaching on one of the suit's seals.

Young and Duke went out the door of their lunar home for three days straight. They were a quarter of a million miles from any kind of support and they were absolutely plastered in lunar dirt. It was so bad that they often wondered if they might not be able to seal their gloves and helmets each day. The regolith acted like an abrasive and scratched everything it came into contact with.

As if that wasn't bad enough the orange juice dispensers were leaking juice inside their helmets making the work environment even more unpleasant.

Ignoring these risks and unpleasantries, Young and Duke plunged into their work with enthusiasm. They drove their lunar rover for kilometers up and down the local terrain stopping at many different locations which they probed and sampled. If they were successful their results would bring about a new understanding of our place in the cosmos.

The Earth is (if you'll pardon the metaphor) like a great big beautiful multi-layered gateaux. The best chefs can make such a wonderful cake if they know what the ingredients are. But after all the mixing and stirring it is all but impossible to know just what ingredients made the cake so special. On the other hand, the Moon is like a table of untouched ingredients. It has been around as long as the Earth and yet it has suffered only a minimal amount of change. The forces of erosion work infinitesimally slowly on the Moon and it is possible for the trained eye to still search for and identify those raw ingredients. If you can identify them, you can understand what makes the Earth the beautiful and unique world that it is.

The landing site at Descartes was chosen because it appeared to have been caused by volcanic activity. Confirmation of such vulcanism would reveal enormous amounts of information about the origins of the Moon — and the Earth. The absence of such volcanic activity would also change many of the conventional theories back here on Earth. As it turned out Descartes was not volcanic but it was *very* old. Nearly 4.5 billion years. Theories of lunar volcanoes which hailed back to the days of Galileo were finally put to rest by Apollo 16.

Young and Duke also set up an ultra-violet observatory on the Moon to study the effects of solar radiation in an environment unimpeded by the Earth's atmosphere. As on previous missions they established a sophisticated science station which would relay information about the lunar environment back to Earth for years to come. The whole time they were on the surface, Ken Mattingly, aboard the orbiting CSM, reeled in enormous amounts of data from orbit with the Scientific Instrument Module.

In hindsight NASA's team was a victim of its own success. If the people back on Earth had not been given such easy access to the information streaming back from Apollo 16 it might have created the same kind of feeding frenzy which surrounds such enigmas as the so-called Area 51. If we hadn't had such perfect TV pictures and such professional well-trained astronauts there might have been more interest.

The simple fact is that Young, Duke, Mattingly were explorers of the first magnitude. They risked their lives in the pursuit of new knowledge to help all of us understand our universe a little better. They didn't travel to the moon in pursuit of gold (like Magellan or Columbus). Nor did they go there to beat the Soviets (that had already been done). They went there for knowledge and for the good of their country. Far more noble motives than those of the explorers of old. The fact that they made it look easy, while *some* of us watched, can never diminish their accomplishments and for me at least the drama is there for those with eyes to see it.

Robert Godwin
(Editor)

NATIONAL AERONAUTICS AND SPACE ADMINISTRATION

Washington, D. C. 20546
Phone: (202) 755-8370

FOR RELEASE: THURSDAY A.M. April 6, 1972
Kenneth C. Atchison (Phone: 202/755-3114)

RELEASE NO: 72-64

APOLLO 16 LAUNCH APRIL 16

Apollo 16 scheduled for an April 16 launch will devote its 12-day duration to gathering additional knowledge about the environment on and around the Moon and about our own planet Earth.

During the three days two Apollo 16 crewmen spend on the lunar surface north of the crater Descartes, they will extend the exploration begun by Apollo 11 in the summer of 1969 and continued through the Apollo 12, 14, and 15 lunar landing missions. In addition to gathering samples of lunar surface material for analysis on Earth, the crew will emplace a fourth automatic scientific station.

An extensive array of scientific experiments in the orbiting command/service module will search out and record data on the physical properties of the Moon and near-lunar space and photographic images to further refine mapping technology. Additionally, the command module pilot will photograph astronomical phenomena in the distant reaches of space.

The Descartes landing site is a grooved, hilly region which appears to have undergone some modification by volcanic processes during formation. The Descartes region is in the southeast quadrant of the visible face of the Moon and will offer an opportunity to examine several young, bright-rayed craters created by impacts in the volcanic terrain.

John W. Young is Apollo 16 mission commander, with Thomas K. Mattingly flying as command module pilot and Charles M. Duke, Jr as lunar module pilot. Young is a US Navy captain, Mattingly a Navy lieutenant commander, and Duke a US Air Force lieutenant colonel.

Young and Duke will climb down from the lunar module onto the lunar surface for three seven-hour periods of exploration and experimentation. A major part of the first EVA will be devoted to establishing the nuclear powered, automatic scientific station — Apollo Lunar Surface Experiment Package (ALSEP) — which will return scientific data to Earth for many months for correlation with data still being returned by the Apollo 12, 14 and 15 ALSEPs.

The second and third EVAs will be devoted primarily to geological exploration and sample gathering in selected areas in the vicinity of the landing site. As in past missions, the crew's observations and comments will be supplemented by panoramic, stereo, and motion picture photographic coverage and also by television coverage. Crew mobility again will be aided by the use of the lunar roving vehicle.

In lunar orbit, Mattingly will operate experiments in the scientific instrument module (SIM) bay for measuring such things as the lunar surface chemical composition, and the composition of the lunar atmosphere. A high-resolution camera and a mapping camera in the SIM bay will add to the imagery and photogrammetry gathered by similar cameras flown on Apollo 15. Mattingly will perform an inflight EVA during transearth coast to retrieve film cassettes from these cameras.

Using hand-held cameras, Mattingly will photograph such phenomena in deep space as the Gegenschein, and looking earthward, photograph the ultraviolet spectra around Earth.

A second subsatellite, similar to the one flown on Apollo 15, will be ejected into lunar orbit to measure the effect of the Earth's magnetosphere upon the Moon and to investigate the solar wind and the lunar gravity field. Apollo 16 is scheduled for launch at 12:54 pm EST April 16 from the NASA Kennedy Space Center's Launch Complex 39, with lunar landing taking place on April 20. The landing crew will remain at Descartes for 73 hours before returning to lunar orbit and for rendezvous with the orbiting command module on April 23. Earth splashdown will occur on April 28 at 3:30 pm EST at 5 degrees north latitude and 158.7 degrees west longitude in the central Pacific just north of Christmas Island. The prime recovery vessel, USS Ticonderoga, an aircraft carrier, will be located near the splashdown point to recover the crew and spacecraft.

Communications call signs to be used during Apollo 16 are "Casper" for the command module and "Orion" for the lunar module. The United States flag will be erected on the lunar surface in the vicinity of the lunar module, and a stainless steel plaque engraved with the landing date and crew signatures will be affixed to the LM front landing gear.

Apollo 16 backup crewmen are civilian Fred W. Haise, Jr., commander; USAF Lt Col Stuart A. Roosa, command module pilot; and USN Captain Edgar D. Mitchell, lunar module pilot.

COUNTDOWN

The Apollo 16 launch countdown will be conducted by a government-industry team working in two control centers at the Kennedy Space Center (KSC).

Overall space vehicle operations will be controlled from Firing Room No.1 in the Complex 39 Launch Control Center. The spacecraft countdown will be run from an Acceptance Checkout Equipment control room in the Manned Spacecraft Operations (MSO) Building.

Extensive checkout of the launch vehicle and spacecraft components are completed before the space vehicle is ready for the final countdown. The prime and backup crews participate in many of these tests, including mission simulations, altitude runs, a flight readiness test and a countdown demonstration test.

The Apollo 16 rollout — the 5.5-kilometer (3.4-nautical mile) trip from the Vehicle Assembly Building (VAB) to the launch pad — took place Dec. 13, 1971. Due to a problem associated with the fuel tank system in the command module reaction control system, the space vehicle was returned to the VAB on Jan. 27. The spacecraft was taken to the MSO Building for changeout of the tanks. After re-mating the spacecraft with the launch vehicle, the KSC team again rolled Apollo 16 back to Pad A on Feb. 9, 1972.

Apollo 16 will be the tenth Saturn V launched from Pad A (eight manned). Apollo 10 was the only launch from Pad B, which will be used again in 1973 for the Skylab program.

The Apollo 16 precount activities will start at T-6 days. The early tasks include electrical connections and pyrotechnic installation in the space vehicle. Mechanical build-up of the spacecraft is completed, followed by servicing of the various gases and cryogenic propellants (liquid oxygen and liquid hydrogen) to the CSM and LM. Once this is accomplished, the fuel cells are activated.

The final countdown begins at T-28 hours when the flight batteries are installed in the three stages and instrument unit of the launch vehicle.

At the T-9 hour mark, a built-in hold of nine hours and 54 minutes is planned to meet contingencies and provide a rest period for the launch crew. A one hour built-in hold is scheduled at T-3 hours 30 minutes.

Following are some of the highlights of the latter part of the count:

T-10 hours, 15 min	Start mobile service structure move to park site.
T-9 hours	Built-in hold for nine hours and 54 minutes. At end of hold, pad is cleared for LV propellant loading.
T-8 hours, 05 min	Launch vehicle propellant loading - Three stages (LOX in first stage, LOX and LH in second and third stages). Continues thru T-3 hours 38 minutes.
T-4 hours, 15 min	Flight crew alerted.
T-4 hours, 00 min	Crew medical examination.
T-3 hours, 30 min	Crew breakfast.
T-3 hours, 30 min	One-hour built-in hold.
T-3 hours, 06 min	Crew departs Manned Spacecraft Operations Building for LC-39 via transfer van.
T-2 hours, 48 min	Crew arrival at LC-39.
T-2 hours, 40 min	Start flight crew ingress.
T-1 hour, 51 min	Space Vehicle Emergency Detection System test (Young participates along with launch team).
T-43 min	Retract Apollo access arm to standby position (12 degrees).
T-42 min	Arm launch escape system. Launch vehicle power transfer test, LM switch to internal power.
T-37 min	Final launch vehicle range safety checks (to 35 minutes).
T-30 min	Launch vehicle power transfer test, LM switch over to internal power.
T-20 min to T-10 min	Shutdown LM operational instrumentation.
T-15 min	Spacecraft to full internal power.
T-6 min	Space vehicle final status checks.
T-5 min, 30 secs	Arm destruct system.
T-5 min	Apollo access arm fully retracted.
T-3 min, 6 secs	Firing command (automatic sequence).
T-50 secs	Launch vehicle transfer to internal power.
T-8.9 secs	Ignition start.
T-2 secs	All engines running.
T-0	Liftoff

NOTE: Some changes in the countdown are possible as a result of experience gained in the countdown test which occurs about two weeks before launch.

LAUNCH WINDOWS

The mission planning considerations for the launch phase of a lunar mission are, to a major extent, related to launch windows. Launch windows are defined for two different time periods: a "daily window" has a duration of a few hours during a given 24-hour period; a "monthly window" consists of a day or days which meet the mission operational constraints during a given month or lunar cycle.

Launch windows are based on flight azimuth limits of 72 to 100° (Earth-fixed heading east of north of the launch vehicle at end of the roll program), on booster and spacecraft performance, on insertion tracking, and on Sun elevation angle at the lunar landing site.

	Launch Windows*		SUN ELEVATION
LAUNCH DATE	OPEN	CLOSE	ANGLE
April 16, 1972	12:54 pm	4:43 pm	11.9°
May 14, 1972	12:17 pm	4:01 pm	6.8°
May 15, 1972	12:30 pm	4:09 pm	6.8°
May 16, 1972	12:38 pm	4:13 pm	18.6°
June 13, 1972**	10:50 am	2:23 pm	13.0°
June 14, 1972**	10:57 am	2:26 pm	13.0°
June 15, 1972**	(———— To Be Determined ————)		

*   April times are Eastern Standard Time; all others are Eastern Daylight Time.
**   Launch window times and Sun elevation angles for June will be refined. The June 15, 1972 launch opportunity is under review.

GROUND ELAPSED TIME UPDATE

It is planned to update, if necessary, the actual ground elapsed time (GET) during the mission to allow the major flight plan events to occur at the pre-planned GET regardless of either a late liftoff or trajectory dispersions that would otherwise have changed the event times.

For example, if the flight plan calls for descent orbit insertion (DOI) to occur at GET 82 hours, 40 minutes and the flight time to the Moon is two minutes longer than planned due to trajectory dispersions at translunar injection, the GET clock will be turned back two minutes during the translunar coast period so that DOI occurs at the pre-planned time rather than at 82 hours, 42 minutes. It follows that the other major mission events would then also be accomplished at the pre-planned times.

Updating the GET clock will accomplish in one adjustment what would otherwise require separate time adjustments for each event. By updating the GET clock, the astronauts and ground flight control personnel will be relieved of the burden of changing their checklists, flight plans, etc.

The planned times in the mission for updating GET will be kept to a minimum and will, generally, be limited to three updates. If required, they will occur at about 53, 97, and 150 hours into the mission. Both the actual GET and the update GET will be maintained in the MCC throughout the mission.

LAUNCH AND MISSION PROFILE

The Saturn V launch vehicle (SA-511) will boost the Apollo 16 spacecraft from Launch Complex 39A at the Kennedy Space Center, Fla., at 12:54 p.m. EST April 16, 1972, on an azimuth of 72 degrees. The first stage (S-IC) will lift the vehicle 68.4 kilometers (37 nautical miles) above the Earth. After separation the booster stage will fall into the Atlantic ocean about 668 km (361 nautical mi.) downrange from Cape Kennedy, approximately nine minutes, 14 seconds after liftoff. The second stage (S-II) will push the vehicle to an altitude of about 173 km (93.6 nautical mi.). After separation the S-II stage will follow a ballistic trajectory which will plunge it into the Atlantic about 4,210 km (2,273 nautical mi.) downrange about 19 minutes, 51 seconds into the mission. The single engine of the third stage (S-IVB) will insert the vehicle and spacecraft into a 173-km (93-nautical mi.) circular Earth parking orbit before it is cut off for a coast period. When reignited, the engine will inject the Apollo spacecraft into a trans-lunar trajectory.

APOLLO 16 vs APOLLO 15 OPERATIONAL DIFFERENCES

ITEM	APOLLO 16	APOLLO-15
LAUNCH AZIMUTH	72 - 100°	80 - 100°
INCLINATION	9°	26°
DOI TRIM MANEUVER	NOT PLANNED	ACCOMPLISHED
HIGH ALTITUDE LANDMARK TRACKING AND TV OF LANDING SITE	NO	YES
PDI	REV 13	REV 14
SURFACE STAY TIME	~73 HOURS	66.9 HOURS
SURFACE REST CYCLE	8 HOURS (REST BEFORE LIFTOFF)	7 - 7 1/2 HOURS (NO REST BEFORE LIFTOFF)
EVA's: DURATION (HRS) (PLANNED)	7 - 7 - 7	7 - 7 - 6
SEVA	NO	YES
TRAVERSE STATION TIMES	9:40 (PLANNED)	7:53 (PLANNED) 5:00 (ACTUAL)
SURFACE ACTIVITIES	ALSEP DEPLOYED FIRST	TRAVERSE FIRST
LUNAR ORBIT PLANE CHANGES	2	1
EARTH RETURN INCLINATION	62°	40°
LANDING	PLAN TO RECOVER PARACHUTES AND RETAIN RCS PROPELLANTS ON BOARD	NEITHER APOLLO 16 TYPE EVENTS WERE PLANNED

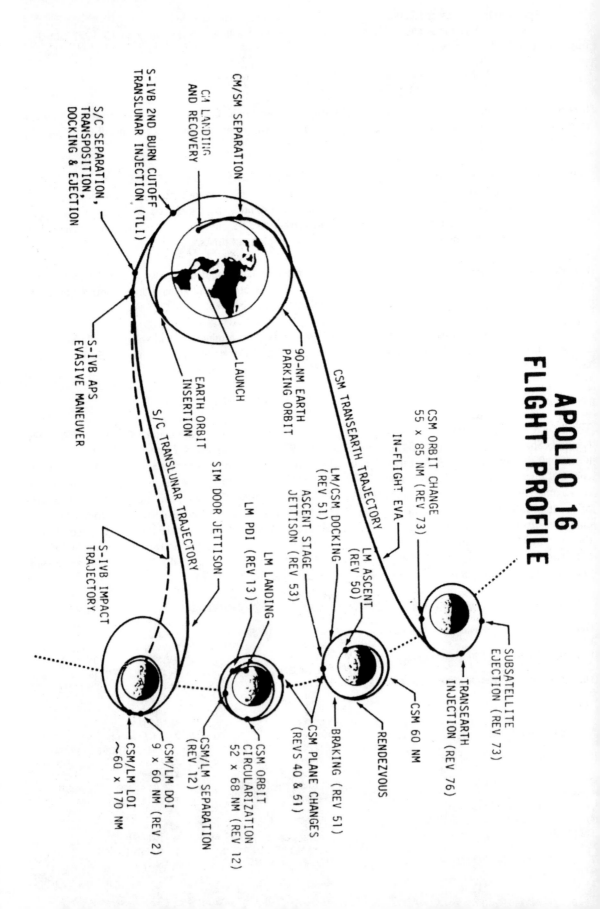

APOLLO 16 FLIGHT PROFILE

S/C SEPARATION, TRANSPOSITION, DOCKING & EJECTION

S-IVB 2ND BURN CUTOFF TRANSLUNAR INJECTION (TLI)

CM LANDING AND RECOVERY

CM/SM SEPARATION

S-IVB APS EVASIVE MANEUVER

90-NM EARTH PARKING ORBIT

LAUNCH

EARTH ORBIT INSERTION

S/C TRANSLUNAR TRAJECTORY

SIM DOOR JETTISON

S-IVB IMPACT TRAJECTORY

CSM/LM LOI ~60 x 170 NM

CSM/LM DOI 9 x 60 NM (REV 2)

CSM/LM SEPARATION (REV 12)

CSM ORBIT CIRCULARIZATION 52 x 68 NM (REV 12)

LM PDI (REV 13)

LM LANDING

BRAKING (REV 51)

RENDEZVOUS

CSM PLANE CHANGES (REVS 40 & 51)

ASCENT STAGE JETTISON (REV 53)

LM/CSM DOCKING (REV 51)

LM ASCENT (REV 50)

CSM 60 NM

CSM TRANSEARTH TRAJECTORY

IN-FLIGHT EVA

CSM ORBIT CHANGE 55 x 85 NM (REV 73)

TRANSEARTH INJECTION (REV 76)

SUBSATELLITE EJECTION (REV 73)

COMPARISON OF APOLLO MISSIONS

	PAYLOAD DELIVERED TO LUNAR SURFACE KG	(LBS)	EVA DURATION (HR:MIN)	SURFACE DISTANCE TRAVERSED (KM)	SAMPLES RETURNED KG	(LBS)
APOLLO 11	104	(225)	2:24	.25	20.7	(46)
APOLLO 12	166	(355)	7:29	2.0	34.1	(75)
APOLLO 14	209	(460)	9:23	3.3	42.8	(94)
APOLLO 15	550	(1210)	18:33	27.9	76.6	(169)
APOLLO 16 (PLANNED)	558	(1228)	21:00	25.7	88.6	(195)

LAUNCH EVENTS

Time Hrs	Min	Sec	Event	Vehicle Wt Kilograms (Pounds)*	Altitude Meters (Feet)*	Velocity Mtrs/Sec (Ft/Sec)*	Range Kilometers (Naut Mi)*
00	00	00	First Motion	2,920,956 (6,439,605)	60 (198)	0 (0)	0 (0)
00	01	20	Maximum Dynamic Pressure	1,853,350 (4,085,937)	12,646 (41,491)	481 (1,577)	5 (3)
00	02	18	S-IC Center Engine Cutoff	1,084,409 (2,390,712)	47,153 (154,703)	1,712 (5,618)	51 (27)
00	02	40	S-IC Outboard Engines Cutoff	843,020 (1,858,541)	67,371 (221,034)	2,369 (7,773)	91 (49)
00	02	42	S-IC/S-II Separation	676,542 (1,491,519)	69,094 (226,685)	2,375 (7,792)	95 (51)
00	02	44	S-II Ignition	676,,542 (1,491,519)	70,645 (231,775)	2,369 (7,771)	98 (53)
00	03	12	S-II Aft Interstage Jettison	644,029 (1,419,840)	95,024 (311,758)	2,474 (8,117)	161 (87)
00	03	18	Launch Escape Tower Jettison	633,419 (1,396,451)	99,428 (326,207)	2,503 (8,213)	174 (94)
00	07	40	S-II Center Engine Cutoff	303,470 (669,037)	174,389 (572,142)	5,201 (17,065)	1,095 (591)
00	09	17	S-II Outboard Engines Cutoff	216,673 (477,682)	174,130 (571,291)	6,568 (21,550)	1,650 (891)
00	09	18	S-II/S-IVB Separation	170,989 (376,967)	174,161 (571,395)	6,571 (21,558)	1,657 (895)
00	09	21	S-IVB First Ignition	170,947 (376,874)	174,235 (571,637)	6,571 (21,559)	1,677 (905)
00	11	44	S-IVB First Cutoff	140,103 (308,873)	172,908 (567,282)	7,400 (24,279)	2,643 (1,427)
00	11	54	Parking Orbit Insertion	140,039 (308,734)	172,914 (567,302)	7,402 (24,284)	2,715 (1,466)
02	33	35	S-IVB Second Ignition	139,052 (306,557)	175,919 (577,162)	7,404 (24,292)	16,203 (8,749)
02	39	19	S-IVB Second Cutoff	65,620 (144,666)	305,315 (1,001,688)	10,441 (34,256)	13,320 (7,192)
02	39	29	Trans-Lunar Injection	65,551 (144,514)	319,199 (1,047,241)	10,433 (34,230)	13,222 (7,139)

* English measurements given in parentheses.

ACTIVITY	TIME	PURPOSE
VISUAL LIGHT FLASH PHENOMENON		TO STUDY THE ORIGIN OF VISUAL LIGHT FLASHES OBSERVED BY ASTRONAUTS ON PREVIOUS MISSIONS.
EYE SHIELDS	TRANSLUNAR & TRANSEARTH	
ALFMED	TRANSLUNAR	
ELECTROPHORETIC SEPARATION	TRANSLUNAR	TO STUDY A PROCESS OF MATERIALS PURIFICATION WHICH MAY PERMIT PRODUCTION SEPARATION OF MATERIALS OF HIGHER PURITY THAN CAN CURRENTLY BE PRODUCED ON EARTH.

APOLLO 16

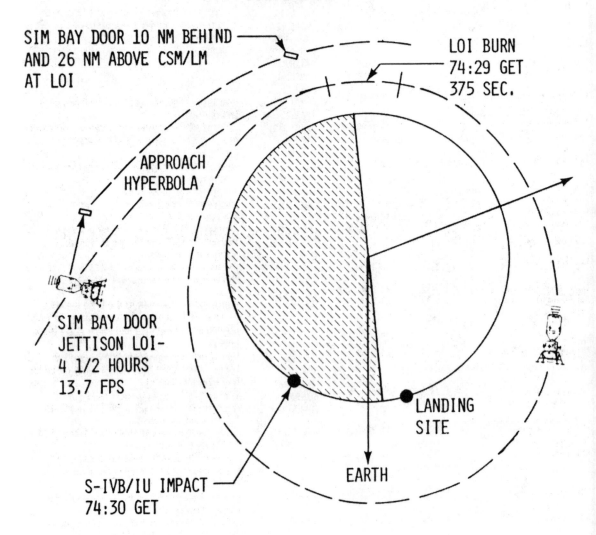

SIM BAY DOOR 10 NM BEHIND AND 26 NM ABOVE CSM/LM AT LOI

LOI BURN 74:29 GET 375 SEC.

APPROACH HYPERBOLA

SIM BAY DOOR JETTISON LOI-4 1/2 HOURS 13.7 FPS

S-IVB/IU IMPACT 74:30 GET

LANDING SITE

EARTH

MISSION EVENTS

Events	GET hrs:min	Date/EST	Velocity change m/sec (ft/sec)	Purpose and resultant orbit
Translunar injection (S-IVB engine start)	2:33	16/3:27 pm	3053 (10,018)	Injection into translunar trajectory with 131 km (71 nm) pericynthion
CSM separation, docking	3:04	16/3:58 pm	—	Mating of CSM and LM
Ejection from SLA	3:59	16/4:53 pm	.3 (1)	Separates CSM-LM from S-IVB/SLA
S-IVB evasive maneuver	4:23	16/5:17 pm	3 (9.8)	Provides separation prior to S-IVB propellant dump and thruster maneuver to cause lunar impact
S-IVB residual propellant dump	4:44	16/6:38 pm		
APS impact burn	5:30	16/7:24 pm		
APS correction burn	9:30	16/11:24 pm		
Midcourse correction 1	TLI+9 hr	17/00:33 am	0*	*These midcourse corrections have a nominal velocity change of 0 m/sec, but will be calculated in real time to correct TLI dispersions;
Midcourse correction 2	TLI+28 hr	17/7:33 pm	0*	
Midcourse correction 3	LOI-22 hr	18/5:23 pm	0*	Trajectory remains within capability of a docked-DPS TEI burn should SPS fail to ignite.
Midcourse correction 4	LOI-5 hr	19/10:23 am	0*	
SIM door jettison	LOX-4.5 hr	19/10:53 am	4.2 (13.7)	
Lunar Orbit insertion	74:29	19/3:23 pm	-855.6 (-2807)	Inserts Apollo 16 into 108x316 km (58x170 nm) elliptical lunar orbit.
S-IVB impacts lunar surface	74:30	19/3:24 pm		Seismic event for Apollo 12, 14 and 15 passive seismometers. Target: 2.2 degrees south latitude by 21.4 degrees west longitude.
Descent orbit insertion	78:36	19/7:30 pm	-62.8 (206)	SPS burn places CSM/LM into 20x109 km (11x59 nm) lunar orbit.
CSM/LM undocking	96:14	20/01:08 pm	—	
CSM circularization burn	97:42	20/2:36 pm	30.3 (99.6)	Inserts CSM into 96x 111 km (52x60 nm) orbit (BPS burn)
LM powered descent	98:35	20/3:29 pm	-2041 (6696)	Three-phase DPS burn to brake LM out of transfer orbit, vertical descent and lunar surface touchdown.
LM lunar surface touchdown	98.46	20/3:41 pm	—	Lunar exploration, deploy ALSEP, collect geological samples, photography.
EVA-1 begins	102:25	20/7:19 pm	—	See separate EVA timelines
EVA-2 begins	124:50	21/5:44 pm	—	See separate EVA timelines
EVA-3 begins	148:25	22/5:19 pm	—	See separate EVA timelines
CSM plane change 1	152:29	22/9:23 pm	48 (159)	Changes CSM orbital plane by 1.7 degree to coincide with LM orbital plane at time of LM ascent from surface
LM ascent	171:45	23/4:39 pm	1,843 (6048)	Boosts ascent stage into 16.6 x 84 km (9 x 45.4 nm)lunar orbit for rendezvous with CSM
Lunar orbit insertion	171:52	23/4:46 pm		
Terminal phase initiate (TPI) LM APS	172:39	23/5:33 pm	15 50	Boosts ascent stage into 81.4 x 114.5 km (44 x 61.9 nm catch-up orbit; LM trails CSM by 59.2 km (32 nm) and 27.7 km (15 nm) below at TPI burn time
Braking: 4 LM RCS burns	173:20	23/6:14 pm	10 33	Line-of-sight terminal phase braking to place LM in 110.7 x 109.8 km (59.8 x 59.3 nm) orbit for final approach, docking.
Docking	173:40	23/6:34 pm	—	CDR and LMP transfer back to CSM
LM jettison, separation	177:31	23/10:25 pm	—	Prevents recontact of CSM with LM ascent stage for remainder of mission
LM ascent stage deorbit (RCS burn)	179:16	24/00:10 am	-70 (-229)	ALSEP seismometers at Apollo 16, 15, 14 and 12 landing sites record impact.
LM impact	179:39	24/00:33 am	—	Impact at about 1691 m/sec (5550 fps) at -3.2 degree angle. 23 km (12 nm) west of Apollo 16 ALSEP.
CSM plane change 2	193:14	24/2:08 pm	86 (282)	Increase lunar surface photo,coverage, changes plane 3 degrees.
CSM shaping burn	216:49	25/1:43 pm	11 (38)	Adjusts CSM orbit for later subsatellite jettison orbit: 102 x 147 km (55 x 85 nm).
Subsatellite jettison	218:02	25/2:56 pm	—	Lunar orbit science experiments
Transearth injection (TEI)	222:21	25/7:15 pm	979 (3212)	Inject CSM into transearth trajectory

Midcourse correction 5	239:21	26/12:17 pm	0	Transearth midcourse corrections will be computed in real-time for entry corridor control and recovery area weather avoidance.
Transearth EVA	241:57	26/2:51 pm	—	Retrieve SM SIM bay film canisters.
Midcourse correction 6	EI-22 hrs	27/5:17 pm	0	
Midcourse correction 7	EI-3 hrs	28/12:17 pm	0	
CM/SM separation	290:08	28/3:02 pm	—	Command module oriented for Earth atmosphere entry
Entry interface (121.9 km, 400,000 ft)	290:23	28/3:17 pm		Command module enters Earth's atmosphere at 11,026 m/s (36,175 fps).
Splashdown	290:36	28/3:30 pm		Landing 2111 km (1140 nm) downrange from entry: splash at 5 degrees north latitude by 158 degrees west longitude.

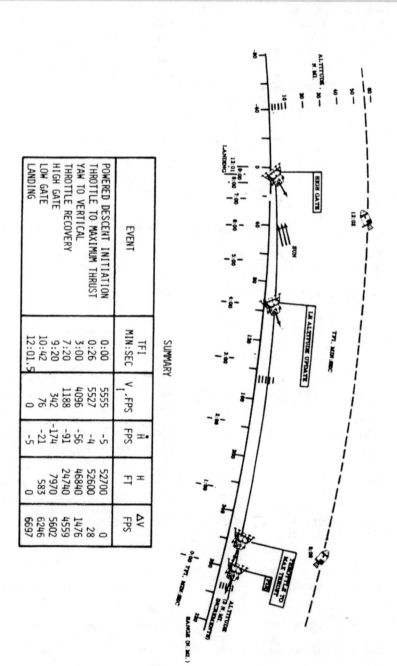

POWERED DESCENT VEHICLE POSITIONS

SUMMARY

EVENT	TFI MIN:SEC	V_I, FPS	$\dot{H}$ FPS	H FT	ΔV FPS
POWERED DESCENT INITIATION	0:00	5555	-5	52700	0
THROTTLE TO MAXIMUM THRUST	0:26	5527	-4	52600	28
YAW TO VERTICAL	3:00	4096	-56	46840	1476
THROTTLE RECOVERY	7:20	1188	-91	24740	4559
HIGH GATE	9:20	342	-174	7970	5602
LOW GATE	10:42	76	-21	583	6246
LANDING	12:01.5	0	-5	0	6697

APPROACH PHASE

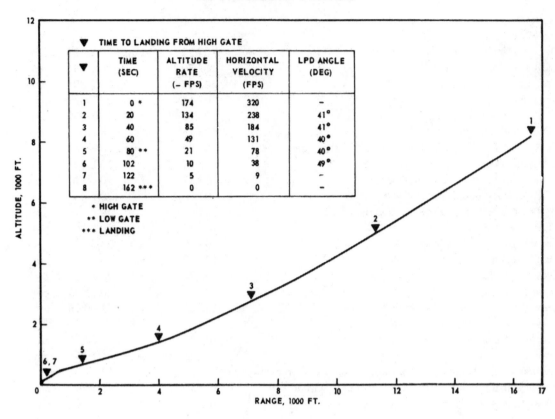

▼	TIME (SEC)	ALTITUDE RATE (− FPS)	HORIZONTAL VELOCITY (FPS)	LPD ANGLE (DEG)
1	0 *	174	320	−
2	20	134	238	41°
3	40	85	184	41°
4	60	49	131	40°
5	80 **	21	78	40°
6	102	10	38	49°
7	122	5	9	−
8	162 ***	0	0	−

▼ TIME TO LANDING FROM HIGH GATE

* HIGH GATE
** LOW GATE
*** LANDING

ALTITUDE, 1000 FT.

RANGE, 1000 FT.

APOLLO 16
CREW POST-LANDING ACTIVITIES

DAYS FROM RECOVERY	DATE	ACTIVITY
SPLASHDOWN	APRIL 28	
R + 1	APRIL 29	DEPART SHIP, ARRIVE HAWAII
R + 2	APRIL 30	DEPART HAWAII, ARRIVE HOUSTON
R + 3 THRU R + 15	MAY 1-12	CREW DEBRIEFING PERIOD

APOLLO 16 ALTERNATE MISSIONS

• EARTH ORBIT MISSION (NO TLI)
   PHOTOGRAPHIC MISSION OVER U.S. OF ABOUT 6 DAYS DURATION. X-RAY AND GAMMA RAY
   SPECTROMETERS WILL ALSO BE OPERATED. CMP EVA TO RETRIEVE FILM CASSETTES.
LUNAR ORBIT MISSIONS
   • APPROXIMATELY 6 DAYS IN LUNAR ORBIT
   • USE OF SIM BAY EXPERIMENTS
   • SUBSATELLITE JETTISON REQUIRED
      TYPES OF CONTINGENCIES
         • CSM ALONE
         • CSM/LM (OPERABLE AND INOPERABLE DPS CASES)
         • FROM LM ABORTS

EVA MISSION EVENTS

Events	GET hrs:min		Date/EST	
Depressurize LM for EVA 1	102:25	Apr	20/7:19	pm
CDR steps onto surface	102:40		20/7:34	pm
LMP steps onto surface	102:45		20/7:39	pm
Crew offloads LRV	102:53		20/7:47	pm
CDR test drives LRV	103:13		20/8:07	pm
LRV parked near MESA	103:15		20/8:09	pm
CDR offloads, deploys far-UV camera/spectroscope	103:24		20/8:18	pm
LMP mounts LCRU, TV on LRV	103:25		20/8:19	pm
Crew loads geology equipment on LRV	103:37		20/8:31	pm
CDR deploys United States flag	103:50		20/8:44	pm
LMP carries ALSEP "barbell" to deployment site	104:02		20/8:56	pm
CDR drives LRV to ALSEP site	104:12		20/9:06	pm
Crew begins ALSEP deploy	104:20		20/9:14	pm
ALSEP deploy complete	106:22		20/11:16	pm
Crew drives to station 1 (Flag Crater)	106:27		20/11:21	pm
Crew arrives at station 1 for rake/ soil sample, crater sample	106:39		20/11:33	pm
Crew drives to station 2 (Spook crater)	107:22		21/00:06	am
Crew arrives at station 2 for crater area sampling, magnetometer site measurement	107:27		21/00:21	am
Crew returns to LM/ALSEP area Station 3	108:23	Apr	21/1:17	am
Crew arrives at station 3 to retrieve deep core. Comprehensive sampling, arm mortar package, perform LRV "Grand Prix".	108:30		21/1:24	am
Crew returns to LM	108:44		21/1:38	am
Crew arrives at LM	108:45		21/1:39	am
LMP deploys solar wind composition experiment	109:02		21/1:56	am
CDR offloads geology equipment, film mags	109:05		21/1:59	am
LMP ingresses LM	109:15		21/2:09	am
CDR ingresses LM	109:22		21/2:16	am
Repressurize LM, end EVA 1	109:25		21/2:19	am
Depressurize LM for EVA 2	124:50	Apr	21/5:44	pm
CDR steps onto surface	125:10		21/5:59	pm
LMP steps onto surface	125:10		21/6:04	pm
Crew completes loading LRV for geology traverse	125:30		21/6:24	pm
Crewmen load geological gear on each other's PLSS	125:32		21/6:26	pm
Crew drives to station 4 (upper slope of Stone Mountain)	125:40		21/6:34	pm
Crew arrives at station 4 for rake soil, Documented, and double core samples, photo panorama, penetrometer reading	126:15		21/7:09	pm
Crew drives to station 5 (on slope of Stone Mountain)	127:11		21/8:07	pm
Crew arrives at station 5 for documented samples, photo panorama	127:19		21/8:13	pm
Crew drives to station 6 (at base of Stone Mountain)	127:59	Apr	21/8:53	pm
Crew arrives at station 6 for documented samples, photo panorama	128:02		21/8:56	pm
Crew drives to station 7 (Stubby Crater, base of Stone Mountain)	128:22		21/9:16	pm
Crew arrives at station 7 for documented samples, photo panorama	128:26		21/9:20	pm
Crew drives to station 8 (rays from South Ray Crater)	128:41		21/9:35	pm

Crew arrives at station 8 for rake soil, documented double core samples, photo panorama	128:46	21/9:40	pm
Crew drives to station 9 (Cayley Plains)	129:45	21/10:39	pm
Crew arrives at Station 9 for surface and core samples, photo panorama	129:50	21/10:44	pm
Crew drives to station 10 (LM/ALSEP area)	130:15	21/11:09	pm
Crew arrives at station 10 for soil mechanics investigation, double core, radial sample, photo panorama	130:36	21/11:30	pm
Crew drives back to LM	131:09	22/00:03	am
Crew arrives at LM	131:10	22/00:04	am
Crew unloads LRV, packs sample containers, clean-up	131:20	22/00:14	am
LMP ingresses LM	131:37	22/00:31	am
CDR ingresses LM	131:48	22/00:42	am
Repressurize LM, end EVA 2	131:50	22/00:44	am
Depressurize LM for EVA 3	148:25	22/5:19	pm
CDR steps onto surface	148:40	22/5:34	pm
LMP steps onto surface	148:46	22/5:40	pm
Crew completes loading LRV, PLSSs for geology traverse	149:02	22/5:56	pm
Crew drives to station 11 (South rim, North Ray Crater)	149:10	22/6:04	pm
Crew arrives at station 11 for polarimetric photos, documented samples, photo panorama	149:54	22/6:48	pm
Crew drives to station 12 (North Ray, Crater, southeast rim, large block area)	150:49	22/7:43	pm
Crew arrives at station 12 for rake/soil and documented samples, photo panorama	150:52	22/7:46	pm
Crew drives to station 13 (outer ejecta blanket, North Ray Crater)	151:47	22/8:41	pm
Crew arrives at station 13 for rock/soil samples, photo panorama	151:52	22/8:46	pm
Crew drives to station 14 (Smoky Mountain)	152:02	22/8:56	pm
Crew arrives at station 14 for rake/soil, double core, documented samples, photo panorama	152:09	22/9:03	pm
Crew drives to station 15 (Cayley Plains)	152:49	22/9:43	pm
Crew arrives at station 15 for magnetometer measurements, rock/soil sample, photo panorama	153:00	22/9:49	pm
Crew drives to station 16 (Dot Crater)	153:10	22/10:04	pm
Crew arrives at station 16 for magnetometer measurement, rock/soil samples, photo panorama	153:19	22/10:13	pm
Crew drives to station 17 (NE rim, Palmetto Crater)	153:29	22./10:23	pm
Crew arrives at station 17for rake/soil and documented samples, Magnetometer readings, photo panorama	153:34	22/10:28	pm
Crew drives back to LM	154:12	22/11:06	pm
Crew arrives at LM	154:30	22/11:24	pm
Crew transfers samples, film packs, etc. stowage in LM; close-out final EVA	154:36	22/11:30	pm
LMP ingresses LM	155:19	23/00:13	am
CDR ingresses LM	155:23	23/00:17	am
Repressurize LM, end EVA 3	155:25	23/00:19	am

FIRST VIEW OF DESCARTES
(HIGH GATE + 8 SECONDS)

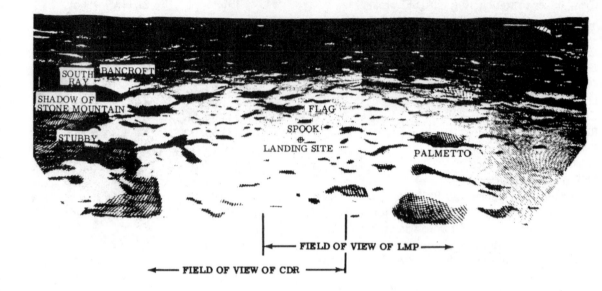

FILM RETRIEVAL FROM THE SIM BAY

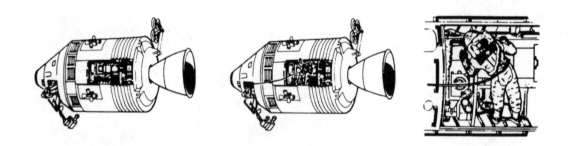

EVA TIME LINE

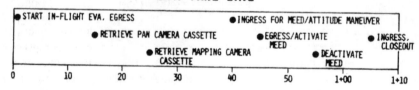

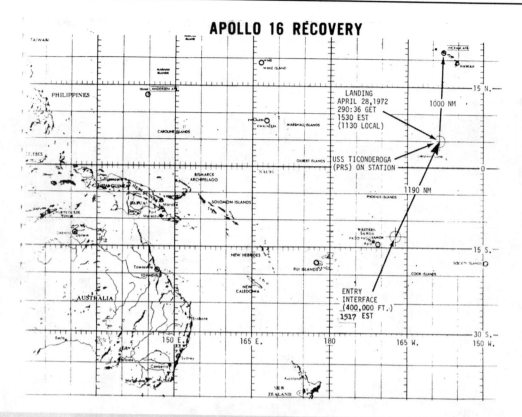

APOLLO 16 RECOVERY

APOLLO 16 TRAVERSE COMPARISON WITH APOLLO 15

	DURATION (HR:MIN)		TRAVERSE STATION TIME (HR:MIN)		LM/ALSEP LRV AREA TIME (HR:MIN)		TRAVERSE DRIVING TIME (HR:MIN)		DISTANCE (KM)	
	PLAN	ACTUAL	PLAN	ACTUAL	PLAN	ACTUAL	PLAN	ACTUAL	PLAN	ACTUAL
EVA I										
APOLLO 15	7:00	6:32	1:16	1:10	4:35	4:14	1:08	1:08	8.2	10.3
APOLLO 16	7:00	-	1:53	-	4:41	-	:26	-	3.2	-
EVA II										
APOLLO 15	7:00	7:12	3:32	2:37	1:29	3:11	1:59	1:24	14.3	12.5
APOLLO 16	7:00	-	4:11	-	1:30	-	1:19	-	9.5	-
EVA III										
APOLLO 15	6:00	4:49	3:05	1:23	1:27	2:50	1:28	:36	10.5	5.1
APOLLO 16	7:00	-	3:36	-	1:40	-	1:44	-	12.6	-
TOTALS										
APOLLO 15	20:00	18:33	7:53	5:00	7:31	10:15	4:35	3:08	33.0	27.9
APOLLO 16	21:00	-	9:40	-	7:51	-	3:29	-	25.2	-

APOLLO 16 MISSION OBJECTIVES

Apollo 16 astronauts will explore the Descartes region, man's first opportunity to explore the lunar highlands. The site is some 2,250 meters (7,400 feet) higher than the Apollo 11 site and is representative of over three fourths of the lunar surface. Preliminary geological analysis of the highlands indicates that the Moon's crust underwent modification early in its history. By studying these modification processes, we hope to achieve a better understanding of the development of this portion of the Moon's surface as well as the development of the Earth's crust, its continents, and ocean basins.

The three basic objectives are to explore and sample the materials and surface features, to set up and activate experiments on the lunar surface which will continue to relay data back to Earth after the crew has returned, and to conduct inflight experiments and photographic tasks. The lunar roving vehicle, used for the first time

DESCARTES LRV TRAVERSES

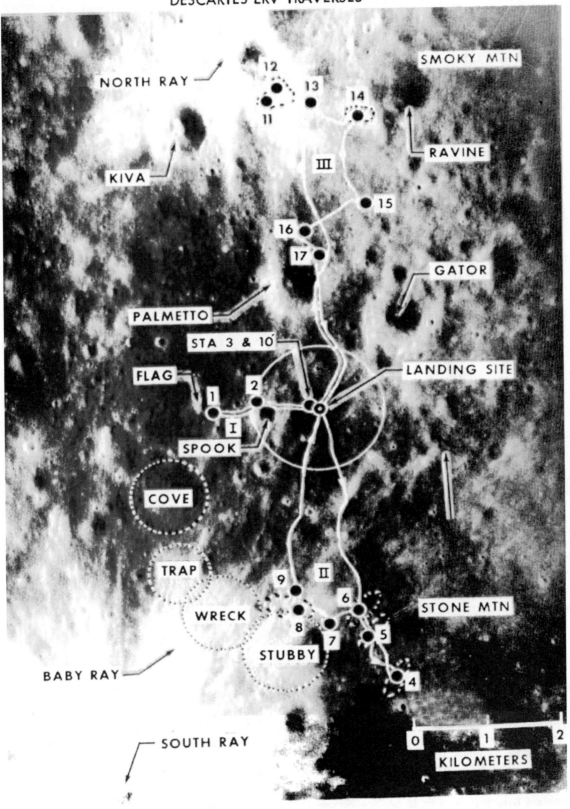

APOLLO 16 LUNAR SURFACE TIMELINE

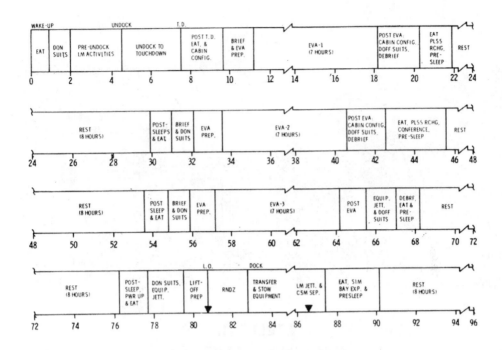

SUMMARY TIME LINE
FROM ALSEP OFFLOAD THROUGH ALSEP DEPLOYMENT

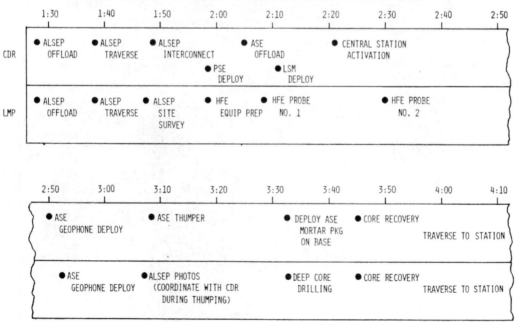

NEAR LM LUNAR SURFACE ACTIVITY

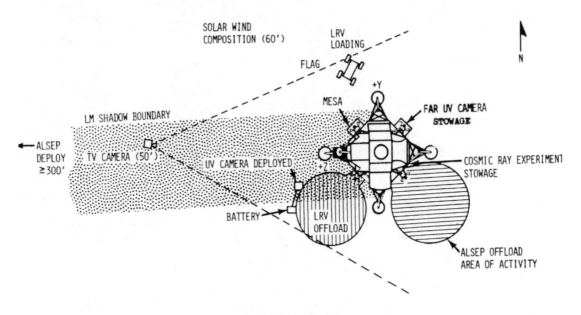

APOLLO 16
WALKING TRAVERSE
(CONTINGENCY)

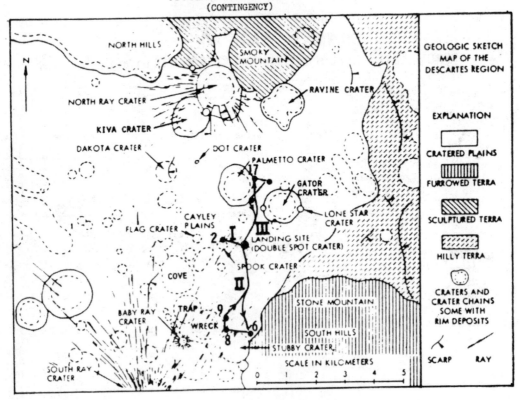

on Apollo 15, will extend the range of the exploration and geological investigations that Young and Duke will make during their three seven-hour EVAs. The Apollo lunar surface package (ALSEP) which the crew will set up and place into operation will, with the Apollo 12, 14 and 15 ALSEPS, become the fourth in a network of lunar surface scientific stations.

The scientific instrument module (SIM) bay in the service module is the heart of the inflight experiment effort on Apollo 16. Quite similar to the SIM bay flown on Apollo 15, the bay contains high-resolution and mapping cameras and scientific sensors for photographing and measuring properties of the lunar surface and the environment around the Moon.

While in lunar orbit, command module pilot Mattingly will have the responsibility for operating the inflight experiments during the time his crewmates are on the lunar surface. During the homeward coast after transearth injection, Mattingly will exit through the command module crew hatch and maneuver hand-over-hand to the SIM bay where he will retrieve film cassettes, pass them back into the cabin for return to Earth and emplace a medical experiment for exposure to the solar-space environment. Mattingly also will take photographs aimed at gathering knowledge of astronomical phenomena such as Gegenschein from lunar orbit and ultraviolet photography of both the Earth and the Moon.

Engineering and operational tasks the Apollo 16 crew will carry out include further evaluation of the lunar roving vehicle and Skylab crew equipment, and use of the SIM bay subsatellite as a navigation tracking aid. Other medical experiments include biostack, ALFMED, and the passive bone mineral measurement.

SCIENTIFIC RESULTS OF APOLLO 11, 12, 14 and 15 MISSIONS

The Apollo 14 and Apollo 15 missions have clearly demonstrated that the major scientific returns from the Apollo program are to be expected from the later Apollo missions. The scientific results from the Apollo 15 mission have advanced the science of understanding the Moon from the stage characterized by interesting and stimulating speculations to one characterized by scientific hypotheses. The hypotheses which lead to specific experiments that, in turn, give answers that rapidly reduce the uncertainty regarding many basic questions concerning the origin and evolution of the Moon. The extended stay times and extended mobility available on the lunar surface and the wide-array of instrumentation carried in the SIM bay of the command and service module have already provided a rich harvest of scientific information from the first J mission. Samples from the first two lunar landings showed us that most of the surface of the Moon dates back to a time when the terrestrial geologic record is just barely decipherable. They show that the lunar maria that covers about 1/3 of the side of the Moon viewed from Earth largely consists of an iron-rich volcanic rock produced from a partially molten shell 100-200 miles beneath the surface of the Moon. The soil samples from these two sites also included a variety of intriguing fragments clearly distinguishable from the mare basalts. These soil samples became the basis of widespread speculation on the composition and origin of the lunar highlands that make up most of the backside and approximately 2/3 of the front side of the Moon. Both the samples and the results of experiments flown in the Apollo 15 CSM have shown that these intriguing bits of rock are indeed representative of the large areas of the Moon that contain the bulk of the early history of the planet.

The measurement of gamma rays, produced by minute amounts of radioactive substances in the lunar soil, and characteristic x-rays, induced by the high energy solar radiation impinging on the sunlit-side of the Moon has shown that much of the northwestern quadrant of the Earth side of the Moon is probably underlain by a very old (probably first formed 4.4-4.5 billion years ago) uranium— and thorium-rich volcanic rock and that most of the backside of the Moon, along with the eastern highland regions on the front side, appears to be made up of an aluminum— and calcium-rich rock (anorthosite) that was first observed in the Apollo 11 soil samples.

The widespread occurrence of both of these rock types tells us a great deal about the early history of the Moon. The aluminum-rich rocks suggest that the primitive Moon had a liquid outer shell that may have been 50-80 miles thick that gave rise to a lunar crust almost simultaneously with the formation of the Moon itself. The old uranium-rich volcanic rocks both limit the thickness of this early liquid layer and tell us something about the composition of the solid interior very early during lunar history.

In particular, they suggest that the Moon was formed from material that condensed at temperatures much higher than previously thought out of the primitive dust cloud that surrounded the early Sun.

In addition to the chemical study of the lunar surface, the Apollo 15 mission returned excellent photographs of more than 10 percent of the surface of the Moon which reveal geological features as small as the size of the LM. Like the rock samples returned from this and previous missions, these photographs will provide a source of data for scientists for generations to come.

A subsatellite left behind by the Apollo 15 mission has measured the present magnetic field over a large percentage of the Moon's surface. These magnetic field measurements, along with studies of the magnetization of lunar rocks, indicate that the Moon's magnetic field 3 billion years ago must have been 100 to 1,000 times stronger than it is today. The subsatellite and altimeter have also shown that the mascons discovered by Lunar Orbiter spacecraft consist of circular plates of very dense rocks which fill the deep circular basins produced by collisions of asteroid-size objects with the Moon.

Most of the experiments deployed on the lunar surface during the Apollo 12, 14, and 15 missions continue to function well, but more important — they have provided information regarding geological processes that are still going on today and have shown us that the interior of the Moon can be investigated with present instrumentation. The impact of the Apollo 15 SIVB was recorded by both the Apollo 14 and Apollo 12 seismometers. Sound waves from this impact penetrated about 80 kilometers (50 miles) into the lunar interior. The relatively high sound velocity found at these depths was a complete surprise to seismologists and lunar scientists. The sound velocity profile in the upper 80 kilometers (50 miles) indicates that the concept of the lunar crust derived from mineralogy and chemistry of rocks was well-founded. Ultrasensitive thermometers implaced in two holes on the lunar surface showed that the amount of heat escaping from the lunar interior exceeds even the highest estimates made by scientists who speculated on the content of radioactive elements found in the Moon.

During the week of January 10, approximately 700 scientists, 15 from foreign countries including three scientists from the Soviet Union, gathered at the Manned Spacecraft Center to present and summarize results obtained by the Apollo 14, Apollo 15, and Luna 16 missions. In addition to the conclusions and ideas already mentioned here, studies of Apollo 14 samples have shown that the largest collision with an asteroidal object recorded on the surface of the Moon took place about 4 billion years ago. This was much later than had been anticipated from various theories dealing with the formation of large planetary objects. It raises the possibility that major collisions of asteroid-sized objects and the Earth continued more than 1 billion years after the planet was formed.

The relatively small Luna 16 samples provided a surprising number of extremely interesting results that indicate that unmanned sampling of the Moon is, indeed, a useful augmentation of the manned program.

As we enter into the homestretch of the Apollo program, it is clear that a successful Apollo 16 and 17 added to the missions already flown will produce a chapter in the history of science that will be a permanent and living testimony to this undertaking.

APOLLO 16 LANDING SITE

A hilly region north of the Descartes crater in a highlands area of the southeastern quadrant of the visible face of the Moon is the landing site chosen for Apollo 16.

The Descartes site appears to have structural characteristics similar to vulcanism sites on Earth, and has two separate volcanic features — Cayley Plains and the Descartes mountains — which will be extensively explored and sampled by the Apollo 16 crew.

The Cayley Plains segment of the landing site is characterized by terrain ranging from smooth to undulating; possibly as a result of fluid volcanic rock flow. The Descartes Mountains, part of the Kant Plateau are

APOLLO 16 TRAVERSES - GEOLOGIC SKETCH

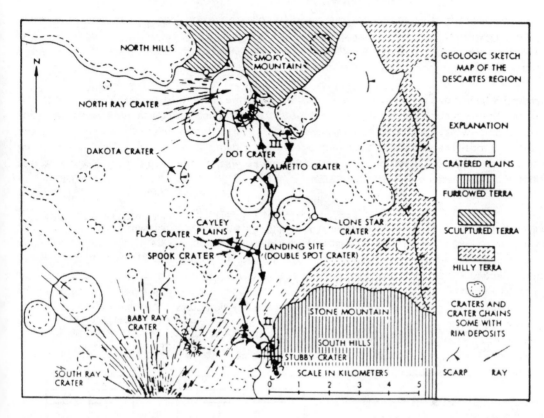

characterized by hilly, furrowed highland plateau material that is thought to have come from a more viscous volcanic flow. Additionally, the Descartes landing site provides an opportunity to study the evolution of young, bright-rayed craters and to extend age-dating to similar craters in other regions of the Moon.

The landing site has two basic areas which will be explored and sampled: Cayley Plains, including North Ray and South Ray craters; Stone Mountain and Smoky Mountain of the South and North Descartes Mountains.

The low crater density in the Cayley Plains suggests an Imbrian age for the rolling, ridged portion of Cayley in the Apollo 16 traverse area. Stone and Smoky mountains, on the other hand, appear to have shapes typical of volcanic formations on Earth — shapes that might be formed by movement of rather viscous material.

North and South Ray craters appear to penetrate deeply into the Cayley formation and reveal the sequence of layering, perhaps through an overlap of both the Cayley and Descartes formations. Taller, subdued craters in the landing site seem to have a characteristic concave bottom, which suggests that the substrata underlying the crater impact were more resistant than in other crater fields on the Moon.

APOLLO LANDING SITES

SITE	LAT.	LONG.	TERRAIN FEATURES	SCIENTIFIC INTEREST
APOLLO 11 (SEA OF TRANQUILITY)	0°41'N.	23°26'E.	MARE: LARGE SUBDUED CRATERS (200-600 METERS DIAMETER), INTERMEDIATE CRATERS (50-200 METERS DIAMETER).	MARE MATERIAL AGE AND COMPOSITION, GEOPHYSICAL DATA.
APOLLO 12 (OCEAN OF STORMS)	3°12'S.	23°24'W	MARE: SAME AS APOLLO 11.	MATERIAL AGE AND COMPOSITION FOR COMPARISON WITH APOLLO 11, SURVEYOR DATA.
APOLLO 14 (FRA MAURO)	3°40'S.	17°28'W	UPLAND: DEEP-SEATED EJECTA BLANKET.	EJECTA AGE, COMPOSITION AND STRUCTURE, GEOPHYSICAL DATA.
APOLLO 15 (HADLEY-APENNINES)	26°04'N.	3°39'E	FLARE/HIGHLAND: V-SHAPED SINUOUS RILLE, RILLE FLOOR BLOCKS; APENNINE HIGHLANDS.	BASIN AND RILLE ORIGIN AND AGE; HIGHLANDS COMPOSITION, GEOPHYSICAL DATA
DESCARTES	10°00'S.	16°00'E	HIGHLAND/HIGHLAND BASIN FILL	VOLCANICS AND PLAINS STRUCTURE, HIGHLAND AGE AND COMPOSITION, VOLCANISM TIME SPAN AND COMPOSITIONAL TRENDS, GEOPHYSICAL DATA.

		APOLLO MISSIONS					
		11	12	14	15	16	17
ALSEP EXPERIMENTS							
S-031	LUNAR PASSIVE SEISMOLOGY	X	X	X	X	X	
S-033	LUNAR ACTIVE SEISMOLOGY			X		X	
S-034	LUNAR TRI-AXIS MAGNETOMETER		X		X	X	
S-035	SOLAR WIND SPECTROMETER		X		X		
S-036	SUPRATHERMAL ION DETECTOR		X	X	X		
S-037	LUNAR HEAT FLOW				X	X	X
S-038	CHARGED PARTICLE LUNAR ENVIRONMENT			X			
S-058	COLD CATHODE GUAGE		X	X	X		
S-202	LUNAR EJECTA AND METEORITES						X
S-203	LUNAR SEISMIC PROFILING						X
S-205	LUNAR ATMOSPHERIC COMPOSITION						X
S-207	LUNAR SURFACE GRAVIMETER						X
M-515	LUNAR DUST DETECTOR	X	X	X	X		
OTHER EXPERIMENTS							
S-059	LUNAR GEOLOGY INVESTIGATION	X	X	X	X	X	X
S-078	LASER RANGING RETRO-REFLECTOR	X		X	X		
S-080	SOLAR WIND COMPOSITION	X	X	X	X	X	
S-151	COSMIC RAY DETECTOR (HELMET)	X					
S-152	COSMIC RAY DETECTOR (SHEETS)					X	
S-184	LUNAR SURFACE CLOSE-UP CAMERA	X	X	F			
S-198	PORTABLE MAGNETOMETER			X		X	
S-199	LUNAR GRAVITY TRAVERSE						X
S-200	SOIL MECHANICS			X	X	X	X
S-201	FAR UV CAMERA/SPECTROSCOPE					X	
S-204	SURFACE ELECTRICAL PROPERTIES						X

F - FACILITY EQUIPMENT

LUNAR SURFACE SCIENCE EXPERIMENT ASSIGNMENTS

Lunar Surface Experiment Scientific Discipline Contribution Correlation

SCIENTIFIC DISCIPLINE / EXPERIMENT	GEOLOGY	GEOPHYSICS	GEOCHEMISTRY	BIOSCIENCES	LUNAR ATMOSPHERE	PARTICLES AND FIELDS	ASTRONOMY
CONTINGENCY SAMPLE COLLECTION	AID IN DETERMINING LUNAR HISTORY BY AGING OF LUNAR SAMPLES		DETERMINE COMPOSITION OF LUNAR SURFACE BY CHEMICAL ANALYSIS OF LUNAR SAMPLES	AID IN DETERMINING POSSIBILITY OF BIOLOGICAL FORMS ON LUNAR SURFACE			
ALSEP PASSIVE SEISMIC (S-031)	AID IN DETERMINING INTERIOR STRUCTURE, TECTONISM AND VOLCANISM	AID IN DETERMINING FREE OSCILLATIONS, TIDES, SECULAR STRAINS, TILT, VELOCITY, ATTENUATION AND DIRECTION OF SEISMIC WAVES					MEASURE METEOROID IMPACTS
ACTIVE SEISMIC (S-033)	AID IN DETERMINING THE TYPE AND CHARACTER AS WELL AS HARDNESS AND BEARING STRENGTH OF LUNAR MATERIALS	AID IN DETERMINING FREE OSCILLATIONS, TIDES, SECULAR STRAINS, TILT, VELOCITY, ATTENUATION AND DIRECTION OF SEISMIC WAVES					MEASURE METEOROID IMPACTS
LUNAR SURFACE MAGNETOMETER (S-034)	AID IN DETERMINING MAGNETIC ANOMALIES, SUBSURFACE FEATURES AND LUNAR HISTORY	AID IN DETERMINING THERMAL STATE OF THE LUNAR INTERIOR				ESTABLISH GROSS ELECTRICAL DIFFUSIVITY. MEASURE MAGNETIC FIELD OF THE MOON	DETERMINE LUNAR RESPONSE TO FLUCTUATIONS IN THE INTERPLANETARY MAGNETIC FIELD
HEAT FLOW (S-037)	AID IN DETERMINING LUNAR EVOLUTION	MEASURE VERTICAL TEMPERATURE GRADIENTS, ABSOLUTE TEMPERATURE OF THE SURFACE TO ESTABLISH VERTICAL THERMAL CONDUCTIVITY	BULK COMPOSITION AND CHEMICAL SORTING MAY BE INFERRED FROM DATA				DETERMINE THERMAL ENVIRONMENT
PORTABLE MAGNETOMETER (S-198)	AID IN DETERMINING LOCAL LUNAR MAGNETIC ANOMALIES, PRESENCE OF MASCONS AND OTHER LOCAL MAGNETIC FIELD SOURCES.	AID IN MEASURING DISTORTION OF EARTH MAGNETIC FIELD BY SOLAR WIND, SHOCK FRONT, BOUNDARY LAYERS, MAGNETOPAUSE.				MEASURE LOCAL MAGNETIC FIELD OF MOON AND MAGNETIC FIELD/SOLAR PLASMA INTERACTION.	DETERMINE LUNAR FIELD FLUCTUATIONS DUE TO DIFFUSION OF INTERPLANETARY MAGNETIC FIELD.
FAR UV CAMERA/ SPECTROSCOPE		PROVIDE INFORMATION ON THE DENSITY, DISTRIBUTION, AND COMPOSITION OF GEOCORONA, SOLAR WIND, AND INTERSTELLAR MEDIUM			DETECT THE POSSIBLE PRESENCE OF A LUNAR ATMOSPHERE OR GAS VENTING FROM THE LUNAR SURFACE		PROVIDE OPPORTUNITY TO DETECT RED-SHIFTED EMISSION FROM INTERGALACTIC HYDROGEN AND OBTAIN AN ACCURATE SURVEY OF FAINT SOURCES OF COSMIC ULTRAVIOLET LIGHT
COSMIC RAY DETECTOR (SHEETS) (S-152)		AID IN DETERMINING THE ORIGIN OF METEORITES (GLASSY BODY OF PROBABLY METEORITIC ORIGIN)	DETERMINE COMPOSITION OF SOLAR WIND PLASMA		AID IN DETERMINING HISTORY OF PLANETARY ATMOSPHERE	MEASURE NEUTRON FLUX ON LUNAR SURFACE. PROVIDE DATA ON COSMIC RAY PARTICLES, BOTH SOLAR AND GALACTIC.	
LUNAR GEOLOGY INVESTIGATION S-059	AID IN DETERMINING LUNAR GEOLOGICAL STRUCTURE AND HISTORY		DETERMINE CHEMICAL COMPOSITION OF LUNAR SAMPLES	LUNAR SAMPLES MAY BE TESTED FOR ABILITY TO SUPPORT LIFE FORMS USED TO DETERMINE POSSIBILITY OF BIOLOGICAL LIFE FORMS ON THE LUNAR SURFACE		MONOPOLE SEARCH	
SOIL MECHANICS S-200	AID IN DETERMINING LUNAR HISTORY. ENABLE DETERMINATION OF COMPOSITIONAL TEXTURAL AND MECHANICAL PROPERTIES OF LUNAR SOIL		ENABLE DETERMINATION OF COMPOSITION OF LUNAR SOIL				
SOLAR WIND COMPOSITION (S-080)			DETERMINE COMPOSITION OF SOLAR WIND PLASMA		AID IN DETERMINING HISTORY OF PLANETARY ATMOSPHERE	PROVIDE INFORMATION ON THE ELEMENTAL AND ISOTOPIC COMPOSITION OF MOBILE GASES AND OTHER ELEMENTS IN THE SOLAR WIND	

APOLLO 16 ALSEP DEPLOYMENT

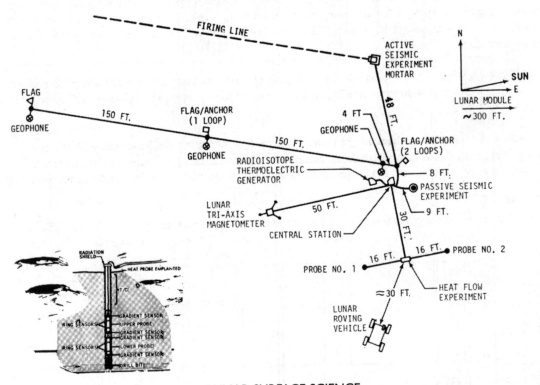

LUNAR SURFACE SCIENCE

The ALSEP array carried on Apollo 16 has four experiments: heat flow, lunar surface magnetometer, passive seismic and active seismic.

Six additional experiments will be conducted on the Descartes landing area: far ultraviolet camera/spectroscope, portable magnetometer, cosmic ray detector (sheets), solar wind composition, lunar geology investigation, and soil mechanics.

Passive Seismic Experiment (PSE): The PSE measures seismic activity of the Moon an gathers and relays to Earth information relating to the physical properties of the lunar crust and interior. The PSE reports seismic data on man-made impacts (LM ascent stage), natural impacts of meteorites, and moonquakes. Dr. Gary Latham of the Lamont-Doherty Geological Observatory (Columbia University) is responsible for PSE design and experiment data analysis.

Three similar PSEs, deployed as a part of the Apollo 12, 14 and 15 ALSEPs, have transmitted to Earth data on lunar surface seismic events since deployment. The Apollo 12, 14, 15 and 16 seismometers differ from the seismometer left at Tranquillity Base in July 1969 by the Apollo 11 crew in that the later PSEs are continuously powered by SNAP-27 radioisotope thermoelectric generators. The Apollo 11 seismometer, powered by solar cells, transmitted data only during the lunar dad and is no longer functioning.

After Apollo 16 trans-lunar injection, the Saturn V's spent S-IVB stage and instrument unit will be aimed to impact the Moon. This will stimulate the passive seismometers left on the lunar surface during previous Apollo missions.

The S-IVB/IU will be commanded to hit the Moon 250 kilometers (135 nautical miles) west of the Apollo 12 ALSEP site, at a target point 2.2 degrees south latitude by 31.4 degrees west longitude, near Lansberg Craters

E and D in the Ocean of Storms.

After the spacecraft is ejected from the launch vehicle, a launch vehicle auxiliary propulsion system (APS) ullage motor will be fired to separate the vehicle a safe distance from the spacecraft. Residual liquid oxygen in the almost spent S-IVB/IU will then be dumped through the engine with the vehicle positioned so the dump will slow it into an impact trajectory. Mid-course corrections will be made with the stage's APS ullage motors if necessary.

The S-IVB/IU will weigh 13,973 kilograms (30,805 pounds) and will be traveling 9,223 kilometers an hour (4,980 nautical mph) at lunar impact. It will provide an energy source at impact equivalent to about 11 tons of TNT.

After Young and Duke have completed their lunar surface operations and rendezvoused with the command module in lunar orbit, the lunar module ascent stage will be jettisoned and later ground-commanded to impact on the lunar surface west of the Apollo 16 landing site at Descartes. The stage will impact the surface at 1.692 m/s (5,550 fps) at a -3.2° angle.

Impacts of these objects of known masses and velocities will assist in calibrating the Apollo 16 PSE readouts as well as in providing comparative readings between the Apollo 12, 14, 15 and 16 seismometers.

ALSEP to Impact Distance Table

	Approximate Distance	
	Kilometers	Nautical miles
Apollo 12 ALSEP (3.03°S, 23.4°W) to:		
Apollo 12 LM A/S Impact	72	39
Apollo 13 S-IVB Impact	137	74
Apollo 14 S-IVB Impact	175	95
Apollo 14 LM A/S Impact	114	62
Apollo 15 S-IVB Impact	354	191
Apollo 15 LM A/S Impact	1130	610
Apollo 16 S-IVB Impact	252	136
Apollo 16 LM A/S Impact	1173	633
Apollo 14 ALSEP (3.67°S, 17.45°W) to:		
Apollo 14 LM A/S Impact	68	37
Apollo 15 S-IVB Impact	184	99
Apollo 15 LM A/S Impact	1048	565
Apollo 16 S-IVB Impact	433	234
Apollo 16 LM A/S Impact	992	536
Apollo 15 ALSEP (26.10°N, 3.65°E) to:		
Apollo 15 LM A/S Impact	92	50
Apollo 16 S-IVB Impact	1347	727
Apollo 16 LM A/S Impact	1129	609
Apollo 16 ALSEP (9.0°S, 15.52°E) to:		
Apollo 16 LM A/S Impact	23	12

Apollo 16 S-IVB Impact 2.3°5, 31°W
Apollo 16 LM A/S Impact 9.48°S, 14.97°E

S-IVB/IU IMPACT

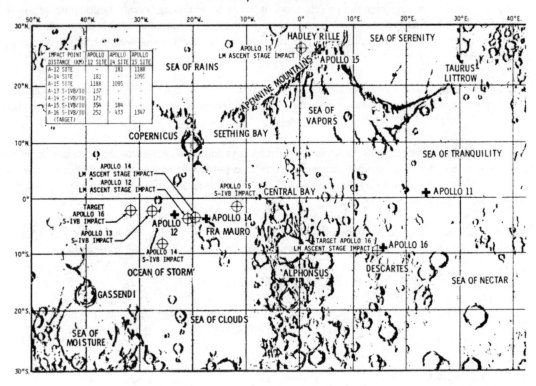

IMPACT POINT DISTANCE (KM)	APOLLO 12 SITE	APOLLO 14 SITE	APOLLO 15 SITE
A-12 SITE	-	181	1188
A-14 SITE	181	-	1095
A-15 SITE	1188	1095	-
A-13 S-IVB/IU	137	-	-
A-14 S-IVB/IU	175	-	-
A-15 S-IVB/IU	554	184	-
A-16 S-IVB/IU (TARGET)	252	433	1347

LUNAR IMPACT TARGET FOR SPENT LM ASCENT STAGE

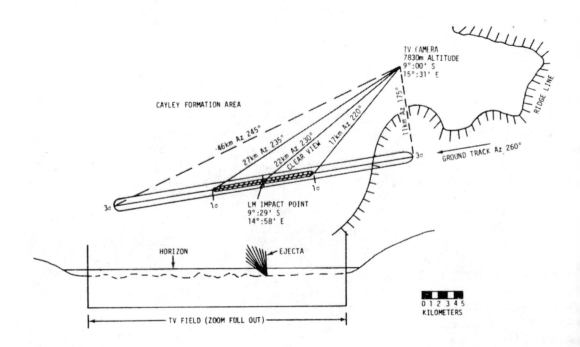

There are three major physical components of the PSE:

1.   The sensor assembly is comprised of 3 long period (LP) and a short period (SP) seismometer, an electrical power and a data subsystem, and a thermal control system. In the LP seismometer, low frequency (approximately 250 to 0.3 second periods) motion of the lunar surface caused by seismic activity is detected by tri-axial, orthogonal displacement amplitude type sensors. In the SP seismometer, the higher frequency (approximately 5 to 0.04 second periods) vertical motion of the lunar surface is detected by a displacement velocity sensor.

2.   The external leveling stool allows manual leveling of the sensor assembly by the crewman to within ±5 degrees. Final internal leveling to within ±3 arc seconds is accomplished by control motors.

3.   The five-foot diameter hat-shaped thermal shroud covers and helps stabilize the temperature of the sensor assembly. The instrument uses thermostatically controlled heaters to protect it from the extreme cold of the lunar night.

The Lunar Surface Magnetometer (LSM): The scientific objective of the magnetometer experiment is to measure the magnetic field at the lunar surface. Charged particles and the magnetic field of the solar wind impact directly on the lunar surface. Some of the solar wind particles are absorbed by the surface layer of the Moon. Others may be deflected around the Moon. The electrical properties of the material making up the Moon determine what happens to the magnetic field when it hits the Moon. If the Moon is a perfect insulator the magnetic field will pass through the Moon undisturbed. If there is material present which acts as a conductor, electric currents will flow in the Moon. A small magnetic field of approximately 35 gammas, one thousandth the size of the Earth's field was recorded at the Apollo 12 site.

Fields recorded by the portable magnetometer on Apollo 14 were about 43 gammas and 103 gammas in two different locations.

(Gamma is a unit of intensity of a magnetic field. The Earth's magnetic field at the Equator, for example, is 35,000 gamma. The interplanetary magnetic field from the Sun has been recorded at 5 to 10 gamma.)

Two possible models are shown in the next drawing. The electric current carried by the solar wind goes through the Moon and "closes" in the space surrounding the Moon (figure a). This current (E) generates a magnetic field (M) as shown. The magnetic field carried in the solar wind will set up a system of electric currents in the Moon or along the surface. These currents will generate another magnetic field which tries to counteract the solar wind field   (figure b). This results in a charge in the total magnetic field measured at the lunar surface.

The magnitude of this difference can be determined by independently measuring the magnetic field in the undisturbed solar wind nearby, yet away from the Moon's surface. The value of the magnetic field change at the Moon's surface can be used to deduce information on the electrical properties of the Moon. This, in turn, can be used to better understand the internal temperature of the Moon and contribute to better understanding of the origin and history of the Moon.

The principal investigator for this experiment is Dr. Palmer Dyal, NASA Ames Research Center, Mountain View, California.

The magnetometer consists of three magnetic sensors aligned in three orthogonal sensing axes, each located at the end of a fiberglass support arm extending from a central structure. This structure houses both the experiment electronics and the electro-mechanical gimbal/flip unit which allows the sensor to be pointed in any direction for site survey and calibration modes. The astronaut aligns the magnetometer experiment to within ±3 degrees east-west using a shadowgraph on the central structure, and to within ±3 degrees of the vertical using a bubble level mounted on the Y sensor boom arm.

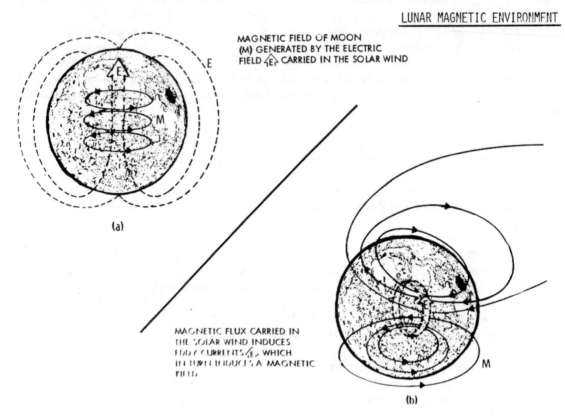

(a)

MAGNETIC FIELD OF MOON
(M) GENERATED BY THE ELECTRIC
FIELD $\langle E \rangle$ CARRIED IN THE SOLAR WIND

MAGNETIC FLUX CARRIED IN
THE SOLAR WIND INDUCES
EDDY CURRENTS $\langle E \rangle$ WHICH
IN TURN INDUCES A MAGNETIC
FIELD

(b)

Size, weight and power are as follows:

Size deployed	102 cm (40 inches) high with 152 cm (60 inches) between sensor heads
Weight (pounds)	8 kilograms (17.5 lbs)

Peak Power Requirements (watts)

Site Survey Mode	11.5
Scientific Mode	6.2
12.3 (night) Calibration Mode	10.8

The magnetometer experiment operates in three modes:

Site Survey Mode — An initial site survey is performed in each of the three sensing modes for the purpose of locating and identifying any magnetic influences permanently inherent in the deployment site so that they will not affect the interpretation of the LSM sensing of magnetic flux at the lunar surface.

Scientific Mode — This is the normal operating mode wherein the strength and direction of the lunar magnetic field are measured continously. The three magnetic sensors provide signal outputs proportional to the incidence of magnetic field components parallel to their respective axes. Each sensor will record the intensity three times per second which is faster'than the magnetic field is expected to change. All sensors have the capability to sense over any one of three dynamic ranges with a resolution of 0.2 gamma.

-100 to +100 gamma
-200 to +200 gamma
-400 to +400 gamma

Calibration Mode — This is performed automatically at 18-hour intervals to determine the absolute accuracy of the magnetometer sensors and to correct any drift from their laboratory calibration.

Magnetic Lunar Sample Returned to the Moon

In the study of returned lunar samples it has been found that there are commonly two components of remanent magnetization. The first of these is quite soft and can be removed by cleaning in alternating fields of about 50 to 100 oersteds. The second component is quite stable and is hardly affected by alternating fields up to 500 oersteds. It is this stable component which is most likely to represent ancient lunar magnetic fields and on which most attention has been focused. It can be simulated in the laboratory by allowing the sample to cool in the presence of a weak field from above its Curie point — about 780 degrees Centigrade (1,400 degrees Fahrenheit) in the case of lunar samples.

The soft component can also be simulated in the laboratory by exposing lunar samples to steady magnetic fields of about 10 to 50 oersteds. The field required to do this is much greater than the Earth's field which is only about 0.5 oersteds. One possibility, therefore, is that this magnetization was acquired on the return journey to Earth either in the spacecraft or in Earth laboratories. Another possibility is that this magnetization is truly of lunar origin, perhaps related to thermal fluctuations in the presence of a weak, solar wind field. It is important to establish the origin of this soft component and two steps are being taken to do this.

At the last lunar portable magnetometer (LPM) station a documented sample (preferably igneous) will be picked up and placed on the LPM after the first reading is taken. It will be documented in place on the magnetometer and the LPM reading retaken. If the intensity of magnetization of the soft component is like that seen in the returned samples it should be measurable at the level of several gammas. On return to Earth this same sample will be placed on the back-up flight model LPM in the shielded room in the LRL and the same procedure repeated. This will tell if additional magnetization has been acquired on the return journey.

The second step consists of taking a lunar sample back to the Moon. The sample chosen is Apollo 12 sample 12002,78.

Weight is 4 grams (0.14 ounces). This sample has been thoroughly tested magnetically and has a very clear stable component of magnetization. When first received in the laboratory it also had a soft component of magnetization which has been removed. Upon return to Earth it will be tested once again to see if it has reacquired the soft component. If it has, the soft remanence is not of lunar origin. The sample will be placed in a bag, sewn inside an interim stowage assembly bag and returned to Earth stowed as on Apollo 12.

Lunar Heat Flow Experiment (HFE): The scientific objective of the heat flow experiment is to measure the steady-state heat flow from the lunar interior. Two predicted sources of heat are:
(1) original heat at the time of the Moon's formation and (2) radioactivity. Scientists believe that heat could have been generated by the infalling of material and its subsequent compaction as the Moon was formed. Moreover, varying amounts of the radioactive elements uranium, thorium and potassium were found in the Apollo 11, 12, 14, and 15 lunar samples which if present at depth, would supply significant amounts of heat. No simple way has been devised for relating the contribution of each of these sources to the present rate of heat flow. In addition to temperature, the experiment is capable of measuring the thermal conductivity of the lunar rock material.

The combined measurement of temperature and thermal conductivity gives the net heat flux from the lunar interior through the lunar surface. Similar measurements on Earth have contributed basic information to our understanding of volcanoes, earthquakes and mountain building processes. In conjuction with the seismic and magnetic data obtained on other lunar experiments the values derived from the heat flow measurements will help scientists to build more exact models of the Moon and thereby give us a better understanding of its origin and history.

The heat flow experiment consists of instrument probes, electronics and emplacement tool and the lunar surface drill. Each of two probes is connected by a cable to an electronics box which rests on the lunar surface. The electronics, which provide control, monitoring and data processing for the experiment, are connected to the ALSEP central station.

Each probe consists of two identical 51 cm (20 in.) long sections each of which contains a "gradient" sensor bridge, a "ring" sensor bridge and two heaters. Each bridge consists of four platinum resistors mounted in a thin-walled fiberglass cylindrical shell. Adjacent areas of the bridge are located in sensors at opposite ends of the 51 cm (20 in.) fiberglass probe sheath. Gradient bridges consequently measure the temperature difference between two sensor locations.

In thermal conductivity measurements at very low values a heater surrounding the gradient sensor is energized with 0.002 watts and the gradient sensor values monitored. The rise in temperature of the gradient sensor is a function of thermal conductivity of the surrounding lunar material. For higher range of values, the heater is energized at 0.5 watts of heat and monitored by a ring sensor. The rate of temperature rise, monitored by the ring sensor is a function of the thermal conductivity of the surrounding lunar material. The ring sensor, approximately 10 cm (4 in.) from the heater, is also a platinum resistor. A total of eight thermal conductivity measurements can be made. The thermal conductivity mode of the experiment will be implemented about 20 days (500 hours) after deployment. This is to allow sufficient time for the perturbing effects of drilling and emplacing the probe in the borehole to decay; i.e., for the probe and casings to come to equilibrium with the lunar Subsurface.

A 9.1 meter (30-ft.) cable connects each probe to the electronics box. The cable contains four evenly spaced thermocouples: at the top of the probe; at 66 cm (26 in.), 114 cm (45 in.) and 168 cm (66 in.). The thermocouples will measure temperature transients propagating downward from the lunar surface. The reference junction temperature for each thermocouple is located in the electronics box. In fact, the feasibility of making a heat flow measurement depends to a large degree on the low thermal conductivity of the lunar surface layer, the regolith. Measurement of lunar surface temperature variations by Earth-based telescopes as well as the Surveyor and Apollo missions show a remarkably rapid rate of cooling. The wide fluctuations in temperature of the lunar surface are to influence only the upper one meter (three ft.) and not the bottom one meter (three ft.) of the borehole.

The astronauts will use the Apollo lunar surface drill (ALSD) to make a lined borehole in the lunar surface for the probes. The drilling energy will be provided by a battery powered rotary percussive power head. The drill borestem rod consists of fiberglass tubular sections reinforced with boron filaments, one section 137 cm (54 in) long and two sections each 71 cm (28 in) long. A closed drill bit, placed on the first drill rod, is capable of penetrating the variety of rock including 95 cm (3 ft.) of vesicular basalt (40 percent porosity). As lunar surface penetration progresses, additional drill rod sections will be connected to the drill string. The drill string is left in place to serve as a hole casing.

An emplacement tool is used by the astronaut to insert the probe to full depth. Alignment springs position the probe within the casing and assure a well-defined radiative coupling between the probe and the borehole. Radiation shields on the hole prevent direct sunlight from reaching the bottom of the hole.

As a part of the field geology experiment, the astronaut will drill a third hole near the HFE, using corestem drill sections and a coring bit, and obtain cores of lunar material for subsequent analysis of thermal properties and composition. Total available core length is 2.5 m (100 in.). The heat flow experiment, design, and data analysis are the responsibility of Dr. Marcus Langseth of the LamontDoherty Geological Observatory.

Active Seismic Experiment: The ASE will produce data on the physical structure and bearing strength of the lunar surface by measuring seismic waves. Two types of man-made seismic sources will be used with the ASE: a crew-actuated pyrotechnic "thumper" and a mortar-like device from which four rocket propelled projectiles can be launched by command from Earth. Naturally produced seismic events will be detected passively by the ASE (the ASE geophones will be turned on remotely for short listening periods). The seismic waves are detected by geophones deployed by the crew. Data on wave penetration, frequency spectra, and velocity to lunar depths of 152 meters (500 ft.) will be obtained and passed to the ALSEP central station for transmittal to the Earth. Dr. Robert Kovach of Stanford University is the Principal Investigator.

The mortar like device will be deployed, aligned and activated about 15 meters (48 ft.) north of the ALSEP central station. The four grenade-like projectiles will be launched sometime after the crew returns and at a time specified by the principal investigator, but no sooner than the third lunation.

The crew will deploy three geophones at 3.5, 45.7 and 91.4 meters (10, 150, 300 ft.) from the ALSEP central station. Enroute back to the central station, the crewman will fire 19 "thumper" charges at 4.5 meters (15 ft.) intervals along the geophone line. The thumper serves as a storage and transport rack for the geophones and their connecting cable.

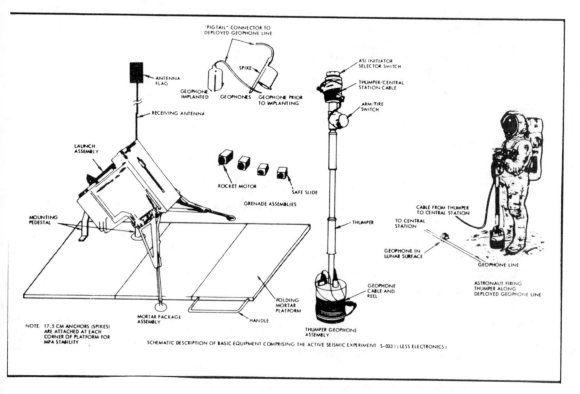

Active Seismic Experiment

The two major components of the ASE are:

1. The thumper-geophone assembly measuring 1.13 m (44.5 in.) when deployed and weighing 3.2 kg. (7 lbs.) including three geophones and cable. Each geophone is 12.2 cm (4.8 in.) high, 4.1 cm (1.6 in.) in diameter and weighs less than 0.4 kg. (1 lb.).

2. The package projectile launch assembly weighs 6.8 kg. (15 lbs.) (including four projectiles) and is 24 cm (9.5 in.) high, 10.2 cm (4 in.) wide and 38 cm (15.6 in.) long. The mortar-like launching device is made of fiberglass and magnesium, and contains firing circuitry and a receiver antenna. The projectile launch assembly is enclosed in a box and consists of four fiberglass launch tubes and four projectiles. The projectiles vary in length and weight according to the propellant and explosive charges. Radio transmitters in each projectile furnish start-and-stop flight time data for telemetry back to Earth. Thus, with the launch angle known, range can be calculated. The geophones provide information on seismic wave travel time. Correlation of this time with range will establish wave velocity through the lunar surface.

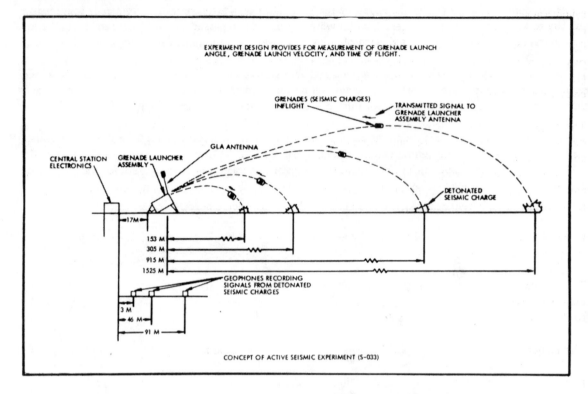

EXPERIMENT DESIGN PROVIDES FOR MEASUREMENT OF GRENADE LAUNCH ANGLE, GRENADE LAUNCH VELOCITY, AND TIME OF FLIGHT.

CONCEPT OF ACTIVE SEISMIC EXPERIMENT (S-033)

Active Seismic Experiment Mortar Mode Concept

ALSEP Central Station: The central station serves as a power distribution and data-handling point for experiments carried on the ALSEP. Central station components are the data subsystem, helical antenna, experiment electronics, and power conditioning unit. The central station is deployed after other experiment instruments are unstowed from the pallet.

The central station data subsystem receives and decodes uplink commands, times and controls experiments, collects and transmits scientific and engineering data downlink, and controls the electrical power subsystem through the power distribution and signal conditioner.

The modified axial-helix S-band antenna receives and transmits a right-hand circularly-polarized signal. The antenna is manually aimed with a two-gimbal azimuth/elevation aiming mechanism.

The ALSEP electrical power subsystem draws electrical power from a SNAP-27 (Systems for Nuclear Auxiliary Power) radioisotope thermoelectric generator.

SNAP-27—Power Source for ALSEP

A SNAP-27 unit, similar to three others deployed on the Moon, will provide power for the ALSEP package. SNAP-27 is one of a series of radioisotope thermoelectric generators, or atomic batteries, developed by the Atomic Energy Commission under its space SNAP program. The SNAP (Systems for Nuclear Auxiliary Power) program is directed at development of generators and reactors for use in space, on land, and in the sea.

SNAP-27 on Apollo 12 marked the first use of a nuclear power system on the Moon. The nuclear generators are required to provide power for, periods of at least one year. Thus far, the SNAP-27 unit on Apollo 12 has operated over two years and the SNAP-27 unit on Apollo 14 has operated a little over a year.

The basic SNAP-27 unit is designed to produce at least 68 watts of electrical power. The SNAP-27 unit is a cylindrical generator, fueled with the radioisotope plutonium-238. It is about 46 cm (18 inches) high and 41 cm (16 inches) in diameter, including the heat radiating fins. The generator, making maximum use of the lightweight material beryllium, weighs about 12.7 kilograms (26 pounds) without fuel. The fuel capsule, made of a superalloy material, is 42 cm (16.5 inches) long and 6.4 cm (2.5 inches) in diameter. It weighs about 7 km (15.5 pounds), of which 3.8 km (8.36 pounds) represent fuel. The plutonium-238 fuel is fully oxidized and is chemically and biologically inert.

The rugged fuel capsule is stowed within a graphite fuel cask from launch through lunar landing. The cask is designed to provide reentry heating protection and containment for the fuel capsule in the event of an aborted mission. The cylindrical cask with hemispherical ends includes a primary graphite heat shield, a secondary beryllium thermal shield, and a fuel capsule support structure. The cask is 58.4 cm (23 inches) long and 20 cm (eight inches) in diameter and weighs about 11 km (24.5 pounds). With the fuel capsule installed, it weighs about 18 km(40 pounds). It is mounted on the lunar module descent stage.

Once the lunar module is on the Moon, an Apollo astronaut will remove the fuel capsule from the cask and insert it into the SNAP-27 generator which will have been placed on the lunar surface near the module.

The spontaneous radioactive decay of the plutonium-238 within the fuel capsule generates heat which is converted directly into electrical energy—at least 68 watts. Units now on the lunar surface are producing 72 to 74 watts. There are no moving parts.

The unique properties of plutonium-238, make it an excellent isotope for use in space nuclear generators. At the end of almost 90 years, plutonium-238 is still supplying half of its original heat. In the decay process, plutonium-238 emits mainly the nuclei of helium (alpha radiation), a very mild type of radiation with a short emission range. Before the use of the SNAP-27 system was authorized for the Apollo program a thorough review was conducted to assure the health and safety of personnel involved in the launch and of the general public. Extensive safety analyses and tests were conducted which demonstrated that the fuel would be safely contained under almost all credible accident conditions.

Soil Mechanics: Mechanical properties of the lunar soil, surface, and subsurface will be investigated through trenching at various locations and through use of the self-recording penetrometer. The self-recording penetrometer measures the characteristics and mechanical properties of the lunar surfac material. The penetrometer consists of a 76 cm (30-in) penetration shaft and recording drum. Three interchangeable penetration cones 1.3, 3.2 and 6.4 cm^2(0.2, 0.5 and 1.0 in^2) cross sections and a 2.5 x 12.7 cm (1 x 5-in.) pressure plate may be attached to the shaft. The crewman forces the penetrometer into the surface and a stylus scribes a force vs. depth plot on the recording drum. The drum can record up to 24 force-depth plots. The upper housing containing the recording drum is detached at the conclusion of the experiment for return to Earth and analysis by the principal investigator. This experiment will be documented with the electric Hasselblad and the 16mm data acquisition cameras.

Soil behavior characteristics are also determined from interactions of LRV wheels, LM footpads, and footprints with the lunar soil.

Lunar Portable Magnetometer; (LPM) The Apollo 16 crew will use the LPM during EVAs 1 and 2 for measuring variations in the lunar magnetic field at several points during the geology traverses. Data gathered will be used to determine the location, strength, and dimensions of the magnetic field, as well as knowledge of the local selenological structure. The LPM experiment was also flown on Apollo 14.

The LPM will be carried on the LRV aft pallet and consists of a flux-gate magnetometer sensor head mounted on a tripod and an electronics data package. The sensor head is connected to the data package by a 15.2-meter (50 ft.) flat cable, and after the crewman aligns the sensor head at least 14 meters (46 ft.) from the data package, he returns to the LRV and relays readouts to Earth by voice.

The sensor head is sensitive to magnetic interference from the crew's portable life support systems, the geology tool, and the LRV, hence the need for emplacing and aligning the sensor well away from them. The mercury-cell powered electronics package has a range of +256 gamma. Readings are displayed in three digital meters — one for each axis (orthogonal X, Y and Z,). At each traverse location for LPM measurements, the crew will call out the meter readings on each axis.

Dimensions of the LPM components are 10.2 x 19 x 12.7 cm (4 x 7 1/2 x 5 inches) for the data package and 8.6 x 14.4 x 6.7 cm (3 3/8 x 5 11/16.x 2 5/8 inches) for the sensor head. The sensor head tripod is 45.7 cm (18 inches) long when retracted and extends to 78.7 cm (31 inches). The principal investigator is Dr. Palmer Dyal, NASA Ames Research Center, Mountain View, Calif.

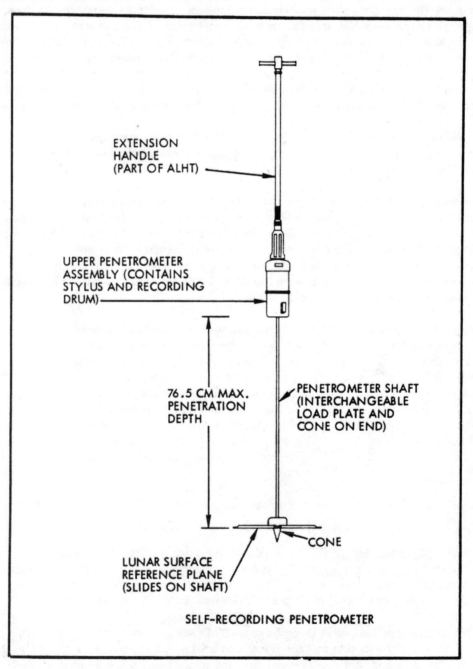

Soil Mechanics Experiment Penetrometer

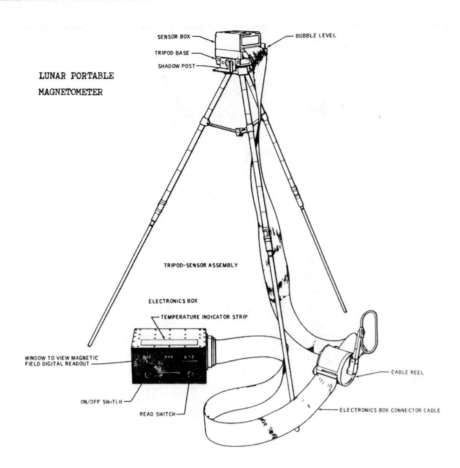

LUNAR PORTABLE MAGNETOMETER

Far Ultraviolet Camera/Spectroscope: Deep-space concentrations of hydrogen in interplanetary, interstellar and intergalactic regions will be mapped by this experiment using an instrument which gathers both photographic images and spectroscopic data in the far ultraviolet spectrum. The experiment will be the first such astronomical observation emplaced on the lunar surface.

Earlier spectrographic searches for hydrogen sources in space from Earth-orbiting astronomical satellites were impaired by the "masking" effect of the Earth's corona. The instrument will be pointed toward such targets as star clouds, nebulae, galaxy clusters and other galactic objects, intergalactic hydrogen, the solar bow cloud and solar wind, lunar atmosphere and any possible lunar volcanic gases, and the Earth's atmosphere and corona. Several astrophysicists have speculated that extremely hot hydrogen exists in intergalactic space and that hydrogen clusters may be detected between galaxies by an instrument such as the far UV camera spectroscope.

The experiment includes a 75mm (3-in) electronographic Schmidt camera with a potassium bromide cathode and 35mm film magazine and transport. Emplaced on a tripod in the LM shadow, the instrument is provided with a battery pack which is placed in the Sun at the end of its connecting cable. The spectroscope is fitted with lithium fluoride and calcium fluoride filters for detecting hydrogen Lyman-alpha radiation in the 1216 Å (angstrom) wavelength.

Measurements will range from 500 to 1550 Å for spectroscopic data and from 1050 to 1550 Å and 1230 to 1550 Å in photographic imagery. Hydrogen gas clouds will be detected through differential measurements of the photoimagery.

Using the elevation/azimuth adjustments on the instrument mount, the crew will align the camera/spectroscope toward specific targets periodically during the three EVAs. Near the end of EVA 3, the camera film cassette will be retrieved for return to Earth.

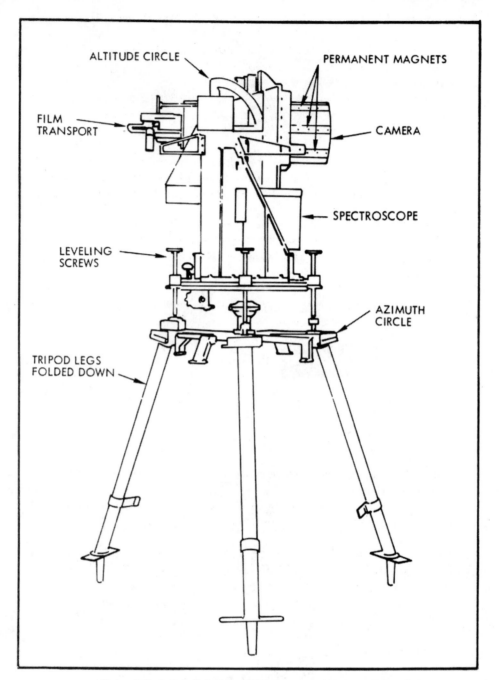

Tripod-Mounted Schmidt Electronographic UV Camera/Spectroscope

Experiment principal investigator is Dr. George R. Carruthers of the E.O. Hurlburt Center for Space Research, Naval Research Laboratory, Washington, D.C.

Solar Wind Composition Experiment (SWC) The scientific objective of the solar wind composition experiment is to determine the elemental and isotopic composition of the noble gases in the solar wind.

As in Apollos 11, 12, 14 and 15, the SWC detector will be deployed on the lunar surface and brought back to Earth by the crew. The detector will be exposed to the solar wind flux for 46 hours compared to 41 hours on Apollo 15, 21 hours on Apollo 14, 18 hours on Apollo 12 and two hours on Apollo 11. The solar wind detector consists of an aluminum and platinum foil .37 square meters (4 square ft.) in area and about 0.5

mils thick rimmed by Teflon for resistance to tearing during deployment. The platinum foil strips have been added to the experiment on Apollo 16 to reduce contamination. This will allow a more accurate determination of solar wind constituents. A staff and yard arrangement will be used to deploy the foil and to maintain the foil approximately perpendicular to the solar wind flux. Solar wind particles will penetrate into the foil, while cosmic rays pass through. The particles will be firmly trapped at a depth of several hundred atomic layers. After exposure on the lunar surface, the foil is rolled up and returned to Earth. Professor Johannes Geiss, University of Berne, Switzerland, is principal investigator. Professor Geiss is sponsored and funded by the Swiss Committee for Space Research.

Cosmic Ray Detector: A four-panel detector array is mounted on the lunar module descent stage for measuring the charge, mass and energy of cosmic ray and solar wind particles impacting the detector array on the way to the Moon and during the stay on the lunar surface. The detector array will be brought back to Earth for analysis.

Energy ranges for the two types of particles to be measured are from 0.5 to 10 thousand electron volts/nucleon and from 0.2 to 200 million electron volts/nucleon. Additionally, various types of glass detectors will be evaluated and lunar surface thermal neutron flux will be measured. Cosmic ray and solar wind particles will impact three portions of the detector during translunar coast, and the fourth panel will be uncovered early in the first lunar surface EVA to gather data on lunar cosmic ray and solar wind particles. The detector will be folded and bagged during the third EVA and returned to Earth.

Detector panel 1 is a sandwich of 31 sheets of Lexan, each 0.025 cm thick, and covered by perforated aluminized Teflon.

Panel 2 is almost identical to panel 1, except that two plastic sheets are pre-irradiated for data analysis calibration.

Panel 3 is made up of 40 layers of 0.2-cm-thick Kodacel cellulose triacetate sheets, overlaid on the upper half by 10 five-micron thick Lexan sheets. Five specimens of glass are imbedded in the panel's lower portion.

Panel 4 is similar in makeup to panel 3, except that the sheets in the lower part of the panel may be shifted. A 0.076-cm-thick aluminum plate, to which are bonded small fragments of mica, glass and natural crystals, covers the lower quarter of the panel. The next highest quarter of the panel is covered by a bonded 2-micron thick foil. A lanyard-actuated slide made of 0.0005-cm thick aluminum foil bonded to a platinum base covers the upper half of panel four. The panel upper half is exposed to the lunar environment when the crewman pulls the lanyard during EVA 1.

Principal investigators for the experiment are Dr. R. L. Fleischer, General Electric Co., Dr. P. B. Price, University of California at Berkeley, and Dr. R. M. Walker, Washington University.

Lunar-Geology Investigation

The fundamental objective of the lunar geology investigation experiment is to provide data in the vicinity of the landing site for use in the interpretation of the geologic history of the Moon. Apollo lunar landing missions offer the opportunity to correlate carefully collected samples with a variety of observational data on at least the upper portions of the mare basin filling and the lunar highlands, the two major geologic subdivisions of the Moon. The nature and origin of the maria and highlands will bear directly on the history of lunar differentiation and differentiation processes. From the lunar bedrock, structure, land forms and special materials, information will be gained about the internal processes of the Moon. The nature and origin of the debris layer (regolith) and the land forms superimposed on the maria and highland regions are a record of lunar history subsequent to their formation. This later history predominately reflects the history of the extralunar environment. Within and on the regolith, there will also be materials that will aid in the understanding of geologic units elsewhere on the Moon and the broader aspects of lunar history.

The primary data for the lunar geology investigation experiment come from photographs, verbal data, and returned lunar samples. Photographs taken according to specific procedures will supplement and illustrate crew comments, record details not discussed by the crew, provide a framework for debriefing, and record a wealth of lunar surface information that cannot be returned or adequately described by any other means.

In any Hasselblad picture taken from the lunar surface, as much as 90 percent of the total image information may be less than 100 feet from the camera, depending on topography and how far the camera is depressed below horizontal.

Images of distant surface detail are so foreshortened that they are difficult to interpret. Therefore, it is important that panoramas be taken at intervals during the traverse and at the farthest excursion of the traverse. This procedure will extend the high resolution photographic coverage to the areas examined and discussed by the astronaut, and will show the regional context of areas of specific interest that have been discussed and photographed in detail.

The polarizing filters will permit the measurement of the degree of polarization and orientation of the plane of polarization contained in light reflected from the lunar surface. Different lunar materials (i.e., fine-grained glass and/or fragments, strongly shocked rocks, slightly shocked rocks and shock-lithified (fragmental material) have different polarimetric functions, in other words, different polarimetric "signatures." Comparison of the polarimetric function of known material, such as returned samples and close-up lunar surface measurements, with materials photographed beyond the traverse of the astronaut will allow the classification and correlation of these materials even though their textures are not resolvable. The polari-metric properties of lunar materials and rock types are a useful tool for correlation and geologic mapping of each landing site, and for extrapolation of geologic data from site to site across the lunar surface.

The "in situ" photometric properties of both finegrained materials and coarse rock fragments will serve as a basis for delineating, recognizing, describing, and classifying lunar materials. The gnomon, with photometric chart attached, will be photographed beside a representative rock and, if practical, beside any rock or fine-grained material with unusual features.

The long focal length (500 mm) lens with the HEDC will be used to provide high resolution data. A 5 to 10 centimeter resolution is anticipated at a distance of 1 to 2 km (0.6 to 1.2 mi.). The high degree of resolution will make it possible to analyze the stratigraphic layering in North Ray Crater walls and potential outcrops and exposed bedrock in distant areas.

Small exploratory trenches, several centimeters deep, are to be dug to determine the character of the regolith down to these depths. The trenches should be dug in the various types of terrain and in areas where the surface characteristics of the regolith are of significant interest as determined by the astronaut crew. The main purpose of the trenches will be to determine the small scale stratigraphy (or lack of) in the upper few inches of the regolith in terms of petrological characteristics and particle size.

The organic control sample, carried in each sample return case (SRC) will be analyzed after the mission in the Lunar Receiving Laboratory to determine the level of contamination in each SRC. This will then be compared to an organic control sample which was removed from the SRC prior to the SRC being shipped to KSC for loading onto the LM.

In order to more fully sample the major geological features of the Apollo 16 landing site, various groupings of sampling tasks are combined and will be accomplished in concentrated areas. This will aid in obtaining vertical as well as lateral data to be obtained in the principal geological settings. Thus, some trench samples, core tube samples and lunar environmental soil samples will be collected in association with comprehensive samples. In addition, sampling of crater rims of widely differing sizes in a concentrated area will give a sampling of the deeper stratigraphic divisions at that site. Repeating this sampling technique at successive traverse stations will show the continuity of the main units within the Cayley formation.

Sampling and photographic techniques used to gather data in the Descartes region include:

* Documented samples of lunar surface material which, prior to gathering, are photographed in color and stereo -using the gnomon and photometric chart for comparison of position and color properties — to show the sample's relation to other surface features.

* Rock and soil samples of Imbrian age rocks from deep layers, and soil samples from the regolith in the immediate area where the rocks are gathered.

* Radial sampling of material on the rim of a fresh crater — material that should be from the deepest strata.

* Photopanoramas for building mosaics which will allow accurate control for landing site map correlation.

* Polarimetric photography for comparison with known materials.

* Double drive tube samples to depths of 60 cm (23.6 in.) for determining the stratigraphy in multi-layer areas.

* Single drive tube samples to depths of 38 cm (15 in.) in the comprehensive sample area and in such target of opportunity areas as mounds and fillets.

* Drill core sample of the regolith which will further spell out the stratigraphy of the area sampled.

* A small exploratory trench, ranging from 6 to 24 cm (2.4 to 9.7 in.) in depth, to determine regolith particle size and small-scale stratigraphy.

* Large equidimensional rocks ranging from 15 to 24 cm (6 to 9.4 in.) in diameter for data on the history of solar radiation. Similar sampling of rocks from 6 to 15 cm (2.4 to 6 in.) in diameter will also be made.

* Fillet sampling which may reveal a relationship between fillet volume and rock size to the length of time the fillet has been in place.

* Vacuum-packed lunar environment soil and rock samples kept biologically pure for postflight gas, chemical and microphysical analysis.

* Special surface samples to determine properties of the upper 10 - 100 microns of the lunar surface.

Apollo Lunar Geology Hand Tools

Sample scale - The scale is used to weigh the loaded sample return containers, sample bags, and other containers to maintain the weight budget for return to Earth. The scale has graduated markings in increments of 5 pounds to a maximum capacity of 80 pounds. The-scale is stowed and used in the lunar module ascent stage.

Tongs - The tongs are used by the astronaut while in a standing position to pick up lunar samples from pebble size to fist size. The tines of the tongs are made of stainless steel and the handle of aluminum. The tongs are operated by squeezing the T-bar grips at the top of the handle to open the tines in addition to picking up samples, the tongs are used to retrieve equipment the astronaut may inadvertantly drop. This tool is 81 cm (32 inches) long overall.

Lunar rake - The rake is used to collect discrete samples of rocks and rock chips ranging from 1.3 cm (one-half inch) to 2.5 cm (one inch) in size. The rake is adjustable for ease of sample collection and stowage. The tines, formed in the shape of a scoop, are stainless steel. A handle, approximately 25 cm (10 inches) long, attaches to the extension handle for sample collection tasks.

Adjustable scoop - The sampling scoop is used to collect soil material or other lunar samples too small for the rake or tongs to pick up. The stainless steel pan of the scoop, which is 5 cm (2 inches) by 11 cm (41/2 inches) by 15 cm (6 inches) has a flat bottom flanged on both sides and a partial cover on the top to prevent loss of contents. The pan is adjustable from horizontal to 55 degrees and 90 degrees from the horizontal for use in scooping and trenching. The scoop handle is compatible with the extension handle.

Hammer - This tool serves three functions; as a sampling tool to chip or break large rocks, as a pick, and as a hammer to drive the drive tubes or other pieces of lunar equipment. The head is made of impact resistant tool steel, has a small hammer face on one end, a broad flat blade on the other, and large hammering flats on the sides. The handle, made of aluminum, is 36 cm (14 inches) long; its lower end fits the extension handle when the tool is used as a hoe.

Extension handle - The extension handle extends the astronaut's reach to permit working access to the lunar surface by adding 76 cm (30 inches) of length to the handles of the scoop, rake, hammer, drive tubes, and other pieces of lunar equipment. This tool is made of aluminum alloy tubing with a malleable stainless steel cap designed to be used as an anvil surface. The lower end has a quick-disconnect mount and lock designed to resist compression, tension, torsion, or a combination of these loads. The upper end is fitted with a sliding "T" handle to facilitate any torqueing operation.

Drive Tubes - These nine tubes are designed to be driven or augured into soil, loose gravel, or soft rock such as pumice. Each is a hollow thin-walled aluminum tube 41 cm (16 inches) long and 4 cm (1.75 inch) diameter with an integral coring bit. Each tube can be attached to the extension handle to facilitate sampling. A deeper core sample can be obtained by joining tubes in series of two or three. When filled with sample, a Teflon cap is used to seal the open end of the tube, and a keeper device within the drive tube is positioned against the top of the core sample to preserve the stratigraphic integrity of the core. Three Teflon caps are packed in a cap dispenser that is approximately a 5.7 cm (2.25 inch) cube.

Gnomon and Color Patch - The gnomon is used as a photographic reference to establish local vertical Sun angle, scale, and lunar color. This tool consists of a weighted staff mounted on a tripod. It is constructed in such a way that the staff will right itself in a vertical position when the legs of the tripod are on the lunar surface. The part of the staff that extends above the tripod gimbal is painted with a gray scale from 5 to 35 percent reflectivity and a color scale of blue, orange, and green. The color patch, similarly painted in gray scale and color scale, mounted on one of the tripod legs provides a larger target for accurately determining colors in color photography.

Apollo Lunar Hand Tool Carrier - This is an aluminum framework with a handle an legs that unfold to provide a steady base. The tools described above may be stowed in special fittings on this framework during lunar surface EVAs. The framework also provides support positions for two sample collection bags and a 20-bag documented sample bag dispenser. The hand tool carrier mounts on the Quad III LRV pallet during launch and stays on that pallet when it is transferred to the aft end of the lunar roving vehicle.

Sample Bags - Several different types of bags are furnished for collecting lunar surface samples. The Teflon documented sample bag (DSB), 19 by 20 cm (.7-1/2 by 8 inches) in size, is prenumbered and packed in a 20-bag dispenser that can be mounted on the hand tool carrier or on a bracket on the Hasselblad camera. Documented sample bags (140) will be available during the lunar surface EVAs. The sample collection bag (SCB), also of Teflon, has interior pockets along one side for holding drive tubes and exterior pockets for the special environmental sample container and for a drive tube cap dispenser. This bag is 17 by 23 by 41 cm (6-3/4 by 9 by 16 inches) in size (exclusive of the exterior pockets) and fits inside the sample return containers. During the lunar surface EVAs this bag is hung on the hand tool, carrier or on the portable life support system tool carrier. Four SCBs will be carried on Apollo 16. The extra sample collection bag (ESCB) is identical to the SCB except that the interior and exterior pockets are omitted. During EVAS it is handled in the same way as an SCB. Four ESCB bags will be carried on the mission.

A sample return bag, 13 by 33 by 57 cm (5 by 13 by 22.5 inches) in size, replaces the third sample return container and is used for the samples collected on the third EVA. It hangs on the LRV pallet during this EVA.

A special type of sample bag, the padded sample bag, is carried on Apollo 16. Although generally similar to the DSB, it has an inner padding of knit Teflon that forms an open-topped box 6 cm (2-1/2 inches) thick and a velcro strap to insure satisfactory closure for return to Earth. Only two of these special purpose bags will be carried on the flight.

Sample Return Container - This container maintains a vacuum environment and padded protection for lunar samples. It is an aluminum box 20 by 28 by 48 cm (8 by 11.5 by 19 inches) in size, with a knife-edge indium-silver seal, a strap latch system for closing, and a lever and pin system to support the container in the LM and CM stowage compartments under all vibration and g-force conditions. The drive tubes, special environmental sample container, and the core sample vacuum container as well as some other specialized pieces of equipment are flown outbound and inbound within the sample return containers, two of which will be carried on Apollo 16.

Special Environmental Sample Containers - This container is used to protect the vacuum environment of selected samples of lunar soil or rocks to be studied in specific experiments upon return to Earth. It is a thin-walled stainless steel can with a knife edge at the top and three-legged press assembly attached to the lid to effect a vacuum-proof seal when used on the lunar surface. Until used upon collection of the sample, the seal surfaces are covered with Teflon protectors that are then discarded.

Core Sample Vacuum Container - The core sample vacuum container is a receptacle or vacuum storage and transport to Earth of a lunar surface drive tube. The core sample vacuum container is about twice as long as the special environmental sample container and includes an insert to grip the drive tube, providing lateral and longitudinal restraint.

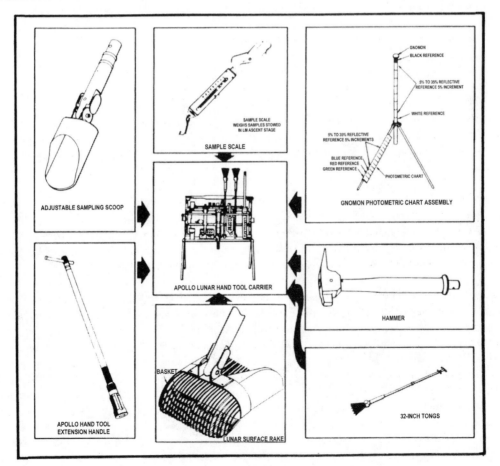

Lunar Geology Hand Tools

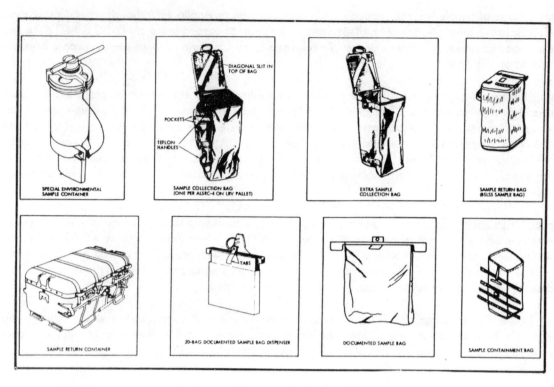

Lunar Geology Sample Containers

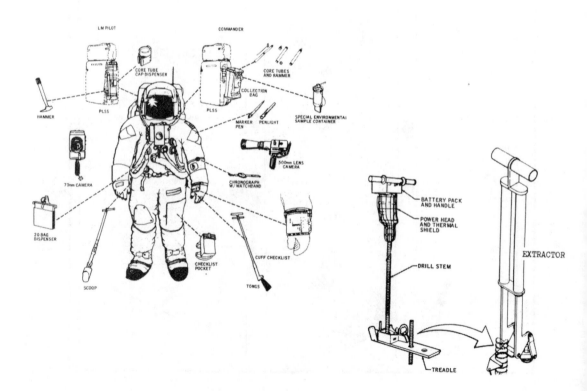

Surface Sampler Tool - The purpose of the surface sampler tool is to capture and return to Earth for analysis a sample from the very top of the lunar surface. The surface sampler tool hopefully will capture 300 to 500 milligrams of material only 100 to 500 microns deep. Each sampler consists of a 1.25-cm (1/2-inch) deep box which attaches to the universal hand tool. The 6.25-by-12.5-cm (2.5-by-5-inch) boxes each contain a plate which is floating in channels inside the box. One plate is covered with a deep pile velvet cloth and the other plate will be covered with either beta cloth or a thin layer of grease. The astronaut will open the spring-loaded door at the box bottom and gently lower the box until the floating plate touches the lunar surface. The lunar surface grains will be trapped in the fabric pile or grease. The astronaut will lift the box, close the lid, and place the box in a numbered bag for return to Earth.

LUNAR ORBITAL SCIENCE

Service Module Sector 1 houses the Scientific Instrument Module (SIM) bay. Eight experiments are carried in the SIM bay: X-ray fluorescence spectrometer, gamma ray spectrometer, alpha-particle spectrometer, panoramic camera, 76-millimeter (3-inch) mapping camera, laser altimeter and a mass spectrometer; a subsatellite carries three integral experiments (particle detectors, magnetometer and S-band transponder) comprising the eighth SIM bay experiment and will be placed into lunar orbit.

Gamma-Ray Spectrometer: On a 7.6 meter (25-foot) extendable boom, the gamma-ray spectrometer measures the chemical composition of the lunar surface in conjunction with the X-ray and alpha particle experiments to gain a compositional "map" of the lunar surface ground track. It detects natural and cosmic rays, induced gamma radioactivity and will operate on the Moon's dark and light sides. Additionally, the experiment will be extended in transearth coast to measure the radiation flux in cislunar space and record a spectrum of cosmological gamma-ray flux. The device can measure energy ranges between 0.1 to 10 million electron volts. The extendable boom is controllable from the command module cabin. Principal investigator is Dr. James R. Arnold, University of California at San Diego.

X-Ray Fluorescence Spectrometer: This geochemical experiment measures the composition of the lunar surface from orbit, and detects X-ray fluorescence caused by solar X-ray interaction with the Moon. It will analyze the sunlit portion of the Moon. The experiment will measure the galactic X-ray flux during transearth coast. The device shares a compartment on the SIM bay lower shelf with the alpha-particle experiment, and the protective door may be opened and closed from the command module cabin. Principal investigator is Dr. Isidore Adler, NASA Goddard Space Flight Center, Greenbelt, Md.

Alpha-Particle Spectrometer: This spectrometer measures monoenergetic alpha particles emitted from the lunar crust and fissures as products of radon gas isotopes in the energy range of 4.7 to 9.3 million electron volts. The sensor is made up of an array of 10 silicon surface barrier detectors. The experiment will construct a "map" of lunar surface alpha-particle emissions along the orbital track and is not constrained by solar illumination. It will also measure deep-space alpha-particle background emissions in lunar orbit and in transearth coast. Protective door operation is controlled from the cabin. Principal investigator is Dr. Paul Gorenstein, American Science and Engineering, Inc., Cambridge, Mass.

| | | APOLLO MISSIONS | | | | |
	11	12	14	15	16	17
SERVICE MODULE EXPERIMENTS						
S-160 GAMMA-RAY SPECTROMETER				X	X	
S-161 X-RAY SPECTROMETER				X	X	
S-162 ALPHA-PARTICLE SPECTROMETER				X	X	
S-164 S-BAND TRANSPONDER (CSM/LM)				X	X	X
S-165 MASS SPECTROMETER			X	X	X	
S-169 FAR UV SPECTROMETER						X
S-170 BISTATIC RADAR			X	X	X	
S-171 IR SCANNING RADIOMETER						X
S-209 LUNAR SOUNDER						X
SUBSATELLITE:						
S-164 S-BAND TRANSPONDER				X	X	
S-173 PARTICLE MEASUREMENT				X	X	
S-174 MAGNETOMETER				X	X	

			APOLLO MISSIONS			
	11	12	14	15	16	17
SM-PHOTOGRAPHIC TASKS:						
24" PANORAMIC CAMERA				X	X	X
3" MAPPING CAMERA				X	X	X
LASER ALTIMETER				X	X	X
COMMAND MODULE EXPERIMENTS						
S-158 MULTISPECTRAL PHOTOGRAPHY		X				
S-176 APOLLO WINDOW METEOROID			X		X	X	X
S-177 UV PHOTOGRAPHY OF EARTH AND MOON				X	X	
S-178 GEGENSCHEIN FROM LUNAR ORBIT			X	X	X	

APOLLO 16 ORBITAL TIMELINE

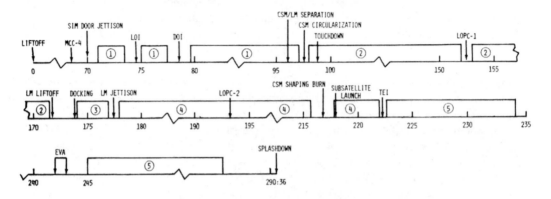

SIM OPERATION PERIODS (NON-CONTINUOUS)

ACTIVITY PERIOD	CONFIGURATION	ɣ-RAY SPEC	α-PARTICLE SPEC	X-RAY SPEC	MASS SPEC	MAP CAMERA	PAN CAMERA	LASER ALTIMETER
①	CSM-LM							
②	CSM							
③	CSM-LM							
④	CSM							
⑤	CSM							
TOTAL TIME (HOURS:MINS IN LUNAR ORBIT)		140:15	140:15	110:53	110:28	19:24	2:44	19:46

LUNAR ORBITAL SCIENCE EXPERIMENT ASSIGNMENTS

Mass Spectrometer: This spectrometer measures the composition and distribution of the ambient lunar atmosphere, identifies active lunar sources of volatiles, and pinpoints contamination in the lunar atmosphere. The sunset and sunrise terminators are of special interest, since they are predicted to be regions of concentration of certain gases. Measurements over at least five lunar revolutions are desired. The mass spectrometer is on a 7.3-meter (24-foot) extendable boom. The instrument can identify species from 12 to 28 atomic mass units (AMU) with the No. 1 ion counter, and 28-66 AMU with the No. 2 counter. Principal investigator is Dr. John H. Hoffman, University of Texas at Dallas.

Panoramic Camera: 610mm (24-inch) SM orbital photo task: The camera gathers mono or stereo high-resolution (2m) photographs of the lunar surface from orbit. The camera produces an image size of 28 x 334 kilometers (17 x 208 nm) with a field of view 11° along the track and 108° cross track. The rotating lens system can be stowed face-inward to avoid contamination during effluent dumps and thruster firings. The 33-kilogram (72-pound) film cassette of 1,650 frames will be retrieved by the command module pilot during a

APOLLO 16
POTENTIAL LUNAR ORBIT COVERAGE

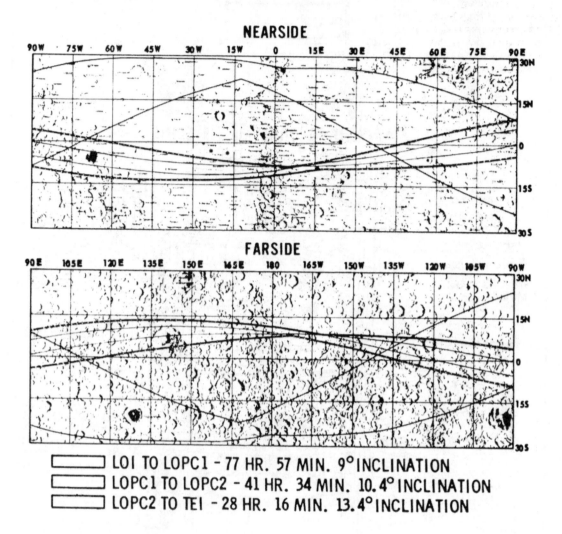

NEARSIDE

FARSIDE

☐ LOI TO LOPC1 - 77 HR. 57 MIN. 9° INCLINATION
☐ LOPC1 TO LOPC2 - 41 HR. 34 MIN. 10.4° INCLINATION
☐ LOPC2 TO TEI - 28 HR. 16 MIN. 13.4° INCLINATION

transearth coast EVA. The camera works in conjunction with the mapping camera and the laser altimeter to gain data to construct a comprehensive map of the lunar surface ground track flown by this mission — about 2.97 million square meters (1.16 million square miles) or 8 percent of the lunar surface.

Mapping Camera 76mm (3-inch): Combines 20-meter resolution terrain mapping photography on five-inch film with 76mm (3-inch) focal length lens with stellar camera shooting the star field on 35mm film simultaneously at 96° from the surface camera optical axis. The stellar photos allow accurate orientation of mapping photography postflight by comparing simultaneous star field photography with lunar surface photos of the nadir (straight down). Additionally, the stellar camera provides pointing vectors for the laser altimeter during darkside passes. The mapping camera metric lens covers a 74° square field of view, or 170 x 170 km (92 x 92 nm) from 111.5 km (60 nm) in altitude. The stellar camera is fitted with a 76mm (3-inch) f/2.8 lens covering a 24° field with cone flats. The 9-kg (20-1b) film cassette containing mapping camera film (3,600 frames) and the stellar camera film will be retrieved during the same EVA described in the panorama camera discussion. The Apollo Orbital Science Photographic Team is headed by Frederick J. Doyle of the U.S. Geological Survey, McLean, Va.

Laser Altimeter: This altimeter measures spacecraft altitude above the lunar surface to within two meters.

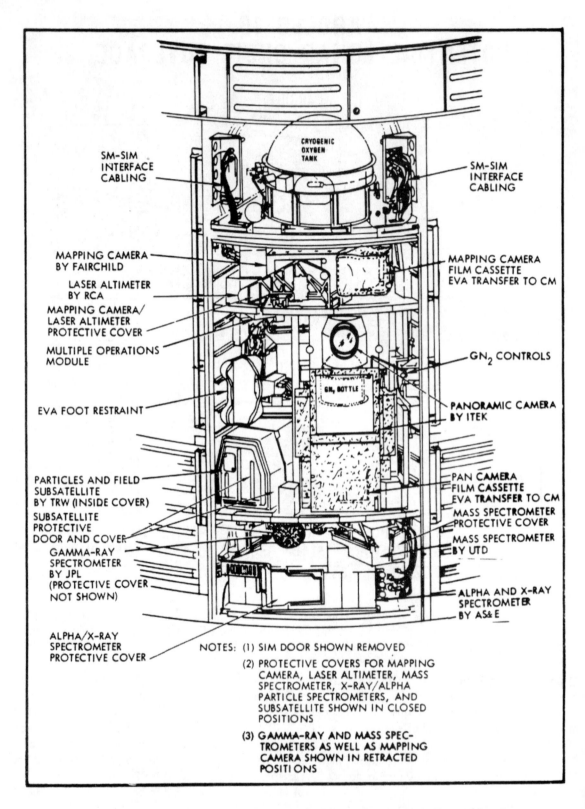

SM-SIM
INTERFACE
CABLING

CRYOGENIC
OXYGEN
TANK

SM-SIM
INTERFACE
CABLING

MAPPING CAMERA
BY FAIRCHILD

MAPPING CAMERA
FILM CASSETTE
EVA TRANSFER TO CM

LASER ALTIMETER
BY RCA

MAPPING CAMERA/
LASER ALTIMETER
PROTECTIVE COVER

MULTIPLE OPERATIONS
MODULE

GN_2 CONTROLS

GN$_2$ BOTTLE

EVA FOOT RESTRAINT

PANORAMIC CAMERA
BY ITEK

PARTICLES AND FIELD
SUBSATELLITE
BY TRW (INSIDE COVER)

PAN CAMERA
FILM CASSETTE
EVA TRANSFER TO CM

SUBSATELLITE
PROTECTIVE
DOOR AND COVER

MASS SPECTROMETER
PROTECTIVE COVER

GAMMA-RAY
SPECTROMETER
BY JPL
(PROTECTIVE COVER
NOT SHOWN)

MASS SPECTROMETER
BY UTD

ALPHA AND X-RAY
SPECTROMETER
BY AS&E

ALPHA/X-RAY
SPECTROMETER
PROTECTIVE COVER

NOTES: (1) SIM DOOR SHOWN REMOVED

(2) PROTECTIVE COVERS FOR MAPPING
CAMERA, LASER ALTIMETER, MASS
SPECTROMETER, X-RAY/ALPHA
PARTICLE SPECTROMETERS, AND
SUBSATELLITE SHOWN IN CLOSED
POSITIONS

(3) GAMMA-RAY AND MASS SPEC-
TROMETERS AS WELL AS MAPPING
CAMERA SHOWN IN RETRACTED
POSITIONS

Mission SIM Bay Science Equipment Installation

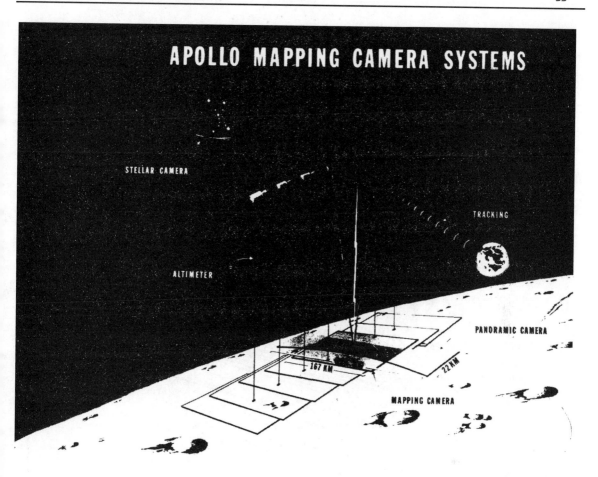

The instrument is boresighted with the mapping camera to provide altitude correlation data for the mapping camera as well as the panoramic camera. When the mapping camera is running, the laser altimeter automatically fires a laser pulse to the surface corresponding to mid-frame ranging for each frame. The laser light source is a pulsed ruby laser operating at 6,943 angstroms, and 200-millijoule pulses of 10 nanoseconds duration. The laser has a repetition rate up to 3.75 pulses per minute. The laser altimeter working group of the Apollo Orbital Science Photographic Team is headed by Dr. William M. Kaula of the UCLA Institute of Geophysics and Planetary Physics.

Subsatellite: The subsatellite is ejected into lunar orbit from the SIM bay and carries three experiments. The subsatellite is housed in a container resembling a rural mailbox and when deployed, is spring-ejected out-of-plane at 1.2 meters per second (4 feet per second) with a spin rate of 140 revolutions per minute. After the satellite booms are deployed, the spin rate is stabilized at about 12 rpm. The subsatellite is 77 centimeters (30 inches)long, has a 35.6 cm (14-inch hexagonal diameter and weighs 40 kg (90 pounds). The folded booms deploy to a length of 1.5 m (five feet). Subsatellite electrical power is supplied by a solar cell array outputting 24 watts for dayside operation and a rechargeable silver-cadmium battery for nightside passes.

Experiments carried aboard the subsatellite are: S-band transponder for gathering data on the lunar gravitational field, especially gravitational anomalies such as the so-called mascons particle shadows/boundary layer for gaining knowledge of the formation and dynamics of the Earth's magnetosphere, interaction of plasmas with the Moon and the physics of solar flares using telescope particle detectors and spherical electrostatic particle: detectors; and subsatellite magnetometer for gathering physical and electrical property data on the Moon and of plasma interaction with the Moon using a biaxial flux-gate magnetometer deployed on one of the three 1.5-m (5-foot) folding booms. Principal investigators for the subsatellite experiments are: particle shadows/boundary layer, Dr. Kinsey A. Anderson, University of California Berkeley; magnetometer, Dr.

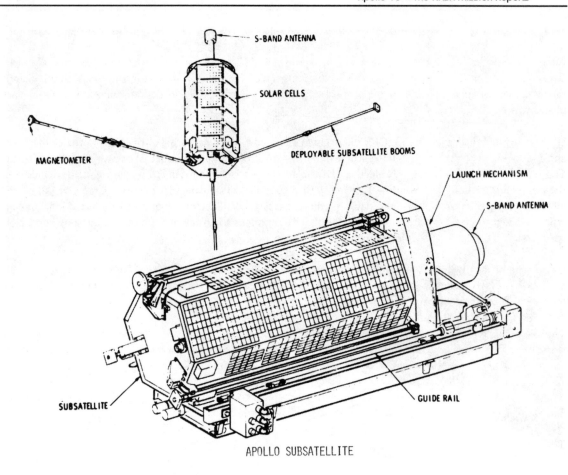

APOLLO SUBSATELLITE

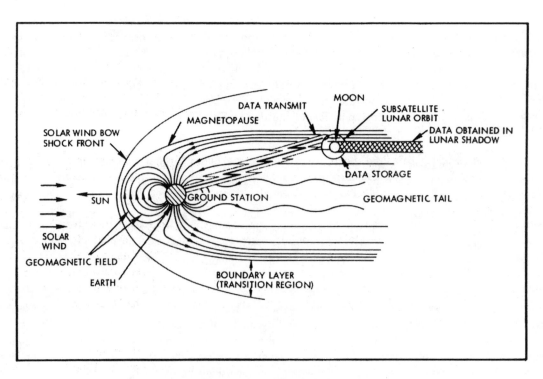

SIM Bay Subsatellite Experiments Concepts

Paul J. Coleman, UCLA; and S-band transponder, Mr. William Sjogren, Jet Propulsion Laboratory.

Other CSM orbital science experiments and tasks not in SIM bay include UV photography-Earth and moon, Gegenschein from lunar orbit, CSM/LM S-band transponder (in addition to that in the subsatellite), bistatic radar, Apollo window meteoroid experiments, microbial response, Skylab contamination, and visual flash phenomenon.

UV Photography-Earth and Moon: This experiment is aimed toward gathering ultraviolet photos of the Earth and Moon for planetary atmosphere studies and investigation of lunar surface short wavelength radiation. The photos will be made with an electric Hasselblad camera bracket-mounted in the right side window of the command module. The window is fitted with a special quartz pane that passes a large portion of the incident UV spectrum. A four-filter pack — three passing UV electromagnetic radiation and one passing visible electromagnetic radiation — is used with a 105mm lens for black and white photography; the visible spectrum filter is used with color film.

Gegenschein from Lunar Orbit: This experiment involves long exposures with a 70mm camera with 55mm f/1.2 lens on high speed black and white film (ASA 6,000). All photos must be made while the command module is in total darkness in lunar orbit.

The Gegenschein is a faint light source covering a 20° field of view along the Earth-Sun line on the opposite side of the Earth from the Sun (anti-solar axis). one theory on the origin of Gegenschein is that particles of matter are trapped at the moulton point and reflect sunlight. Moulton point is a theoretical point located 15,131,000 km (817,000 nm) from the Earth along the anti-solar axis where the sum of all gravitational forces is zero. From lunar orbit, the moulton point region can be photographed from about 15 degrees off the Earth-Sun axis, and the photos should show whether Gegenschein results from the moulton point theory or stems from zodiacal light or from some other source. The experiment was conducted on Apollo 14 and 15. The principal investigator is Lawrence Dunkelman of the Goddard Space Flight Center.

CSM/LM S-Band Transponder: The objective of this experiment is to detect variations in lunar gravity along the lunar surface track. These anomalies in gravity result in minute perturbations of the spacecraft motion and are indicative of magnitude and location of mass concentrations on the Moon. The Spaceflight Tracking and Data Network(STDN) and the Deep Space Network (DSN) will obtain and record S-band doppler tracking measurements from the docked CSM/LM and the undocked CSM while in lunar orbit; S-band doppler tracking measurements of the LM during non-powered portions of the lunar descent; and S-band doppler tracking measurements of the LM ascent stage during non-powered portions of the descent for lunar impact. The CSM and LM S-band transponders will be operated during the experiment period. The experiment was conducted on Apollo 14 and 15.

S-band doppler tracking data have been analyzed from the Lunar Orbiter missions and definite gravity variations were detected. These results showed the existence of mass concentrations (mascons) in the ringed maria. Confirmation of these results has been obtained with Apollo tracking data.

With appropriate spacecraft orbital geometry much more scientific information can be gathered on the lunar gravitational field. The CSM and/or LM in low-altitude or bits can provide new detailed information on local gravity anomalies. These data can also be used in conjunction with high-altitude data to possibly provide some description on the size and shape of the perturbing masses. Correlation of these data with photographic and other scientific records will give a more complete picture of the lunar environment and support future lunar activities. Inclusion of these results is pertinent to any theory of the origin of the Moon and the study of lunar subsurface structure.

There is also the additional benefit of obtaining better navigational capabilities for future lunar missions in that an improved lunar gravity model will be known. William Sjogren, Jet Propulsion Laboratory, Pasadena, California is principal investigator.

Bistatic Radar Experiment (BRE): The downlink bistatic radar experiment seeks to measure the electromagnetic properties of the lunar surface by monitoring that portion of the spacecraft telemetry and communications beacons which are reflected from the Moon.

The CSM S-band telemetry beacon (f = 2.2875 gigahertz), the VHF voice communications link (f = 259.7 megahertz), and the spacecraft omni-directional and high gain antennas are used in the experiment. The spacecraft is oriented so that the radio beacon is incident on the lunar surface and is successively reoriented so that the angle at which the signal intersects the lunar surface is varied. The radio signal is reflected from the surface and is monitored on Earth. The strength of the reflected signal will vary as the angle at which it intersects the surface is varied.

The electromagnetic properties of the surface can be determined by measuring the reflected signal strength as a function of angle of incidence on the lunar surface. The angle at which the reflected signal strength is a minimum is known as the Brewster angle and determines the dielectric constant. The reflected signals can also be analyzed for data on lunar surface roughness and surface electrical conductivity.

The S-band signal will primarily provide data on the surface. However, the VHF signal is expected to penetrate the gardened debris layer (regolith) of the Moon and be reflected from the underlying rock strata. The reflected VHF signal will then provide information on the depth of the regolith over the Moon.

The S-band BRE signal will be monitored by the 210-foot antenna at the Goldstone, California site and the VHF portion of the BRE signal will be monitored by the 150-foot antenna at the Stanford Research Institute in California. The experiment was flown on Apollos 14 and 15.

Lunar bistatic radar experiments were also performed using the telemetry beacons from the unmanned Lunar Orbiter I in 1966 and from Explorer 35 in 1967. Taylor Howard, Stanford University, is the principal investigator:

Apollo Window Meteoroid: This is a passive experiment in which command module windows are scanned under high magnification pre- and postflight for evidence of meteoroid cratering flux of one-trillionth gram or larger. Such particle flux may be a factor in degradation of surfaces exposed to space environment. Principal investigator is Burton Cour-Palais, NASA Manned Spacecraft Center.

MEDICAL EXPERIMENTS AND TESTS

Apollo 16 will carry four medical experiments and tests, two of which are passive and require no crew manipulation and two in which the crew participates. The experiments are: microbial response in space environment, visual light flash phencmenon, biostack and bone mineral measurement.

Microbial Response in Space Environment: A rectangular container with some 60 million microbial passengers aboard will be attached to the command module hatch camera boom for a 10 minute period during the CMP's transearth EVA. The experiment will measure the effects of reduced oxygen pressure, vacuum, zero-g and solar ultraviolet irradiation upon five strains of bacteria, fungi and viruses. Specific strains chosen for the experiment are Rhodotorula rubra, bacillus thuringiensis, bacillus subtilis, aeromanas proteolytica and chaetomium globosum. None of these bacteria are harmful to man.

The 11.6 x 11.6 x 24.8 cm (4.5 x4.5 x 9.75 inch) container—microbial ecology evaluation device (ICED) — contains three separate trays, each with 280 chambers for temperature sensors, ultraviolet measuring solutions, and the microorganisms. About two-thirds of the microorganisms will be in a dry state while the remainder will be in water suspension. The MEED will be opened and pointed toward the Sun for the 10 — minute test period, then capped and returned to Earth for analysis.

Microbial experiments run on Geminis 9 and 12, Biosatellite 2 and in the Soviet Union's Vostok spacecraft suggest that perhaps zero-g coupled with a reduced partial pressure of oxygen has an effect upon growth and mutation rates of microorganisms. Experiment principal investigators are: Dr. Paul Volz, Eastern Michigan University, Ypsilanti, Mich.; Dr. Bill G. Foster, Texas A&M University, College Station, Tex.; and Dr. John Spizien, Scripps Clinic and Research Foundation, La Jolla, Calif.

<u>Visual Light Flash Phenomenon:</u> Mysterious specks of light penetrating closed eye have been reported by crewmen of every Apollo lunar mission since Apollo 11. Usually the light specks and streaks are observed in a darkened command module cabin while the crew is in a rest period. Averaging two flashes a minute, the phenomena was observed in previous missions in translunar and transearth coast and in lunar orbit. Two theories have been proposed on the origin of the flashes. One theory is that the flashes stem from visual phosphenes induced by cosmic rays. The other theory is that Cerenkov radiation by high-energy atomic particles either enter the eyeball or ionize upon collision with the retina or cerebral cortex.

The Apollo 16 crew will run a controlled experiment during translunar and transearth coast in an effort to correlate light flashes to incident primary cosmic rays. One crewman will wear an emulsion plate device on his head called the Apollo light flash moving emulsion detector (ALFMED), while his crewmates wear eye shields. The ALFMED emulsion plates cover the front and sides of the wearer's head and will provide data on time, strength, and path of high-energy atomic particles penetrating the emulsion plates. This data will be correlated with the crewman's verbal reports on flash observations during the tests.

Dr. R. E. Benson of the NASA Manned Spacecraft Center Preventive Medicine Division is the principal investigator.

<u>Biostack:</u> The German biostack experiment is a passive experiment requiring no crew action. Selected biological material will be exposed to high-energy heavy ions in cosmic radiation and the effects analyzed postflight. Heavy ion energy measurements cannot be gathered from ground-based radiation sources.

The experiment results will add to the knowledge of how these heavy ions may present a hazard for man during long space flights.

Alternate layers of biological materials and radiation track detectors are hermetically sealed in a cylindrical aluminum container 12.5 cm in diameter by 9.8 cm high (3.4 x 4.8 inches) and weighing 2.4 kg (5.3 lbs). The biostack container will be mounted in the command module in a position to minimize shielding to cosmic radiation. This container has no organisms harmful to man.

After command module recovery, the container will be returned to the principal investigator, Dr. Horst Bucker, University of Frankfurt, Frankfurt am main, Federal Republic of Germany, whose participation is funded and sponsored by the German Ministry for Education and Science. Dr. Bucker will be assisted in his investigations by scientists at the University of Strasbourg, France, and at the Aerospace Medicine Research Center in Paris. The experiment is endorsed by the working Party on Space Biophysics of the Council of Europe's parliamentary Committee on Science and Technology.

<u>Bone Mineral Measurement:</u> Mineral changes in human bones caused by reduced gravity are measured in this passive experiment. X-ray absorption techniques pre- and post flight will measure bone mineral content of the radius, ulna, and os calcis (heel) for comparisons.

Principal investigator is Dr. J. M. Vogel, U.S. Public Health Service Hospital, San Francisco, Calif.

ENGINEERING OPERATIONAL TESTS AND DEMONSTRATION

Four tests aimed at gathering data for improving operations in future space flight missions will be flown aboard Apollo 16. An investigation into external spacecraft contamination and an evaluation of the Skylab food packaging system will provide data useful to next year's Skylab series of long-duration missions. Tests of an improved gas/water separator for removing hydrogen gas bubbles from potable water generated as a fuel cell byproduct and evaluation of a new type of fecal collection bag also will be conducted. An electrophoresis separation demonstration will also be conducted.

Skylab Contamination Study: Since John Glenn reported seeing fireflies outside a tiny window of his Mercury spacecraft Friendship 7 a decade ago, space crews have noted light-scattering particles that hinder visual observations as well as photographic tasks. Opinions on the origin of the clouds of particles surrounding spacecraft have ranged from ice crystals generated by water vapor dumped from the spacecraft to a natural phenomenon of particles inhabiting the space environment.

The phenomenon could be of concern in the Skylab missions during operation of the solar astronomy experiments. The Apollo 16 tests will attempt to further identify the sources of contamination outside the spacecraft. The crew will use a combination of photography and visual observations to gather the data, which in turn will serve as baselines for predicting contamination in the vicinity of the Skylab space station and for devising methods for minimizing contaminate levels around the Skylab vehicle.

Skylab Food Package: Since the Skylab orbital workshop must carry enough food to last nine men for 140 days, the methods of packaging the food must be somewhat different than the methods used in packaging food for the relatively short Apollo missions.

Packed in with the Apollo 16 crew's regular fare of meals will be several Skylab food items for the crew to evaluate. Among the types of packaging to be checked are two sizes of snap-top cans containing foods ranging from dried peaches and puddings to peanuts, wet-pack spoon-bowl foods, postage-stamp size salt dispensers, and plastic-bellows drink containers.

The Apollo 16 crew will take still and motion pictures of the food packages in use and make subjective comments on ease of handling and preparation of the food.

Improved Gas/Water Separator: During transearth coast the Apollo 16 crew will unstow an improved gas/water separator and use it on the food preparation water spigot for one meal period. Command module potable water is a byproduct of the fuel cell powerplants and contains hydrogen gas bubbles which cause crew discomfort when ingested. Gas/water separators flown previously have had leaks around the seals and have not adequately separated gas bubbles from the water.

The improved separator is expected to deliver bubblefree drinking and food preparation water.

Improved Fecal Collection Bag: Defecation bags used in previous Apollo missions are virtually identical to the Gemini "blue bags". During previous missions, the crews have experienced some difficulty in attaching the bag to the body, in post-defecation sealing, and personal hygiene clean-up. The improved fecal collection bag has been redesigned for better fit and sealing.

Electrophoretic Separation Demonstration: The Apollo 16 astronauts will use the natural weightless environment in space flight to again demonstrate an improved process in materials purification which may permit separation of materials of higher purity than can now be produced on Earth.

The space demonstration is called electrophoretic separation, referring to the movement of electrically charged particles under the influence of an electric field.

Most organic molecules become electrically charged when they are placed in a solution and will move through such a solution if an electric field is applied. Since molecules of different size and shape move at different speeds, the faster molecules in a mixture will "outrun" the slower ones as they move from one end of the solution to the other.

Thus, like particles separate into bands as they drift towards the attracting electrode. Separation is accomplished by removing the desired bands of particles.

On Earth the purification or separation process is generally performed with filters or other supporting devices because materials of any significant size rapidly sink or settle in the solution before separation occurs.

Additionally, light materials remaining are easily mixed by convective currents, thus reducing purity.

The Apollo 14 electrophoretic separation demonstration experienced some technical problems which prohibited obtaining information on two of the three samples. Results with the third sample (a mixture of red and blue dyes) indicated the feasibility of obtaining sharper separations in space than on Earth. The Apollo 16 demonstration has been modified to overcome the problems encountered with the Apollo 14 demonstration and will use polystyrene particles to simulate separation of large biological particles. The Apollo 16 demonstration will allow a more accurate evaluation of the separation process.

If successful, the demonstration could show that one of the practical uses for future manned space stations may be to economically produce vaccine and medical preparations of very high purity.

LUNAR ROVING VEHICLE

The lunar roving vehicle (LRV) will transport two astronauts on three exploration traverses of the Moon's Descartes region during the Apollo 16 mission. The LRV will also carry tools, scientific and communications equipment, and lunar samples.

The four-wheel, lightweight vehicle greatly extends the lunar area that can be explored by man. It is the first manned surface transportation system designed to operate on the Moon, and it represents a solution to challenging new problems without precedent in Earth-bound vehicle design and operation.

The LRV must be folded into a small package within a wedge-shaped storage bay of the lunar module descent stage for transport to the Moon. After landing, the vehicle must be unfolded from its stowed position and deployed on the surface. It must then operate in an almost total vacuum under extremes of surface temperatures, low gravity, and on unfamiliar terrain.

The first lunar roving vehicle, used on the Apollo 15 lunar mission, was driven for three hours during its exploration traverses, covering a distance of 27.8 kilometers (17.3 statute miles) at an average speed of 9.2 kilometers an hour (5.7 miles an hour).

General Description

The LRV is 3.1 meters long (10.2 feet); has a 1.8 meter (six-foot) tread width; is 1.14 meters high (44.8 inches); and has a 2.3-meter wheel base (7.5 feet).

Each wheel is powered by a small electric motor. The maximum speed reached on the Apollo 15 mission was about 13 km/hr (eight mph).

Two 36-volt batteries provide vehicle power, and either battery can run all systems. The front and rear wheels have separate steering systems; if one fails it can be disconnected and the LRV will operate with the other system.

Weighing about 208 kilograms (457 pounds), Earth weight when deployed on the Moon, the LRV can carry a total payload of about 490 kilograms (1,080 pounds), more than twice its own weight. The payload includes two astronauts and their portable life support systems (about 363 kilograms; 800 pounds), 45.4 kilograms (100 pounds) of communications equipment, 54.5 kilograms (120 pounds) of scientific equipment and photographic gear, and 27.2 kilograms (60 pounds) of lunar samples.

The LRV is designed to operate for a minimum of 78 hours during the lunar day. It can make several exploration sorties to a cumulative distance of 92 kilometers (57 miles). The maximum distance the LRV will be permitted to range from the lunar module will be approximately 9.7 kilometers (six miles), the distance the crew could safely walk back to the LM in the unlikely event of a total LRV failure. This walkback distance limitation is based upon the quantity of oxygen and coolant available in the astronaut's portable life support

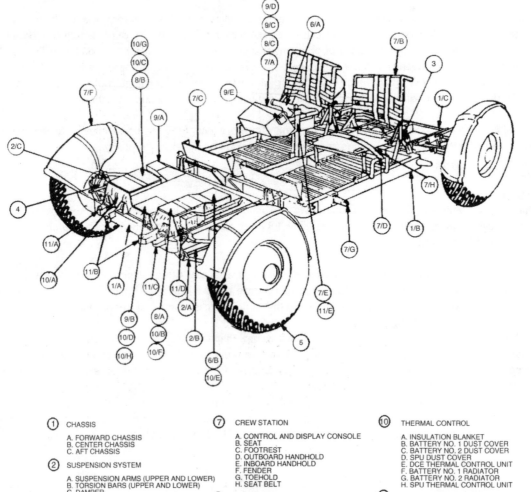

① CHASSIS

    A. FORWARD CHASSIS
    B. CENTER CHASSIS
    C. AFT CHASSIS

② SUSPENSION SYSTEM

    A. SUSPENSION ARMS (UPPER AND LOWER)
    B. TORSION BARS (UPPER AND LOWER)
    C. DAMPER

③ STEERING SYSTEM (FORWARD AND AFT)

④ TRACTION DRIVE

⑤ WHEEL

⑥ DRIVE CONTROL

    A. HAND CONTROLLER
    B. DRIVE CONTROL ELECTRONICS (DCE)

⑦ CREW STATION

    A. CONTROL AND DISPLAY CONSOLE
    B. SEAT
    C. FOOTREST
    D. OUTBOARD HANDHOLD
    E. INBOARD HANDHOLD
    F. FENDER
    G. TOEHOLD
    H. SEAT BELT

⑧ POWER SYSTEM

    A. BATTERY #1
    B. BATTERY #2
    C. INSTRUMENTATION

⑨ NAVIGATION

    A. DIRECTIONAL GYRO UNIT (DGU)
    B. SIGNAL PROCESSING UNIT (SPU)
    C. INTEGRATED POSITION INDICATOR (IPI)
    D. SUN SHADOW DEVICE
    E. VEHICLE ATTITUDE INDICATOR

⑩ THERMAL CONTROL

    A. INSULATION BLANKET
    B. BATTERY NO. 1 DUST COVER
    C. BATTERY NO. 2 DUST COVER
    D. SPU DUST COVER
    E. DCE THERMAL CONTROL UNIT
    F. BATTERY NO. 1 RADIATOR
    G. BATTERY NO. 2 RADIATOR
    H. SPU THERMAL CONTROL UNIT

⑪ PAYLOAD INTERFACE

    A. TV CAMERA RECEPTACLE
    B. LCRU RECEPTACLE
    C. HIGH GAIN ANTENNA RECEPTACLE
    D. AUXILIARY CONNECTOR
    E. LOW GAIN ANTENNA RECEPTACLE

LRV WITHOUT STOWED PAYLOAD

systems. This area contains about 292 square kilometers (1.13 square miles) available for investigation, 10 times the area that can be explored on foot.

The vehicle can negotiate obstacles 30.5 centimeters (one foot) high, and cross crevasses 70 centimeters wide (28 inches). The fully loaded vehicle can climb and descend slopes as steep as 25 degrees, and park on slopes up to 35 degrees. Pitch and roll stability angles are at least 45 degrees, and the turn radius is three meters (10 feet).

Both crewmen sit so the front wheels are visible during normal driving. The driver uses an on-board dead reckoning navigation system to determine direction and distance from the lunar module, and total distance traveled at any point during a traverse.

The LRV has five major systems: mobility, crew station, navigation, power, and thermal control. Secondary systems include the deployment mechanism, LM attachment equipment, and ground support equipment.

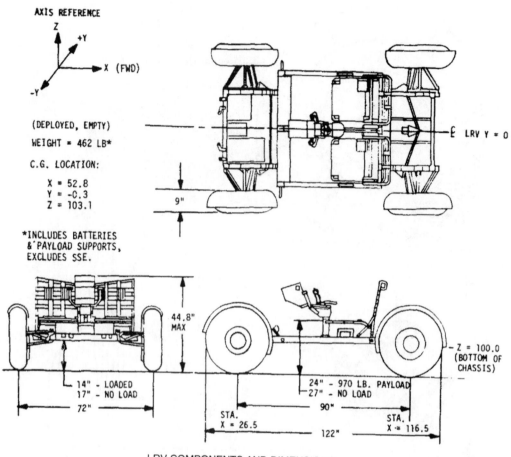

LRV COMPONENTS AND DIMENSIONS

The aluminum chassis is divided into three sections that support all equipment and systems. The forward and aft sections fold over the center one for stowage in the LM. The forward section holds both batteries, part of the navigation system, and electronics gear for the traction drive and steering systems. The center section holds the crew station with its two seats, control and display console, and hand controller. The floor of beaded aluminum panels can support the weight of both astronauts standing in lunar gravity. The aft section holds the scientific payload.

Auxiliary LRV equipment includes the lunar communications relay unit (LCRU) and its high and low gain antennas for direct communications with Earth, the ground commanded television assembly (GCTA), a motion picture camera, scientific equipment, tools, and sample stowage bags.

Mobility System

The mobility system is the major LRV system, containing the wheels, traction drive, suspension, steering, and drive control electronics subsystems.

The vehicle is driven by a T-shaped hand controller located on the control and display console post between the crewmen. Using the controller, the astronaut maneuvers the LRV forward, reverse, left and right.

Each LRV wheel has a spun aluminum hub and a titanium bump stop (inner frame) inside the tire (outer frame). The tire is made of a woven mesh of zinc-coated piano wire to which titanium treads are riveted in a chevron pattern around the outer circumference. The bump stop prevents excessive inflection of the mesh tire during heavy impact. Each wheel weighs 5.4 kilograms (12 pounds) on Earth and is designed to be driven

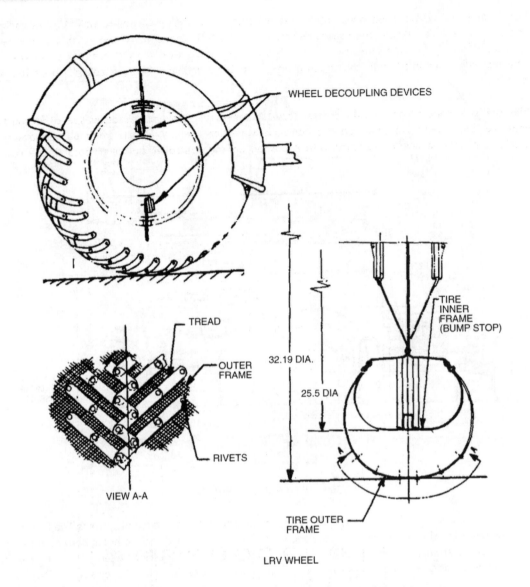

WHEEL DECOUPLING DEVICES

TREAD

OUTER
FRAME

RIVETS

VIEW A-A

32.19 DIA.

25.5 DIA

TIRE
INNER
FRAME
(BUMP STOP)

TIRE OUTER
FRAME

LRV WHEEL

at least 180 kilometers (112 miles). The wheels are 81.3 centimeters (32 inches) in diameter, and 22.9 centimeters (nine inches) wide.

A traction drive attached to each wheel has a motor harmonic drive gear unit, and a brake assembly. The harmonic drive reduces motor speed at an 80-to-1 rate for continuous operation at all speeds without gear shifting. The drive has an odometer pickup (measuring distance traveled) that sends data to the navigation system. Each motor develops 0.18 kilowatt (1/4-horsepower) and operates from a 36-volt input.

Each wheel has a mechanical brake connected to the hand controller. Moving the controller rearward de-energizes the drive motors and forces brake shoes against a drum, stopping wheel hub rotation. Full rear movement of the controller engages and locks a parking brake.

The chassis is suspended from each wheel by two parallel arms mounted on torsion bars and connected to each traction drive. Tire deflection allows a 35.6-centimeter (14-inch) ground clearance when the vehicle is fully loaded, and 43.2 centimeters (17 inches) when unloaded.

Both front and rear wheels have independent steering systems that allow a "wall-to-wall" turning radius of 3.1 meters (122 inches), exactly the vehicle length. If either set of wheels has a steering failure, its steering system can be disengaged and the traverse can continue with the active steering assembly. Each wheel can also be manually uncoupled from the traction drive and brake to allow "free wheeling" about the drive housing.

Pushing the hand controller forward increases forward speed; rear movement reduces speed. Forward and reverse are controlled by a knob on the controller's vertical stem. With the knob pushed down, the controller can only be pivoted forward; with it pushed up, the controller can be pivoted to the rear for reverse.

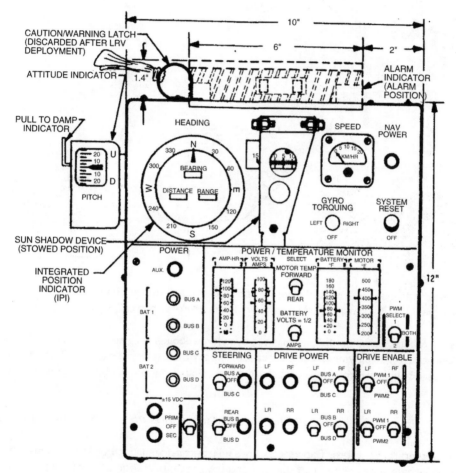

LRV CREW STATION COMPONENTS - CONTROL AND DISPLAY CONSOLE

Crew Station

The crew station consists of the control and display console, seats, seat belts, an armrest, footrests, inboard and outboard handholds, toeholds, floor panels, and fenders.

The control and display console is separated into two main parts: The top portion holds navigation system displays; the lower portion contains monitors and controls. Attached to the upper left side of the console is an attitude indicator that shows vehicle pitch and roll.

At the console top left is a position indicator. Its outer circumference is a large dial that shows vehicle heading (direction) with respect to lunar north. Inside the dial are three digital indicators that show bearing and range to the LM and distance traveled by the LRV. In the middle of the console upper half is a Sun compass device that is used to update the LRV's navigation system. Down the left side of the console lower half are control

HAND CONTROLLER OPERATION:

T-HANDLE PIVOT FORWARD - INCREASED DEFLECTION FROM NEUTRAL INCREASES FORWARD SPEED.

T-HANDLE PIVOT REARWARD - INCREASED DEFLECTION FROM NEUTRAL INCREASES REVERSE SPEED.

T-HANDLE PIVOT LEFT - INCREASED DEFLECTION FROM NEUTRAL INCREASES LEFT STEERING ANGLE.

T-HANDLE PIVOT RIGHT - INCREASED DEFLECTION FROM NEUTRAL INCREASES RIGHT STEERING ANGLE.

T-HANDLE DISPLACED REARWARD - REARWARD MOVEMENT INCREASES BRAKING FORCE. FULL 3 INCH
REARWARD APPLIES PARKING BRAKE. MOVING INTO BRAKE
POSITION DISABLES THROTTLE CONTROL AT 15° MOVEMENT
REARWARD.

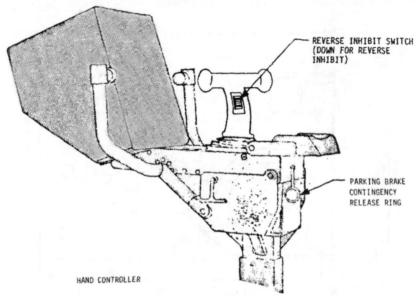

REVERSE INHIBIT SWITCH
(DOWN FOR REVERSE
INHIBIT)

PARKING BRAKE
CONTINGENCY
RELEASE RING

HAND CONTROLLER

switches for power distribution, drive and steering, and monitors for power and temperature. A warning flag atop the console pops up if a drive motor goes above limits in either battery.

The LRV seats are tubular aluminum frames spanned by nylon webbing. They are folded flat during launch and erected by crewmen after deployment. The seat backs support the astronaut portable life support systems. Nylon webbing seat belts, custom fitted to each crewman, snap over the outboard handholds with metal hooks.

The armrest, located directly behind the LRV hand controller, supports the arm of the driving crewman.

The footrests, attached to the center floor section, are adjusted before launch to fit each crewman. Inboard handholds help crewmen get in and out of the LRV, and have receptacles for a 16 mm camera and the low gain antenna of the LCRU.

The lightweight, fiberglass fenders keep lunar dust from being thrown on the astronauts, their equipment, sensitive vehicle parts, and from obstructing vision while driving. Front and rear fender sections are retracted during flight and extended by crewmen after LRV deployment on the lunar surface.

Navigation System

The navigation system is based on the principle of starting a sortie from a known point, recording speed, direction and distance traveled, and periodically calculating vehicle position.

The system has three major components: a directional gyroscope to provide vehicle headings; odometers on each wheel's traction drive unit to give speed and distance data; and a signal processing unit (a small, solid-state computer) to determine heading, bearing, range, distance traveled, and speed.

All navigation system readings are displayed on the control console. The system is reset at the beginning of each traverse by pressing a system reset button that moves all digital displays and internal registers to zero.

The directional gyroscope is aligned by measuring the inclination of the LRV (using the attitude indicator) and measuring vehicle orientation with respect to the Sun (using the Sun compass). This information is relayed to ground controllers and the gyro is adjusted to match calculated values read back to the crew.

Each LRV wheel revolution generates odometer magnetic pulses that are sent to the console displays.

Power System

The power system consists of two 36-volt, non-rechargeable batteries and equipment that controls and monitors electrical power. The batteries are in magnesium cases, use plexiglass monoblock (common cell walls) for internal construction, and have silver-zinc plates in potassium hydroxide electrolyte. Each battery has 23 cells and a 121-ampere-hour capacity.

Both batteries are used simultaneously with an approximately equal load during LRV operation. Each battery can carry the entire electrical load; if one fails, its load can be switched to the other.

The batteries are activated when installed on the LRV at the launch pad about five days before launch. During LRV operation all mobility system power is turned off if a stop exceeds five minutes, but navigation system power remains on throughout each sortie. The batteries normally operate at temperatures of 4.4 to 51.7 degrees C. (40-125 degrees F.).

An auxiliary connector at the LRV's forward end supplies 150 watts of 36-volt power for the lunar communications relay unit.

Thermal Control

The basic concept of LRV thermal control is heat storage during vehicle operation and radiation cooling when it is parked between sorties. Heat is stored in several thermal control units and in the batteries. Space radiators are protected from dust during sorties by covers that are opened at the end of each sortie when battery temperatures cool to about 7.2 degrees C. (45 degrees F.), the covers automatically close.

A multi-layer insulation blanket protects forward chassis components. Display console instruments are mounted to an aluminum plate isolated by radiation shields and fiber glass mounts. Console external surfaces are coated with thermal control paint and the face plate is anodized, as are handholds, footrests, tubular seat sections, and center and aft floor panels.

Stowage and Deployment

Space support equipment holds the folded LRV in the lunar module during transit and deployment at three attachment points with the vehicle's aft end pointing up.

Deployment is essentially manual. One crewman releases a cable attached to the top (aft end) of the folded LRV as the first step in the deployment. This cable is held taut during deployment, until all four LRV wheels are on the lunar surface.

One of the crewmen then ascends the LM ladder part way and pulls a D-ring on the side of the descent stage. This releases the LRV, and lets the vehicle swing out at the top about 12.7 centimeters (five inches) until it is stopped by two steel cables. Descending the ladder, the crewman walks to the LRV's right side, takes the end of a deployment tape from a stowage bag, and pulls the tape hand-over-hand. This unreels two support cables that swivel the vehicle outward from the top. As the aft chassis is unfolded, the aft wheels

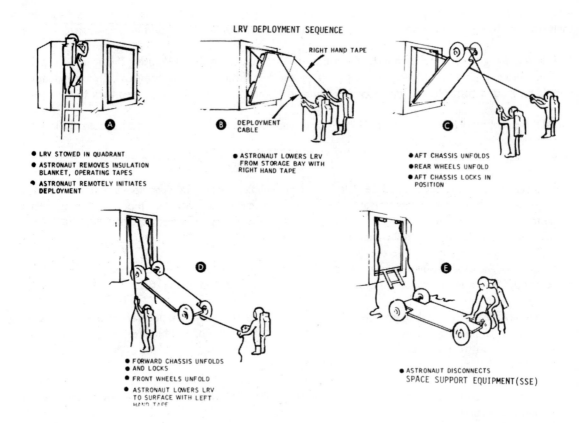

LRV DEPLOYMENT SEQUENCE

A
- LRV STOWED IN QUADRANT
- ASTRONAUT REMOVES INSULATION BLANKET, OPERATING TAPES
- ASTRONAUT REMOTELY INITIATES DEPLOYMENT

B DEPLOYMENT CABLE
RIGHT HAND TAPE
- ASTRONAUT LOWERS LRV FROM STORAGE BAY WITH RIGHT HAND TAPE

C
- AFT CHASSIS UNFOLDS
- REAR WHEELS UNFOLD
- AFT CHASSIS LOCKS IN POSITION

D
- FORWARD CHASSIS UNFOLDS AND LOCKS
- FRONT WHEELS UNFOLD
- ASTRONAUT LOWERS LRV TO SURFACE WITH LEFT HAND TAPE

E
- ASTRONAUT DISCONNECTS SPACE SUPPORT EQUIPMENT(SSE)

automatically unfold and deploy, and all latches are engaged. The crewman continues to unwind the tape, lowering the LRV's aft end to the surface, and the forward chassis and wheels spring open and into place.

When the aft wheels are on the surface, the crewman removes the support cables and walks to the vehicle's left side. There he pulls a second tape that lowers the LRV's forward end to the surface and causes telescoping tubes to push the vehicle away from the LM. The two crewmen then deploy the fender extensions, set up the control and display console, unfold the seats, and deploy other equipment.

One crewman will board the LRV and make sure all controls are working. He will back the vehicle away slightly and drive it to the LM quadrant that holds the auxiliary equipment. The LRV will be powered down while the crewmen load auxiliary equipment aboard the vehicle.

LUNAR COMMUNICATIONS RELAY UNIT (LCRU)

The range from which an Apollo crew can operate from the lunar module during EVAS while maintaining contact with the Earth is extended over the lunar horizon by a suitcase-size device called the lunar communications relay unit (LCRU). The LCRU acts as a portable relay station for voice, TV and telemetry directly between the crew and Mission Control Center instead of through the lunar module communications system. First use of the LCRU was on Apollo 15.

Completely self-contained with its own power supply and erectable hi-gain S-Band antenna, the LCRU may be mounted on a rack at the front of the lunar roving vehicle (LRV) or handcarried by a crewman. In addition to providing communications relay, the LCRU receives ground-command signals for the ground commanded television assembly (GCTA) for remote aiming and focusing the lunar surface color television camera. The GCTA is described in another section of this press kit.

Between stops with the lunar roving vehicle, crew voice is beamed Earthward by a wide beam-width helical S-Band antenna. At each traverse stop, the crew must boresight the high-gain parabolic antenna toward Earth

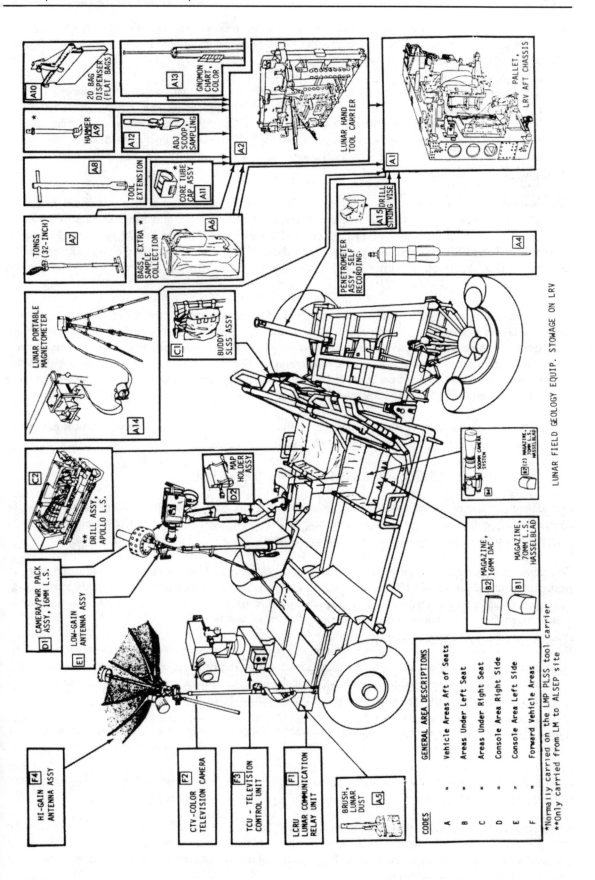

LUNAR FIELD GEOLOGY EQUIP. STOWAGE ON LRV

A10 20 BAG DISPENSER (FLAT BAGS)

A13 GNOMON CHART, COLOR

A2 LUNAR HAND TOOL CARRIER

A1 PALLET, LRV AFT CHASSIS

* HAMMER A9

A12 ADJ SCOOP SAMPLING

A8 TOOL EXTENSION

* CORE TUBE CAP ASSY. A11

A15 DRILL STRING VISE

A7 TONGS (32-INCH)

A6 BAGS, EXTRA SAMPLE COLLECTION *

PENETROMETER ASSY., SELF RECORDING. A4

LUNAR PORTABLE MAGNETOMETER

C1 BUDDY SLSS ASSY

A14

C2 ** DRILL ASSY, APOLLO L.S.

D2 MAP HOLDER ASSY

500MM CAMERA SYSTEM B4

B3 (2) MAGAZINE, 70MM L.S. HASSELBLAD

D1 CAMERA/PWR PACK ASSY, 16MM L.S.

E1 LOW-GAIN ANTENNA ASSY

B2 MAGAZINE, 16MM DAC

B1 MAGAZINE, 70MM L.S. HASSELBLAD

F4 HI-GAIN ANTENNA ASSY

F2 CTV - COLOR TELEVISION CAMERA

F3 TCU - TELEVISION CONTROL UNIT

F1 LCRU LUNAR COMMUNICATION RELAY UNIT

A5 BRUSH, LUNAR DUST

CODES	GENERAL AREA DESCRIPTIONS
A	= Vehicle Areas Aft of Seats
B	" Areas Under Left Seat
C	" Areas Under Right Seat
D	" Console Area Right Side
E	" Console Area Left Side
F	" Forward Vehicle Areas

*Normally carried on the LMP PLSS tool carrier
**Only carried from LM to ALSEP site

before television signals can be transmitted. VHF signals from the crew portable life support system (PLSS) transceivers are converted to S-band by the LCRU for relay to the ground, and conversely, from S-Band to VHF on the uplink to the EVA crewmen.

The LCRU measures 55.9 x 40.6 x 15.2 cm (22 x 16 x 6 inches) not including antennas, and weighs 25 Earth kg (55 Earth pounds) (9.2 lunar pounds). A protective thermal blanket around the LCRU can be peeled back to vary the amount of radiation surface which consists of 1.26 m² (196 square inches) of radiating mirrors to reflect solar heat. Additionally, wax packages on top of the LCRU enclosure stabilize the LCRU temperature by a melt-freeze cycle. The LCRU interior is pressurized to 7.5 psia differential. (one-half atmosphere).

Internal power is provided to the LCRU by a 19-cell silver-zinc battery with a potassium hydroxide electrolyte. The battery weighs 4.1 kg (nine Earth pounds)(1.5 lunar pounds) and measures 11.8 x 23.9 x 11.8 cm (4.7 x 9.4 x 4.65 inches). The battery is rated at 400 watt hours, and delivers 29 volts at a 3.1-ampere current load. The LCRU may also be operated from the LRV batteries. The nominal plan is to operate the LCRU using LRV battery power during EVA-1 and EVA-2. The LCRU battery will provide the power during EVA-3.

Three types of antennas are fitted to the LCRU system: a low-gain helical antenna for relaying voice and data when the LRV is moving and in other instances when the high-gain antenna is not deployed; a .9 m (three-foot) diameter parabolic rib-mesh high-gain antenna for relaying a television signal; and a VHF omni-antenna for receiving crew voice and data from the PLSS transceivers. The high-gain antenna has an optical sight which allows the crewman to boresight on Earth for optimum signal strength. The Earth subtends one-half degree angle when viewed from the lunar surface.

The LCRU can operate in several modes: mobile on the LRV, fixed base such as when the LRV is parked, or handcarried in contingency situations such as LRV failure.

TELEVISION AND GROUND COMMANDED TELEVISION ASSEMBLY

Two different color television cameras will be used during the Apollo 16 mission. One, manufactured by Westinghouse, will be used in the command module. It will be fitted with a 5 cm (2 in) black and white monitor to aid the crew in focus and exposure adjustment.

The other camera, manufactured by RCA, is for lunar surface use and will be operated from three different positions-mounted on the LM MESA, mounted on a tripod and connected to the LM by a 30.5 m (100 ft) cable, and installed on the LRV with signal transmission through the lunar communication relay unit rather than through the LM communications system as in the other two positions.

While on the LRV, the camera will be mounted on the ground commanded television assembly (GCTA). The camera can be aimed and controlled by astronauts or it can be remotely controlled by personnel located in the Mission Control Center. Remote command capability includes camera "on" and "off", pan, tilt, zoom, iris open/closed (f2.2 to f22) and peak or average automatic light control.

The GCTA is capable of tilting the TV camera upward 85 degrees, downward 45 degrees, and panning the camera 340 degrees between mechanical stops. Pan and tilt rates are three degrees per second.

The TV lens can be zoomed from a focal length of 12.5mm to 75mm corresponding to a field of view from three to nine degrees.

At the end of the third EVA, the crew will park the LRV about 91.4 m (300 ft) east of the LM so that the color TV camera can cover the LM ascent from the lunar surface. Because of a time delay in a signal going the quarter million miles out to the Moon, Mission Control must anticipate ascent engine ignition by about two seconds with the tilt command.

The GCTA and camera each weigh approximately 5.9 kg (13 lb). The overall length of the camera is 46 cm (18.0 in) its width is 17 cm (6.7 in), and its height is 25 cm (10 in). The GCTA and LCRU are built by RCA.

APOLLO 16 TELEVISION EVENTS

DATE		TIME (GET)	TIME (EST)	DURATION (HRS:MIN)	EVENT
APRIL	16	3:05	1559	0:19	TD & E
APRIL	20	102:25	1919	6:48	EVA-1
APRIL	21	124:50	1744	6:30	EVA-2
APRIL	22	148:25	1719	6:40	EVA-3
APRIL	23	171:40	1634	0:14	LM LIFT-OFF
		173:20	1814	0:06	RENDEZVOUS
		173:44	1838	0:07	DOCKING
APRIL	26	241:55	1449	1:10	CMP EVA
TBD					TRANSEARTH COAST

PHOTOGRAPHIC EQUIPMENT

Still and motion pictures will be made of most spacecraft maneuvers and crew lunar surface activities. During lunar surface operations, emphasis will be on documenting placement of lunar surface experiments, documenting lunar samples, and on recording in their natural state the lunar surface features.

Command module lunar orbit photographic tasks and experiments include high-resolution photography to support future landing missions, photography of surface features of special scientific interest and astronomical phenomena such as solar corona, Gegenschein, zodiacal light, libration points, and galactic poles.

Camera equipment stowed in the Apollo 16 command module consists of one 70mm Hasselblad electric camera, a 16mm Maurer motion picture camera, and a 35mm Nikon F single-lens reflex camera. The command module Hasselblad electric camera is normally fitted with an 80mm f/2.8 Zeiss Planar lens, but a bayonet-mount 250mm lens can be fitted for long-distance Earth/Moon photos. A 105mm f/4.3 Zeiss UV Sonnar is provided for the ultraviolet photography experiment.

The 35mm Nikon F is fitted with a 55mm f/1.2 lens for the Gegenschein and dim-light photographic experiments.

The Maurer 16mm motion picture camera in the command module has lenses of 10, 18 and 75mm focal length available. Accessories include a right-angle mirror, a power cable and a sextant adapter which allows the camera to film through the navigation sextant optical system.

Cameras stowed in the lunar module are two 70mm Hasselblad data cameras fitted with 60mm Zeiss Metric lenses, an electric Hasselblad with 500mm lens and two 16mm Maurer motion picture cameras with 10mm lenses. One of the Hasselblads and one of the motion picture cameras are stowed in the modular equipment stowage assembly (MESA) in the LM descent stage.

The LM Hasselblads have crew chest mounts that fit dovetail brackets on the crewman's remote control unit, thereby leaving both hands free. One of the LM motion picture cameras will be mounted in the right-hand window to record descent, landing, ascent and rendezvous. The 16mm camera stowed in the MESA will be carried aboard the lunar roving vehicle to record portions of the three EVAs. Descriptions of the 24-inch panoramic camera and the 3-inch mapping/stellar camera are in the orbital science section of this press kit.

TV AND PHOTOGRAPHIC EQUIPMENT

NOMENCLATURE	CSM AT LAUNCH	LM AT LAUNCH	CM TO LM	LM TO CM	CM AT ENTRY
TV, COLOR, ZOOM LENS (MONITOR WITH CM SYSTEM)	1	1			1
CAMERA, DATA ACQUISITION, 16 MM	1	1			1
LENS - 10 MM	1	1			1
- 18 MM	1				1
- 75 MM	1				1
FILM MAGAZINES	13				13
CAMERA, 35 MM NIKON	1				1
LENS - 55 MM	1				1
CASSETTE, 35 MM	9				9
CAMERA, 16 MM, BATTERY OPERATED (LUNAR SURFACE)		1			
LENS - 10 MM		1			
FILM MAGAZINES	8		8	8	8
CAMERA, HASSELBLAD, 70 MM ELECTRIC	1				1
LENS - 80 MM	1				1
- 250 MM	1				1
- 105 MM UV (4 BAND-PASS FILTERS)	1				1
FILM MAGAZINES	7				7
FILM MAGAZINE, 70 MM UV	1				1
CAMERA, HASSELBLAD ELECTRIC DATA (LUNAR SURFACE)		2			
LENS - 60 MM		2			
FILM MAGAZINES	11		11	11	11
POLARIZING FILTER		1			
CAMERA, 24-IN. PANORAMIC (IN SIM)	1				
FILM MAGAZINE (EVA TRANSFER)	1				1
CAMERA, LUNAR SURFACE ELECTRIC		1			
LENS - 500 MM		1			
FILM MAGAZINES	2		2	2	2
CAMERA, 3-IN MAPPING STELLAR(SIM)	1				
FILM MAGAZINE CONTAINING 5-IN. MAPPING AND 35 MM STELLAR FILM (EVA TRANSFER)	1				1
CAMERA, ULTRAVIOLET, LUNAR SURFACE		1			
FILM MAGAZINE, UV, LS		1		1	1

ASTRONAUT EQUIPMENT

Space Suit

Apollo crewmen wear two versions of the Apollo space suit: the command module pilot version (A-7.LB-CMP) for intravehicular operations in the command module and for extravehicular operations during SIM bay film retrieval during transearth coast; and the extravehicular version (A-7LB—EV) worn by the commander and lunar module pilot for lunar surface EVAs.

The A-7LB-EV suit differs from Apollo suits flown prior to Apollo 15 by having a waist joint that allows greater mobility while the suit is pressurized — stooping down for setting up lunar surface experiments, gathering samples and for sitting on the lunar roving vehicle.

From the inside out, an integrated thermal meteroid suit cover layer worn by the commander and lunar module pilot starts with rubber-coated nylon and progresses outward with layers of nonwoven Dacron, aluminized Mylar film and Beta marquisette for thermal radiation protection and thermal spacers, and finally with a layer of nonflammable Teflon-coated Beta cloth and an abrasion-resistant layer of Teflon fabric - a total of 18 layers.

Both types of the A-7LB suit have a pressure retention portion called a torso limb suit assembly consisting of neoprene coated nylon and an outer structural restraint layer.

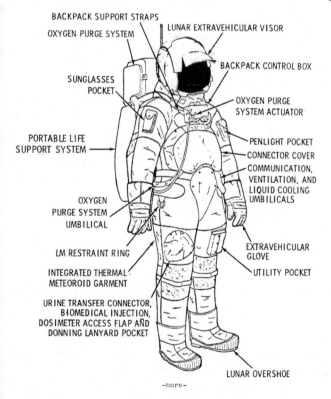

BACKPACK SUPPORT STRAPS
OXYGEN PURGE SYSTEM
LUNAR EXTRAVEHICULAR VISOR
BACKPACK CONTROL BOX
SUNGLASSES POCKET
OXYGEN PURGE SYSTEM ACTUATOR
PORTABLE LIFE SUPPORT SYSTEM
PENLIGHT POCKET
CONNECTOR COVER
COMMUNICATION, VENTILATION, AND LIQUID COOLING UMBILICALS
OXYGEN PURGE SYSTEM UMBILICAL
LM RESTRAINT RING
EXTRAVEHICULAR GLOVE
INTEGRATED THERMAL METEROID GARMENT
UTILITY POCKET
URINE TRANSFER CONNECTOR, BIOMEDICAL INJECTION, DOSIMETER ACCESS FLAP AND DONNING LANYARD POCKET
LUNAR OVERSHOE

-more-

EXTRAVEHICULAR MOBILITY UNIT

The space suit with gloves, and dipped rubber convolutes which serve as the pressure layer liquid cooling garment, portable life support system (PLSS), oxygen purge system, lunar extravehicular visor assembly, and lunar boots make up the extravehicular mobility unit (EMU). The EMU provides an extravehicular crewman with life support for a seven-hour mission outside the lunar module without replenishing expendables.

Lunar extravehicular visor assembly: The assembly consists of polycarbonate shell and two visors with thermal control and optical. coatings on them. The EVA visor is attached over the pressure helmet to provide impact, micrometeoroid, thermal and ultraviolet-infrared light protection to the EVA crewmen.

After Apollo 12, a sunshade was added to the outer portion of the LEVA in the middle portion of the helmet rim.

Extravehicular gloves: Built of an outer shell of Chromel-R fabric and thermal insulation the gloves provide protection when handling extremely hot and cold objects. The finger tips are made of silicone rubber to provide more sensitivity.

Constant-wear garment: A one-piece constant-wear garment, similar to "long johns" is worn as an undergarment for the space suit in intravehicular-and on CSM EV operations, and with the inflight coveralls. The garment is porous-knit cotton with a waist-to neck zipper for donning. Biomedical harness attach points are provided.

<u>Liquid-cooling garment</u> - The knitted nylon-spandex garment includes a network of plastic tubing through which cooling water from the PLSS is circulated. It is worn next to the skin and replaces the constant-wear garment during lunar surface EVA.

<u>Portable life support system (PLSS)</u> - The backpack supplies oxygen at 3.7 psi and cooling water to the liquid cooling garment. Return oxygen is cleansed of solid and gas contaminants by a lithium hydroxide and activated charcoal canister. The PLSS includes communications and telemetry equipment, displays and controls, and a power supply. The PLSS is covered by a thermal insulation jacket, (two stowed in LM).

<u>Oxygen purge system (OPS)</u> - Mounted atop the PLSS, the oxygen purge system provides a contingency 30-75 minute supply of gaseous oxygen in two bottles pressurized to 5,880 psia, (a minimum of 30 minutes in the maximum flow rate and 75 minutes in the low flow rate). The system may also be worn separately on the front of the pressure garment assembly torso for contingency EVA transfer from the LM to the CSM or behind the neck for CSM EVA. It serves as a mount for the VHF antenna for the PLSS, (two stowed in LM).

<u>Coveralls</u> - During periods out of the space suits, crewmen wear two-piece Teflon fabric inflight coveralls for warmth and for pocket stowage of personal items.

<u>Communications carriers</u> - "Snoopy hats" with redundant microphones and earphones are worn with the pressure helmet; a light-weight headset is worn with the inflight coveralls.

<u>Water Bags</u> - .9 liter (1 qt.) drinking water bags are attached to the inside neck rings of the EVA suits. The crewman can take a sip of water from the 6-by-8-inch bag through a 1/8-inch-diameter tube within reach of his mouth. The bags are filled from the lunar module potable water dispenser.

<u>Buddy Secondary Life Support System</u> - A connecting hose system which permits a crewman with a failed PLSS to share cooling water in the other crewman's PLSS. The BSLSS lightens the load on the oxygen purge system in the event of a total PLSS failure in that the OPS would supply breathing and pressurizing oxygen while the metabolic heat would be removed by the shared cooling water from the good PLSS. The BSLSS will be stowed on the LRV.

<u>Lunar Boots</u> - The lunar boot is a.thermal and abrasion protection device worn over the inner garment and boot assemblies. It is made up of layers of several different materials beginning with Teflon coated beta cloth for the boot liner to Chromel R metal fabric for the outer shell assembly. Aluminized Mylar, Nomex felt, Dacron, Beta cloth and Beta marquisette Kapton comprise the other layers. The lunar boot sole is made of high-strength silicone rubber.

<u>Personal Hygiene</u>

Crew personal hygiene equipment aboard Apollo 16 includes body cleanliness items, the waste management system, and one medical kit.

Packaged with the food are a toothbrush and a two-ounce tube of toothpaste for each crewman. Each man-meal package contains a 3.5-by-4-inch wet-wipe cleansing towel. Additionally, three packages of 12-by-12-inch dry towels are stowed beneath the command module pilot's couch. Each package: contains seven towels. Also stowed under the command module pilot's couch are seven tissue dispensers containing 53 three-ply tissues each.

Solid body wastes are collected in plastic defecation bags which contain a germicide to prevent bacteria and gas formation. The bags are sealed after use, identified, and stowed for return to Earth for post-flight analysis.

Urine collection devices are provided for use while wearing either the pressure suit or the inflight coveralls. The urine is dumped overboard through the spacecraft urine dump valve in the CM and stored in the LM. On Apollo 16 urine specimens will be returned to Earth for analysis.

Survival Kit

The survival kit is stowed in two rucksacks in the righthand forward equipment bay of the CM above the lunar module pilot.

Contents of rucksack No. 1 are: two combination survival lights, one desalter kit, three pairs of sunglasses, one radio beacon, one spare radio beacon battery and spacecraft connector cable, one knife in sheath, three water containers, two containers of Sun lotion, two utility knives, three survival blankets and one utility netting.

Rucksack No. 2: one three-man life raft with CO2 inflater, one sea anchor, two sea dye markers, three sunbonnets, one mooring lanyard, three manlines and two attach brackets.

The survival kit is designed to provide a 48-hour postlanding (water or land) survival capability for three crewmen between 40 degrees North and South latitudes.

Medical Kits

The command module crew medical supplies are contained in two kits. Included in the larger medical accessories kit are antibiotic ointment, skin cream, eye drops, nose drops, spare biomedical harnesses, oral thermometer and pills of the following types: 18 pain, 12 stimulant, 12 motion sickness, 24 diarrhea, 60 decongestant, 21 sleeping, 72 aspirin and 60 each of two types of antibiotic. A smaller command module auxiliary drug kit contains 80 and 12 of two types of pills for treatment of cardiac arrythymia and two injectors for the same symptom.

The lunar module medical kit contains eye drops, nose drops, antibiotic ointment, bandages and the following pills: 4 stimulant, 4 pain, 8 decongestant, 8 diarrhea, 12 aspirin and 6 sleeping. A smaller kit in the LM contains 8 and 4 units of injectable drugs for cardiac arrythymia and 2 units for pain suppression.

Crew Food System

The Apollo 16 crew selected menus for their flight from the largest variety of foods ever available for a U.S. manned mission. However, some menu constraints have been imposed by a series of medical requirements. This has limited to some degree the latitude of crew menu selection. For the first time since the flight of Gemini 7 in December 1965, the preflight, inflight, and postflight diets are being precisely controlled to facilitate interpretation of the most extensive series of medical tests to be performed on an Apollo mission.

Menus were designed upon individual crew member physiological requirements in the unique conditions of weightlessness and one-sixth gravity on the lunar surface. Daily menus provide 2750, 2500, and 2650 calories per day for the Commander, Command Module Pilot, and Lunar Module Pilot, respectively.

Food items are assembled into meal units and identified as to crew member and sequence of consumption. Foods stored in a "pantry" section may be used as substitutions for nominal meal items so long as the nutrient intake for a 24-hour period is not altered significantly.

There are various types of food used in the menus. These include freeze-dried rehydratables in spoon-bowl packages; thermostabilized foods (wet packs) in flexible packages and metal easy-open cans; intermediate moisture foods; dry bite-size cubes, and beverages.

Water for drinking and rehydrating food is obtained from two sources in the command module — a portable dispenser for drinking water and a water spigot at the food preparation station which supplies water at about 145 degrees and 55 degrees Fahrenheit. The potable water dispenser provides a continuous flow of water as long as the trigger is held down, and the food preparation spigot dispenses water in one-ounce increments. A continuous flow water dispenser similar to the one in the command module is used aboard the lunar module for cold water reconstitution of food stowed aboard the lunar module.

Water is injected into a food package and the package is kneaded and allowed to sit for several minutes. The bag top is then cut open and the food eaten with a spoon. After a meal, germicide tablets are placed in each bag to prevent fermentation and gas formation. The bags are then rolled and stowed in waste disposal areas in the spacecraft.

Several prototype food packages which are intended for use in Skylab missions in 1973 will be tested by the Apollo 16 crew. These include an improved beverage package design, a salt dispenser, and an improved rehydratable package design. Functional aspects of the package and the behavior of liquid food during extended periods of weightlessness will be observed.

The in-suit drink device, which was used for water previously only on the Apollo 15 mission, will contain a specially formulated beverage powder which will be reconstituted with water prior to extravehicular activities on the lunar surface. This beverage will provide the crewman with fluid, electrolytes (especially potassium), and energy during each of the 3 scheduled EVAs. As on Apollo 15, the crewmen on the lunar surface will have an in-suit food bar to snack on.

New foods for the Apollo 16 mission are limited to thermostabilized ham steak, rehydratable grits, and an intermediate moisture cereal bar.

APOLLO 16 MENU – BLUE

MEAL	Day 1*, 5**, 9, 13**		Day 2, 10		Day 3, 11			Day 4, 8***, 12	
A	Peaches +	WP	Fruit Cocktail	R	Peaches	RSB	Mixed Fruit +	WP
	Scrambled Eggs	RSB	Sausage Patties	R	Scrambled Eggs	RSB	Ham Steak	WP
	Bacon Squares (8)	IMB	Spiced Fruit Cereal	RSB	Bacon Squares (8)	IMB	Cornflakes	RSB
	Grits	RSB	Orange Juice	R	Grits	RSB	White Bread (1)~ Jelly++	WP
	Orange Juice	R	Cocoa	R	Orange Juice	R	Orange Juice	R
	Cocoa	R			Cocoa	R	Cocoa	R
	(+Peaches-Day 13	RSB)					(+Fruit Cocktail - Day 12	R)
							(++ Delete on Day 12)	
B	Chicken & Rice Soup	RSB	Corn Chowder	RSB	Lobster Bisque	RSB	Pea Soup	RSB
	Hamburger &		Turkey & Gravey	WP	Bread Rye (2)	WP	Meatballs w/Sauce	WP
	White Bread (1)	WP	Vanilla Pudding	WP	Tuna Spread	WP	Lemon Pudding+	WP
	Pears	IMB	White Bread (1) &		Cherry Food Bar (2)	IMB	Sugar Cookies (4)	DB
	Inst. Breakfast	R	Peanut Butter	WP	Graham Cracker-Cubes(6)	DB	Peaches	IMB
	Cereal Bar	DB	Apple Food Bar (2)	IMB	Citrus Beverage	R	Orange-Grapefruit Drink	R
	Citrus Beverage	R	Orange Drink	R			(+Pork and Scalloped	
							Potatoes—Day 8 & 12	RSB)
C	Cr. Tomato Soup	RSB	Cr. Potato Soup	RSB	Romaine Soup	RSB	Beef & Gravy	WP
	Spaghetti/Meat Sauce	RSB	Frankfurters (4)		Beef Steak	WP	Chicken, Stew	RSB
	Peach Ambrosia	RSB			Chicken & Rice	RSB	Butterscotch Pudding	RSB
	Apricot Cereal.		Chocolate Pudding		Pin. Fruitcake (4)	DB	Chocolate Bar	DB
	Cubes (4)	DB	Orange-Grapefruit		Pecans (6)	DB	Gingerbread (4)	DB
	Pecans (6)	DB	Drink	R	Grape Drink	R	Citrus Beverage	R
	Cocoa	R						

DB	= Dry Bite
IMB	= Intermediate Moisture Bite
R	= Rehydratable
RSB	= Rehydratable Spoon Bowl
WP	= Wet Pack
RC	= Rehydratable Can
SBD	= Skylab Beverage Dispenser

3B SKYLAB MEAL	
Turkey & Rice Soup	RC
Rye Bread (2)	
Chicken Spread (1/3)	WP
Peaches	WP
Peanuts	CAN
Orange Beverage	SBD

* Meal C only
** Meal A only
*** Meal B and C only

APOLLO 16 - LM11 MENU (Blue Velcro LMP, Charles M. Duke)

MEAL	DAY 5		DAY 6		DAY 7		DAY 8	
B	Cr. of Tomato Soup	RSB A	Peaches (39 gms)	IMB A	Peaches (39 gins)	IMB A	Peaches (39 gms)	IMB
	Rye Bread (2)		Ham Steak	WP	Beef Steak	WP	Ham Steak	WP
	Tuna Spread	WP	Scrambled Eggs	RSB	Bacon Squares (8)	IMB	Scrambled Eggs	RSB
	Apple Food Bar (2)	IMB	Cinn. Toasted Bread		Spiced Fruit Cereal	RSB	Cereal Bar	DB
	Chocolate Bar	DB	Cubes (6)	DB	Instant Breakfast	R	Apricot Cereal Cubes (6)	
DB								
	Orange-Grapefruit		Instant Breakfast	R	Orange-Grapefruit		Orange Beverage	R
	Beverage	R	orange-Grapefruit		Beverage	R	Cocoa R	
			Beverage R		Cherry Food Bar (2)	IMB		
			Lemon Food Bar (2)	IMB				
C	Shrimp Cocktail	RSB B	Pea Soup	RSB B	Romaine Soup	RSB		
	Turkey & Gravy	WP	Salmon Salad	RSB	Tuna Salad	RSB		
	Chocolate Pudding	RSB	Frankfurters (4)	WP	Meatballs w/Sauce	WP		
	Graham Cracker		Peach Ambrosia	RSB	Chicken &-Rice	RSB		
	Cubes (6)	DB	Pears	IMB	Butterscotch Pudding	RSB		
	Cocoa	R	Cereal Bar	DB	Gingerbread (6)	DB		
	Citrus Beverage	R	Citrus Beverage	R	Citrus Beverage	R		
			Cocoa	R	Cocoa	R		

APOLLO XVI

PANTRY STOWAGE ITEMS

Line

P/N: 14-0123

Items 17W & 17X

BEVERAGES	QTY	SOUPS/SALADS MEATS	QTY
Cocoa	6	Salmon Salad	3
Coffee (B)	16	Tuna Salad	3
Instant Breakfast	12	Shrimp Cocktail	3
Grapefruit Drink	6	Romaine Soup	3
Orange Beverage	6	Potato Soup	3
Orange-Grapefruit Beverage	6	Pea Soup	3
Orange Juice	12		
Orange-Pineapple Drink	6	Spaghetti w/Meat Sauce	3
		Chicken Stew	3
BREAKFAST ITEMS		SANDWICH SPREADS	
Bacon Squares (8)	6	Peanut Butter	3
Spiced Fruit Cereal	3	Jelly	3
Cornflakes	3	Ham Salad	1
Scrambled Eggs	6		
Grits	3	Catsup*	7
Peach Ambrosia	3	Mustard*	7
Sausage Patties	3		
SNACK ITEMS		ACCESSORIES	
Pecans (6)	3	Wet Skin Cleaning Towels	9
Apricots (IMB)(38o5 gm)	6	Contingency Feeding System	
Peaches (IMB) (39 gm)	8	3 Food Restrainer Pouches	
Pears (IMB) (42 gm)	6	3 Beverage Packages	
Apricot Food Bar (1)(26 gm)	9	1 Valve Adapter (pontube)	
Apple Food Bar (1) (26 gm)	9	Germicidal Tablets	
Lemon Food Bar (1) (26 gm)	9	Index Card	
Cherry Food Bar (1) (26 gm)	9		
Cereal Bar	6		
Chocolate Bar	3		
Sugar Cookies (4)	3		
Graham Crackers (6)	3		
Cheese Cracker Cubes (4)	3		

*Stowage locations TBD

APOLLO 16 FLAGS, LUNAR MODULE PLAQUE

The United States flag to be erected on the lunar surface measures 78 by 125 cm (30 by 48 inches) and will be deployed on a two-piece aluminum tube 2.1 meters (eight feet) long. The nylon flag will be stowed in the lunar module descent stage modularized equipment stowage assembly.

Also carried on the mission and returned to Earth will be 25 United States flags, 50 individual state flags, flags of United States territories, flags of other national states which are generally accepted as independent in the world community, and flags of the United Nations and other international organizations. These flags are 10 by 15 cm (four by six inches).

An 18 by 23 cm (seven by nine inch) stainless steel plaque, similar to that flown on Apollo 15, will be fixed to the LM front leg. The plaque has on it the words "Apollo 16" with "April 1972" and the signatures of the three crewmen located beneath.

SATURN V LAUNCH VEHICLE

The Saturn V launch vehicle (SA-511) assigned to the Apollo 16 mission is very similar to the vehicles used for the missions of Apollo 8 through Apollo 15.

First Stage
The five first stage (S-IC) F-1 engines develop about 34 million newtons (7.7 million pounds) of thrust at launch. Major stage components are the forward skirt, oxidizer tank, intertank structure, fuel tank, and thrust structure. Propellant to the five engines normally flows at a rate of about 13,300 kilograms (29,300 pounds; 3,390 gallons) a second. One engine is rigidly mounted on the stage's centerline; the outer four engines are mounted on a ring at 90-degree angles around the center engine. These outer engines are gimbaled to control the vehicle's attitude during flight.

Second Stage
The five second stage (S-II) J-2 engines develop a total of about 5.15 million newtons (1.15 million pounds) of thrust during flight. Major components are the forward skirt, liquid hydrogen and liquid oxygen tanks (separated by an insulated common bulkhead), a thrust structure, and an interstage section that connects the first and second stages. The engines are mounted and used in the same arrangement as the first stage's F-1 engines: four outer engines can be gimbaled; the center one is fixed.

Third Stage
Major components of the third stage (S-IVB) are a single J-2 engine, aft interstage and skirt, thrust structure, two propellant tanks with a common bulkhead, and forward skirt. The gimbaled engine has a maximum thrust of 1.01 million newtons (230,000 pounds), and can be restarted in Earth orbit.

Instrument Unit
The instrument unit (IU) contains navigation, guidance and control equipment to steer the Saturn V into Earth orbit and translunar trajectory. The six major systems are structural, environmental control, guidance and control, measuring and telemetry, communications, and electrical.

The IU's inertial guidance platform provides space-fixed reference coordinates and measures acceleration during flight. if the platform should fail during boost, systems in the Apollo spacecraft are programmed to provide launch vehicle guidance. After second stage ignition, the spacecraft commander can manually steer the vehicle if its guidance platform is lost.

Propulsion
The Saturn V has 31 propulsive units, with thrust ratings ranging from 311 newtons (70 pounds) to more than 6.8 million newtons (1.53 million pounds). The large main engines burn liquid propellants; the smaller units

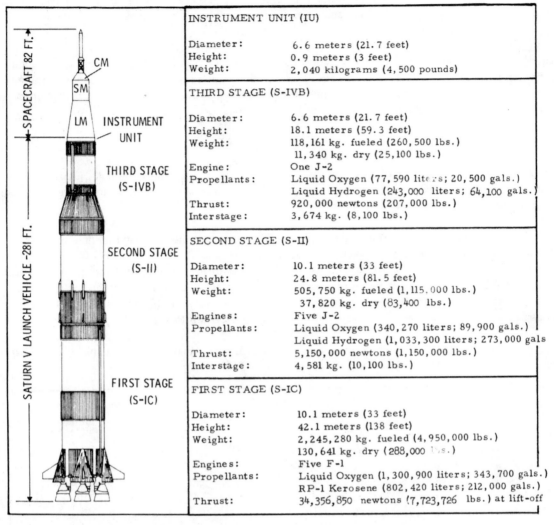

INSTRUMENT UNIT (IU)	
Diameter:	6.6 meters (21.7 feet)
Height:	0.9 meters (3 feet)
Weight:	2,040 kilograms (4,500 pounds)

THIRD STAGE (S-IVB)	
Diameter:	6.6 meters (21.7 feet)
Height:	18.1 meters (59.3 feet)
Weight:	118,161 kg. fueled (260,500 lbs.)
	11,340 kg. dry (25,100 lbs.)
Engine:	One J-2
Propellants:	Liquid Oxygen (77,590 liters; 20,500 gals.)
	Liquid Hydrogen (243,000 liters; 64,100 gals.)
Thrust:	920,000 newtons (207,000 lbs.)
Interstage:	3,674 kg. (8,100 lbs.)

SECOND STAGE (S-II)	
Diameter:	10.1 meters (33 feet)
Height:	24.8 meters (81.5 feet)
Weight:	505,750 kg. fueled (1,115.000 lbs.)
	37,820 kg. dry (83,400 lbs.)
Engines:	Five J-2
Propellants:	Liquid Oxygen (340,270 liters; 89,900 gals.)
	Liquid Hydrogen (1,033,300 liters; 273,000 gals
Thrust:	5,150,000 newtons (1,150,000 lbs.)
Interstage:	4,581 kg. (10,100 lbs.)

FIRST STAGE (S-IC)	
Diameter:	10.1 meters (33 feet)
Height:	42.1 meters (138 feet)
Weight:	2,245,280 kg. fueled (4,950,000 lbs.)
	130,641 kg. dry (288,000 lbs.)
Engines:	Five F-1
Propellants:	Liquid Oxygen (1,300,900 liters; 343,700 gals.)
	RP-1 Kerosene (802,420 liters; 212,000 gals.)
Thrust:	34,356,850 newtons (7,723,726 lbs.) at lift-off

NOTE: Weights and measures given above are for the nominal vehicle configuration for Apollo.   The figures may vary slightly due to changes before launch to meet changing conditions. Weights of dry stages and propellants do not equal total weight because frost and miscellaneous smaller items are not included in chart.

SATURN V LAUNCH VEHICLE

use solid or hypergolic (self-igniting) propellants.

The five F-1 engines give the first stage a thrust range of from 34,356,850 newtons (7,723,726 pounds) at liftoff to 40,576,160 newtons (9,121,883 pounds) at center engine cutoff. Each F-1 engine weighs almost nine metric tons (10 short tons), is more than 5.5 meters long (18 feet), and has a nozzle exit diameter of nearly 4.6 meters (14 feet). Each engine uses almost 2.7 metric tons (3 short tons) of propellant a second.

J-2 engine thrust on the second and third stages averages 960,000 newtons (216,000 pounds) and 911,000 newtons (205,000 pounds) respectively during flight. The 1,590-kilogram (3,500 pound) engine uses high-energy, low-molecular-weight liquid hydrogen as propellant.

The first stage has eight solid-fuel retro-rockets that fire to separate the first and second stages. Each rocket produces a thrust of 237,000 newtons (75,800 pounds) for 0.54 seconds.

Four retrorockets, located in the third stage's aft interstage, separate the second and third stages. Two jettisonable ullage rockets settle propellants before engine ignition. Six smaller engines in two auxiliary propulsion system modules on the third stage provide three-axis attitude control.

<u>Significant Vehicle Changes</u>
Saturn vehicle SA-511 is similar in configuration to the Apollo 15 launch vehicle. The first stage (S-IC) has eight retrorocket motors, double the number on the SA-510 vehicle, because flight evaluation of the Apollo 15 mission revealed that the separation distance between the first and second stages was less than predicted. Eight retrorockets will give a greater safety margin should one motor fail during separation.

APOLLO SPACECRAFT

The Apollo spacecraft consists of the command module, service module, lunar module, a spacecraft lunar module adapter (SLA), and a launch escape system. The SLA houses the lunar module and serves as a mating structure between the Saturn V instrument unit and the SM.

<u>Launch Escape System (LES)</u> — The function of the LES is to propel the command module to safety in an aborted launch. It has three solid-propellant rocket motors: a 658,000-newton (147,000-pound)-thrust launch escape system motor, a 10,750-newton (2,400-pound)-thrust pitch control motor, and a 141,000 newton (31,500-pound)-thrust tower jettison motor. Two canard vanes deploy to turn the command module aerodynamically to an attitude with the heatshield forward. The system is 10 meters (33 feet) tall and 1.2 meters (four feet) in diameter at the base, and weighs 4,158 kilograms (9,167 pounds).

<u>Command Module (CM)</u> — The command module is a pressure vessel encased in heat shields, cone-shaped, weighing 5,543.9 kg (12,874 lb.) at launch.

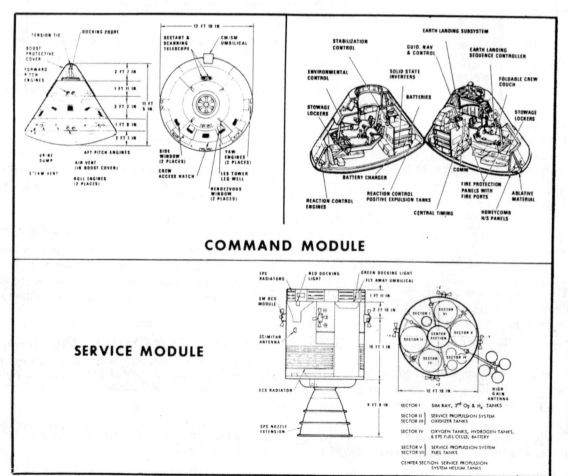

COMMAND MODULE

SERVICE MODULE

The command module consists of a forward compartment which contains two reaction control engines and components of the Earth landing system; the crew compartment or inner pressure vessel containing crew accommodations, controls and displays, and many of the spacecraft systems; and the aft compartment housing ten reaction control engines, propellant tankage, helium tanks, water tanks, and the CSM umbilical cable. The crew compartment contains 6m^3 (210 ft.^{3}) of habitable volume.

Heat-shields around the three compartments are made of brazed stainless steel honeycomb with an outer layer of phenolic epoxy resin as an ablative material.

The CSM and LM are equipped with the probe-and-drogue docking hardware. The probe assembly is a powered folding coupling and impact attentuating device mounted in the CM tunnel that mates with a conical drogue mounted in the LM docking tunnel. After the 12 automatic docking latches are checked following a docking maneuver, both the probe and drogue are removed to allow crew transfer between the CSM and LM.

Service Module (SM) — The Apollo 16 service module will weigh 24,514 kg (54,044 lb.) at launch, of which 18,415 kg (40,594 lb.) is propellant for the 91,840-newton (20,500 pound)-thrust service propulsion engine: (fuel: 50/50 hydrazine and unsymmetrical dimethyl-hydrazine; oxidizer: nitrogen tetroxide). Aluminum honeycomb panels 2.54 centimeters (one inch) thick form the outer skin, and milled aluminum radial beams separate the interior into six sections around a central cylinder containing service propulsion system (SPS) helium pressurant tanks. The six sectors of the service module house the following components: Sector I — oxygen tank 3 and hydrogen tank 3, J-mission SIM bay; Sector II — space radiator, +Y RCS package, SPS oxidizer storage tank; Sector III — space radiator, +Z RCS package, SPS oxidizer storage tank; Sector IV - three fuel cells, two oxygen tanks, two hydrogen tanks, auxiliary battery; Sector V — space radiator, SPS fuel sump tank, -Y RCS package; Sector VI — space radiator, SPS fuel storage tank, -Z RCS package.

Spacecraft-LM adapter (SLA)Structure — The spacecraft-LM adapter is a truncated cone 8.5 m (28 ft.) long tapering from 6.7 m (21.6 ft.) in diameter at the base to 3.9 m (12.8 ft.) at the forward end at the service module mating line. The SLA weighs 1,841 kg (4,059 lb.) and houses the LM during launch and the translunar injection maneuver until CSM separation, transposition, and LM extraction. The SLA quarter panels are jettisoned at CSM separation.

Lunar Module (LM)
The lunar module is a two-stage vehicle designed for space operations near and on the Moon. The lunar module stands 7 m (22 ft. 11 in.) high and is 9.5 m (31 ft.) wide (diagonally across landing gear). The ascent and descent stages of the LM operate as a unit until staging, when the ascent stage functions as a single spacecraft for rendezvous and docking with the CM.

Ascent Stage — Three main sections make up the ascent stage: the crew compartment, midsection, and aft equipment bay. Only the crew compartment and midsection are pressurized 337.5 grams per square centimeter (4.8 pounds per square inch gauge). The cabin volume is 6.7 cubic meters (235 cubic feet). The stage measures 3.8 m (12 ft. 4 in.) high by 4.3 m (14 ft. 1 in.) in diameter. The ascent stage has six substructural areas: crew compartment, midsection, aft equipment bay, thrust chamber assembly cluster supports, antenna supports, and thermal and micrometeoroid shield.

The cylindrical crew compartment is 2.35 m (7 ft. 10 in.) in diameter and 1.47 m (3 ft. 6 in.) deep. Two flight stations are equipped with control and display panels, armrests, body restraints, landing aids, two front windows, an overhead docking window, and an alignment: optical telescope in the center between the two flight stations. The habitable volume is 4.5 m^3 (160 ft^3).

A tunnel ring atop the ascent stage meshes with the command module docking latch assemblies. During docking, the CM docking ring and latches are aligned by the LM drogue and the CSM probe.

The docking tunnel extends downward into the midsection 40 cm (16 in.). The tunnel is 81 cm (32 in.) in

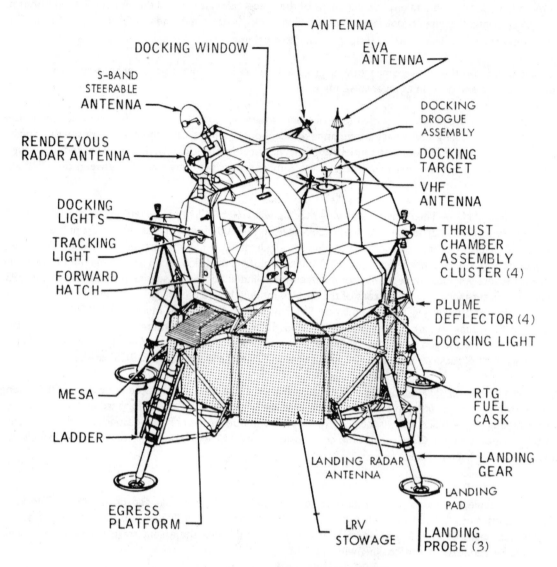

LUNAR MODULE

diameter and is used for crew transfer between the CSM and LM. The upper hatch on the inboard end of the docking tunnel opens inward and cannot by opened. without equalizing pressure on both hatch surfaces.

A thermal and micrometeoroid shield of multiple layers of Mylar and a single thickness of thin aluminum skin encases the entire ascent stage structure.

Descent Stage — The descent stage center compartment houses the descent engine, and descent propellant tanks are housed in the four bays around the engine. Quadrant II contains ALSEP. The radioisotope thermoelectric generator (RTG) is externally mounted. Quadrant IV contains the MESA. The descent stage measures 3.2 m (10 ft. 7 in.) high by 4.3 m (14 ft. 1 in.) in diameter and is encased in the Mylar and aluminum alloy thermal and micrometeoroid shield. The LRV is stowed in Quadrant I.

The LM egress platform or "porch" is mounted on the forward outrigger just below the forward hatch. A ladder extends down the forward landing gear strut from the porch for crew lunar surface operations.

The landing gear struts are released explosively and are extended by springs. They provide lunar surface landing impact attenuation. The main struts are filled with crushable aluminum honeycomb for absorbing compression loads. Footpads 0.95 m (37 in.) in diameter at the end of each landing gear provide vehicle support on the lunar surface.

Each pad (except forward pad) is fitted with a 1.7-m (68-in.) long lunar surface sensing probe which upon contact with the lunar surface signals the crew to shut down the descent engine.

The Apollo LM has a launch weight of 16,429 kg (36,218 lb.). The weight breakdown is as follows:

	kilograms	pounds
1. Ascent stage, dry*	2,134	4,704
2. APS propellants (loaded)	2,378	5,242
3. Descent stage, dry	2,759	6,083
4. DPS propellants (loaded)	8,872	19,558
5. RCS propellants (loaded)	286	631
	16,429 kg	36,218 lbs

* Includes water and oxygen; no crew.

NATIONAL AERONAUTICS AND SPACE ADMINISTRATION
WASHINGTON, D. C. 20546

BIOGRAPHICAL DATA

NAME: John W. Young (Captain, USN) Apollo 16 Mission Commander NASA Astronaut

BIRTHPLACE AND DATE: Born in San Francisco, California, on September 24, 1930. His parents, Mr. and Mrs. William H. Young, reside in Orlando, Florida.

PHYSICAL DESCRIPTION: Brown hair; green eyes, height: 5 feet 9 inches; weight: 165 pounds.

EDUCATION: Graduated from Orlando High School. Orlando, Florida; received a Bachelor of Science degree in Aeronautical Engineering from the Georgia Institute of Technology in 1952; recipient of an Honorary Doctorate of Laws degree from Western State University College of Law in 1969, and an Honorary Doctorate of Applied Science from Florida Technological University in 1970.

MARITAL STATUS: Married to the former Susy Feldman of St. Louis, Missouri.

CHILDREN: Sandy, April 30, 1957; John, January 17, 1959 (by a previous marriage).

RECREATIONAL INTERESTS: He plays handball, runs and works out in the full pressure suit to stay in shape.

ORGANIZATIONS: Fellow of the American Astronautical Society, Associate Fellow of the Society of Experimental Test Pilots, and Member of the American Institute of Aeronautics and Astronautics.

SPECIAL HONORS: Awarded the NASA Distinguished Service Medal, two NASA Exceptional Service Medals, the MSC Certificate of Commendation (1970), the Navy Astronaut Wings, the Navy Distinguished Service Medals, and three Navy Distinguished Flying Crosses.

EXPERIENCE: Upon graduation from Georgia Tech, Young entered the U.S. Navy in 1952; he holds the rank of Captain in that service.

He completed test pilot training at the U.S. Naval Test Pilot School in 1959, and was then assigned as a test pilot at the Naval Air Test Center until 1962. Test projects in which he participated include evaluations of the F8D "Crusader" and F4B "Phantom" fighter weapons systems, and in 1962, he set world time-to-climb records to 3,000 and 25,000 meter altitudes in the Phantom. Prior to his assignment to NASA, he was maintenance officer of All-Weather-Fighter Squadron 143 at the Naval Air Station, Miramar, California. He has logged more than 5,900 hours flying time, including more than 4,900 hours in jet aircraft, and completed three space flights totaling 267 hours and 42 minutes.

Captain Young was selected as an astronaut by NASA in September 1962. He served as pilot with command pilot Gus Grissom on the first manned Gemini flight — a 3-orbit mission, launched on March 23, 1965, during which the crew accomplished the first manned spacecraft orbital trajectory modifications and lifting reentry, and flight tested all systems in Gemini 3. After this flight, he was backup pilot for Gemini 6. On July 18, 1966, Young occupied the command pilot seat for the Gemini 10 mission and, with Michael Collins as pilot, effected a successful rendezvous and docking with the Agena target vehicle. The large Agena main engine was subsequently ignited, propelling the docked combination to a record altitude of approximately 475 miles above the earth, the first manned operation of a large rocket engine in space. They later performed a completely optical rendezvous (without radar) on a second, passive Agena which had been placed in orbit during the Gemini 8 mission. After the rendezvous, Young flew formation on the slowly rotating passive Agena while Collins performed extra-vehicular activity to it and recovered a micrometeorite detection experiment, accomplishing an in-space retrieval of the detector which had been orbiting the earth for three months. The flight was concluded, after 3 days and 44 revolutions, with a precise splashdown in the West Atlantic, 2.6 miles from the recovery ship USS GUADALCANAL.

He was then assigned as the backup command module pilot for Apollo 7.

Young was command module pilot for Apollo 10, May 18-26, 1969, the comprehensive lunar-orbital qualification test of the Apollo lunar module. He was accompanied on the 248,000 nautical mile lunar mission by Thomas P. Stafford (spacecraft commander) and Eugene A. Cernan (lunar module pilot). In achieving all mission objectives, Apollo 10 confirmed the operational performance, stability, and reliability of the command-service module/lunar module configuration during translunar coast, lunar orbit insertion, and lunar module separation and descent to within 8 nautical miles of the lunar surface. The latter maneuvers employed all but the final landing phase and facilitated extensive evaluations of the lunar module landing radar devices and propulsion systems and lunar module and command-service module rendezvous radars in the subsequent lunar rendezvous. In addition to discovering unexpectedly that the lunar gravitational field targeted the lunar module over 4 miles south of the Apollo 11 lunar landing site, Apollo 10 photographed and accurately located this site for the lunar landing.

Captain Young served as backup spacecraft commander for Apollo 13.

He was assigned as spacecraft commander for the Apollo 16 flight March 3, 1971.

NATIONAL AERONAUTICS AND SPACE ADMINISTRATION
WASHINGTON, D. C. 20546

BIOGRAPHICAL DATA

NAME: Thomas K. Mattingly II (Lieutenant Commander, USN) Apollo 16 Command Module Pilot NASA Astronaut

BIRTHPLACE AND DATE: Born in Chicago, Illinois, March 17, 1936. His parents, Mr. and Mrs. Thomas K. Mattingly, now reside in Hialeah, Florida.

PHYSICAL DESCRIPTION: Brown hair; blue eyes; height: 5 feet 10 inches; weight: 140 pounds.

EDUCATION: Attended Florida elementary and secondary schools and is a graduate of Miami Edison High School, Miami, Florida; received a Bachelor of Science degree in Aeronautical Engineering from Auburn University in 1958.

MARITAL STATUS: Married to the former Elizabeth Dailey of Hollywood, California.

RECREATIONAL INTERESTS: Enjoys water skiing and playing handball and tennis.

ORGANIZATIONS: Member of the American Institute of Aeronautics and Astronautics and the U.S. Naval Institute.

SPECIAL HONORS: Presented the MSC Certificate of Commendation (1970).

EXPERIENCE: Prior to reporting for duty at the Manned Spacecraft Center, he was a student at the Air Force Aerospace Research Pilot School.

He began his Navy career as an Ensign in 1958 and received his wings in 1960. He was then assigned to VA-35 and flew AIH aircraft aboard the USS SARATOGA from 1960 to 1963. In July 1963, he served in VAH-11 deployed aboard the USS FRANKLIN D. ROOSEVELT where he flew the A3B aircraft for two years. He logged 4,200 hours of flight time — 2,300 hours in jet aircraft. Lt. Commander Mattingly is one of the 19 astronauts selected by NASA in April 1966. He served as a member of the astronaut support crews for the Apollo 8 and 9 missions. He was then designated command module pilot for Apollo 13 but was removed from flight status 72 hours prior to the scheduled launch due to exposure to the german measles. Mattingly was replaced on the flight of Apollo 13, April 13-17, 1970, by backup command module pilot John L. Swigert, Jr. He was designated to serve as command module pilot for the Apollo 16 flight March 3, 1971.

NATIONAL AERONAUTICS AND SPACE ADMINISTRATION
WASHINGTON, D.C. 20546

BIOGRAPHICAL DATA

NAME: Charles Moss Duke, Jr. (Lieutenant Colonel, USAF) Apollo 15 Lunar Module Pilot NASA Astronaut

BIRTHPLACE AND DATE: Born in Charlotte, North Carolina, on October 3, 1935. His parents, Mr. and Mrs. Charles M. Duke, make their home in Lancaster, South Carolina.

PHYSICAL DESCRIPTION: Brown hair; brown eyes; height: 5 feet 11 1/2 inches; weight: 155 pounds.

EDUCATION: Attended Lancaster High School in Lancaster, South Carolina, and was graduated valedictorian from the Admiral Farragut Academy in St. Petersburg, Florida; received a Bachelor of Science degree in Naval sciences from the U.S. Naval Academy in 1957 and a Master of Science degree in Aeronautics and Astronautics from the Massachusetts Institute of Technology in 1964.

MARITAL STATUS: Married to the former Dorothy Meade Claiborne of Atlanta, Georgia; her parents are Dr. and Mrs. T. Sterling Claiborne of Atlanta.

CHILDREN: Charles M., March 8, 1965; Thomas C., May 1, 1967.

RECREATIONAL INTERESTS: Hobbies include hunting, fishing, reading, and playing golf.

ORGANIZATIONS: Member of the Air Force Association, the Society of Experimental Test Pilots, the Rotary Club, the American Legion, and the American Fighter Pilots Association.

SPECIAL HONORS: Awarded the MSC Certificate of Commendation (1970).

EXPERIENCE: When notified of his selection as an astronaut, Duke was at the Air Force Aerospace Research Pilot School as an instructor teaching control systems and flying in the F-104, F-101, and T-33 aircraft. He was graduated from the Aerospace Research Pilot School in September 1965 and stayed on there as an instructor.

He is an Air Force Lt. Colonel and was commissioned in 1957 upon graduation from the Naval Academy. Upon entering the Air Force, he went to Spence Air Base, Georgia, for primary flight training and then to Webb Air Force Base, Texas, for basic flying training, where in 1958 he became a distinguished graduate. He was again a distinguished graduate at Moody Air Force Base, Georgia, where he completed advanced training in F-86L aircraft. Upon completion of this training he was assigned to the 526th Fighter Interceptor Squadron at Ramstein Air Base, Germany, where he served three years as a fighter interceptor pilot. He has logged 3,000 hours flying time, which includes 2,750 hours in jet aircraft. Lt. Colonel Duke is one of the 19 astronauts selected by NASA in April 1966. He served as a member of the astronaut support crew for the Apollo 10 flight and as backup lunar module pilot for the Apollo 13 flight. He was designated March 3, 1971 to serve as lunar module pilot for the Apollo 16 mission.

NATIONAL AERONAUTICS AND SPACE ADMINISTRATION WASHINGTON D.C. 20546

BIOGRAPHICAL DATA

LANE: Fred Wallace Haise, Jr. (Mr.) Apollo 16 Backup Commander NASA Astronaut

BIRTHPLACE AND DATE: Born in Biloxi, Mississippi, on November 14, 1933; his mother, Mrs. Fred W. Haise, Sr., resides in Biloxi.

PHYSICAL DESCRIPTION: Brown hair; brown eyes, height: 5 feet 9 1/2 inches; weight: 155 pounds.

EDUCATION: Graduated from Biloxi High School, Biloxi, Mississippi; attended Perkinston Junior College (Association of Arts); received a Bachelor of Science degree with honors in Aeronautical Engineering from the University of Oklahoma in 1959, and an Honorary Doctorate of Science from Western Michigan University in 1970.

MARITAL STATUS: Married to the former Mary Griffin Grant of Biloxi, Mississippi. Her parents, Mr. and Mrs. William J. Grant, Jr., reside in Biloxi.

CHILDREN: Mary M., January 25, 1956; Frederick T., May 13., 1958; Stephen W., June 30, 1961; Thomas J., July 6, 1970.

ORGANIZATIONS: Fellow of the American Astronautical Society, and member of the Society of Experimental Test Pilots, Tau Beta Pi, Sigma Gamma Tau, and Phi Theta Kappa.

SPECIAL HONORS: Awarded the Presidential Medal for Freedom (1970), the NASA Distinguished Service Medal, the AIAA Haley Astronautics Award for 1971, the American Astronautical Society Flight Achievement Award for 1970, the City of New York Gold Medal in 1970, the City of Houston Medal for Valor in 1970, the Jeff Davis Award (1970), the Mississippi Distinguished Civilian Service Medal (1970), the American Defense Ribbon, the Society of Experimental Test Pilots Ray E. Tenhoff Award for 1966, and the A.B. Honts Trophy as the outstanding graduate of Class 64A from the Aerospace Research Pilot School in 1964.

EXPERIENCE: Haise was a research pilot at the NASA Flight Research Center at Edwards, California, before coming to Houston and the Manned Spacecraft Center; and from September 1959 to March 1963, he was a research pilot at the NASA Lewis Research Center in Cleveland, Ohio. During this time he authored the following papers which have been published: a NASA TND, entitled "An Evaluation of the Flying Qualities of Seven General-Aviation Aircraft"; NASA TND 3380, "Use of Aircraft for Zero Gravity Environment, May 1966"; SAE Business Aircraft Conference Paper, entitled "An Evaluation of General Aviation Aircraft Flying Qualities, March 30 - April 1, 1966"; and a paper delivered at the Tenth Symposium of the Society of Experimental Test Pilots, entitled "A Quantitative/Qualitative Handling Qualities Evaluation of Seven General-Aviation Aircraft, 1966."

He was the Aerospace Research Pilot School's outstanding graduate of Class 64A and served with the U.S. Air Force from October 1961 to August 1962 as a tactical fighter pilot and as chief of the 164th Standardization-Evaluation Flight of the 164th Tactical Fighter; Squadron at Mansfield, Ohio. From March 1957 to September 1959, Haise was a fighter interceptor pilot with the 185th Fighter Interceptor Squadron in the Oklahoma Air National Guard.

He also served as a tactics and all weather flight instructor in the U.S. Navy Advanced Training Command at NAAS Kingsville, Texas, and was assigned as a U.S. Marine Corps fighter pilot to VMF-533 and 114 at MCAS Cherry Point, North Carolina, from March 1954 to September 1956.

His military career began in October 1952 as a Naval Aviation Cadet at the Naval Air Station in Pensacola, Florida.

He has accumulated 6,700 hours flying time, including 3,300 hours in jets.

Mr. Haise is one of the 19 astronauts selected by NASA in April 1956, He served as backup lunar module pilot for the Apollo 8 and 11 missions.

Haise was lunar module pilot for Apollo 13, April 11-17, 1970. Apollo 13 was programmed for ten days and was committed to our first landing in the hilly, upland Fra Mauro region of the Moon; however, the original flight plan was modified enroute to the Moon due to a failure of the service module cryogenic oxygen system which occurred at approximately 55 hours into the flight. Haise and fellow crewmen, James A. Lovell (spacecraft commander) and John L. Swigert (command module pilot), working closely with Houston ground controllers, converted their lunar module "Aquarius" into an effective lifeboat. Their emergency activation and operation of lunar module systems conserved both electrical power and water in sufficient supply to assure their safety and survival while in space and for the return to Earth.

In completing his first space flight, Mr. Haise logged a total of 142 hours and 54 minutes in space.

He was designated as backup spacecraft commander for the Apollo 16 mission March 3, 1971.

NATIONAL AERONAUTICS AND SPACE ADMINISTRATION
WASHINGTON D.C. 20546

BIOGRAPHICAL DATA

NAME: Stuart Allen Roosa (Lieutenant Colonel, USAF) Apollo 16 Backup Command Module Pilot, NASA Astronaut

BIRTHPLACE AND DATE: Born August 16, 1933, in Durango, Colorado. His parents, Mr. and Mrs. Dewey Roosa, now reside in Tucson, Arizona.

PHYSICAL DESCRIPTION: Red hair; blue eyes, height: 5 feet 10 inches; weight: 155 pounds.

EDUCATION: Attended Justice Grade School and Claremore High School in Claremore, Oklahoma; studied at Oklahoma State University and the University of Arizona and was graduated with honors and a Bachelor of Science degree in Aeronautical Engineering from the University of Colorado; presented an Honorary Doctorate of Letters from the University of St. Thomas (Houston, Texas) in 1971.

MARITAL STATUS: His wife is the former Joan C. Barrett of Tupello, Mississippi; and her mother, Mrs. John T. Barrett, resides in Sessums, Mississippi.

CHILDREN: Christopher A., June 29, 1959; John D., January 2, 1961; Stuart A., Jr., March 12, 1962; Rosemary D., July 23, 1963.

RECREATIONAL INTERESTS: His hobbies are hunting, boating, and fishing.

ORGANIZATIONS: Associate Member of the Society of Experimental Test Pilots.

SPECIAL HONORS: Presented the NASA Distinguished Service Medal, the MSC Superior Achievement Award (1970), the Air Force Command Pilot Astronaut Wings, the Air Force Distinguished Service Medal, the Arnold Air Society's John F. Kennedy Award (1971), and the City of New York Gold Medal in 1971.

EXPERIENCE: Roosa, a Lt. Colonel in the Air Force, has been on active duty since 1953. Prior to joining NASA, he was an experimental test pilot at Edwards Air Force Base, Calif. — an assignment he held from September 1965 to May 1966, following graduation from the Aerospace Research Pilots School. He was a maintenance flight test pilot at Olmstead Air Force Base, Pennsylvania, from July 1962 to August 1964, flying F-101 aircraft. He served as Chief of Service Engineering (AFLC) at Tachikawa Air Base for two years following graduation from the University of Colorado under the Air Force Institute of Technology Program. Prior to this tour of duty, he was assigned as a fighter pilot at Langley Air Force Base, Virginia, where he flew the F-84F and P-100 aircraft.

He attended Gunnery School at Del Rio and Luke Air Force Bases and is a graduate of the Aviation Cadet Program at Williams Air Force Base, Arizona, where he received his flight training and commission in the Air Force. Since 1953, he has acquired 4,600 flying hours — 4,100 hours in jet aircraft. Lt. Colonel Roosa is one of the 19 astronauts selected by NASA in April 1966. He was a member of the astronaut support crew for the Apollo 9 flight. He completed his first space flight as command module pilot on Apollo 14, January 31-February 9, 1971. With him on man's third lunar landing mission were Alan B. Shepard (spacecraft commander) and Edgar D. Mitchell (lunar module pilot).

Maneuvering their lunar module, "Antares," to a landing in the hilly upland Fra Mauro region of the moon, Shepard and Mitchell subsequently deployed and activated various scientific equipment and experiments and proceeded to collect almost 100-pounds of lunar samples for return to earth. Throughout this 33-hour period of lunar surface activities, Roosa remained in lunar orbit aboard the command module, "Kittyhawk," to conduct a variety of assigned photographic and visual observations. Apollo 14 achievements include: first use of Mobile Equipment Transporter (MET); largest payload placed in lunar orbit; longest distance traversed on the lunar surface; largest payload returned from the lunar surface; longest lunar surface stay time (33 hours); longest lunar surface EVA (9 hours and 17 minutes); first use of shortened lunar orbit rendezvous techniques; first use of colored TV with new vidicon tube on lunar surface; and first extensive orbital science period conducted during CSM solo operations. In completing his first space flight, Roosa logged a total of 216 hours and 42 minutes. He was designated to serve as backup command module pilot for Apollo 16 on March 3, 1971.

NATIONAL AERONAUTICS AND SPACE ADMINISTRATION
WASHINGTON D.C. 20546

BIOGRAPHICAL DATA

NAME: Edgar Dean Mitchell (Captain, USN) Apollo 16 Backup Lunar Module Pilot NASA Astronaut

BIRTHPLACE AND DATE: Born in Hereford, Texas, on September 17, 1930, but considers Artesia, New,Mexico, his hometown. His mother, Mrs. J. T. Mitchell, now resides in Tahlequah, Oklahoma.

PHYSICAL DESCRIPTION: Brown hair; green eyes; height: 5 feet 11 inches; weight: 180 pounds.

EDUCATION: Attended primary schools in Roswell, New Mexico, and is a graduate of Artesia High School in Artesia, New Mexico; received a Bachelor of Science degree in Industrial Management from the Carnegie Institute of Technology in 1952, a Bachelor of Science degree in Aeronautical Engineering from the U.S. Naval Postgraduate School in 1961, and a Doctorate of Science degree in Aeronautics/Astronautics from the Massachusetts Institute of Technology in 1964; presented an Honorary Doctorate of Science from New Mexico State University in 1971, and an Honorary Doctorate of Engineering from Carnegie-Mellon University in 1971.

MARITAL STATUS: Married to the former Louise Elizabeth Randall of Muskegon, Michigan. Her mother, Mrs. Winslow Randall, now resides in Pittsburgh, Pennsylvania.

CHILDREN: Karlyn L., August 12, 1953; Elizabeth R., March 24, 1959.

RECREATIONAL INTERESTS: He enjoys handball and swimming, and his hobbies are scuba diving and soaring.

ORGANIZATIONS: Member of the American Institute of Aeronautics and Astronautics; the Society of Experimental Test Pilots; Sigma Xi; and Sigma Gamma Tau.

SPECIAL HONORS: Presented the Presidential Medal of Freedom (1970), the NASA Distinguished Service Medal, the MSC Superior Achievement Award (1970), the Navy Astronaut Wings, the Navy Distinguished Service Medal, the City of New York Gold Medal (1971), and the Arnold Air Society's John F. Kennedy Award (1971).

EXPERIENCE: Captain Mitchell's experience includes Navy operational flight, test flight, engineering, engineering management, and experience as a college instructor. Mitchell came to the Manned Spacecraft Center after graduating first in his class from the Air Force Aerospace Research Pilot School where he was both student and instructor.

He entered the Navy in 1952 and completed his basic training at the San Diego Recruit Depot in May 1953, after completing instruction at the Officers' Candidate School at Newport, Rhode Island, he was commissioned as an ensign. He completed flight training in July 1954 at Hutchinson, Kansas, and subsequently was assigned to Patrol Squadron 29 deployed to Okinawa. From 1957 to 1958, he flew A3 aircraft while assigned to Heavy Attack Squadron Two deployed aboard the USS BON HOMME RICHARD and USS TICONDEROGA; and he was a research project pilot with Air Development Squadron Five until 1959. His assignment from 1964 to 1965 was as Chief, Project Management Division of the Navy Field Office for Manned Orbiting Laboratory.

He has accumulated 4,000 hours flight time -1,900 hours in jets. Captain Mitchell was in the group selected for astronaut training in April 1966. He served as a member of the astronaut support crew for Apollo 9 and as backup lunar module pilot for Apollo 10. He completed his first space flight as lunar module pilot on Apollo 14, January 31 - February 9, 1971. With him on man's third lunar landing mission were Alan B. Shepard (spacecraft commander) and Stuart A. Roosa (command module pilot).

Maneuvering their lunar module, "Antares," to a landing in the hilly upland Fra Mauro region of the Moon, Shepard and Mitchell subsequently deployed and activated various scientific equipment and experiments and collected almost 100 pounds of lunar samples for return to Earth. Other Apollo 14 achievements include: first use of Mobile Equipment Transporter (MET); largest payload placed in lunar orbit; longest distance traversed on the lunar surface; largest payload returned from the lunar surface; longest lunar surface stay time (33 hours); longest lunar surface EVA (9 hours and 17 minutes); first use of shortened lunar orbit rendezvous techniques; first use of colored TV with new vidicon tube on lunar surface; and first extensive orbital science period conducted during CSM solo operations.

In completing his first space flight, Mitchell logged a total of 216 hours and 42 minutes.

He was designated as backup lunar module pilot for Apollo 16 on March 3, 1971.

SPACEFLIGHT TRACKING & DATA SUPPORT NETWORK

NASA's worldwide Spaceflight Tracking and Data Network (STDN) will provide communication with the Apollo astronauts, their launch vehicle and spacecraft. It will also maintain the communications link between Earth and the Apollo experiments left on the lunar surface and track the particles and fields subsatellite injected into lunar orbit during Apollo 16.

The STDN is linked together by the NASA Communication Network (NASCOM) which provides for all information and data flow.

In support of Apollo 16, the STDN will employ 11 ground tracking stations equipped with 9.1-meter (30-foot) and 25.9-meter (85-foot) antennas, an instrumented tracking ship, and four instrumented aircraft. This portion of the STDN was known formerly as the Manned Space Flight Network. For Apollo 16, the network will be augmented by the 64-meter (210-foot) antenna system at Goldstone, California (a unit of NASA's Deep Space Network).

The STDN is maintained and operated by the NASA Goddard Space Flight Center, Greenbelt, Md., under the direction of NASA's Office of Tracking and Data Acquisition. Goddard will become an emergency control center if the Houston Mission Control Center is impaired for an extended time.

NASA Communications Network (NASCOM): The tracking network is linked together by the NASA Communications Network. All information flows to and from MCC Houston and the Apollo spacecraft over this communications system.

The NASCOM consists of more than 3.2 million circuit kilometers (1.7 million nm), using satellites, submarine cables, land lines, microwave systems, and high frequency radio facilities. NASCOM control center is located at Goddard. Regional communication switching centers are in Madrid; Canberra, Australia; Honolulu; and Guam.

Intelsat communications satellites will be used for Apollo 16. One satellite over the Atlantic will link Goddard with Ascension Island and the vanguard tracking ship. Another Atlantic satellite will provide a direct link between Madrid and Goddard for TV signals received from the spacecraft. One satellite positioned over the mid-Pacific will link Carnarvon, Canberra, Guam and Hawaii with Goddard through the Jamesburg, California ground station. An alternate route of communications between Spain and Australia is available through another Intelsat satellite positioned over the Indian Ocean if required.

Mission Operations: Prelaunch tests, liftoff, and Earth orbital flight of the Apollo 16 are supported by the Apollo subnet station at Merritt Island Fla., 6.4 km (3.5 nm) from the launch pad.

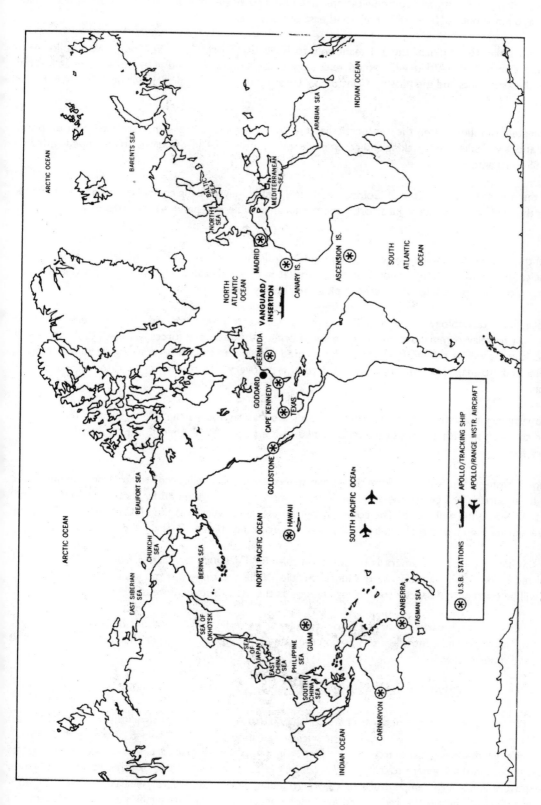

MANNED SPACE FLIGHT TRACKING NETWORK

During the critical period of launch and insertion of the Apollo 16 into Earth orbit, the USNS Vanguard provides tracking, telemetry, and communications functions. This single sea-going station of the Apollo subnet will be stationed about 1610 km (870 nm) southeast of Bermuda.

When the Apollo 16 conducts the TLI maneuver to leave Earth orbit for the Moon, two Apollo range instrumentation aircraft (ARIA) will record telemetry data from Apollo and relay voice communications between the astronauts and the Mission Control Center at Houston. These aircraft will be airborne between Australia and Hawaii.

Approximately one hour after the spacecraft has been injected into a translunar trajectory, three prime MSFN stations will take over tracking and communication with Apollo. These stations are equipped with 25.9 meter (85-foot) antennas.

Each of the prime stations, located at Goldstone, Madrid, and Honeysuckle is equipped with dual systems for tracking the command module in lunar orbit and the lunar module in separate flight paths or at rest on the Moon.

For reentry, two ARIA (Apollo Range Instrumented Aircraft) will be deployed to the landing area to relay communications between Apollo and Mission Control at Houston. These aircraft also will provide position information on the Apollo after the blackout phase of reentry has passed.

An applications technology satellite (ATS) terminal has been placed aboard the recovery ship USS Ticonderoga to relay command control communications of the recovery forces, via NASA's ATS satellite: Communications will be relayed from the deck mounted terminal to the NASA tracking stations at Mojave, California and Rosman, N.C. through Goddard to the recovery control centers located in Hawaii and Houston.

Prior to recovery, the astronauts aeromedical records are transmitted via the ATS satellite to the recovery ship for comparison with the physical data obtained in the post flight examination performed aboard the recovery ship.

Television Transmissions: Television from the Apollo spacecraft during the journey to and from the moon and on the lunar surface will be received by the three prime stations, augmented by the 64-meter (210-foot) antennas at Goldstone and Parkes. The color TV signal must be converted at the MSC Houston. A black and white version of the color signal can be released locally from the stations in Spain and Australia.

Before the lunar surface TV camera is mounted on the LRV TV signals originating from the Moon will be transmitted to the MSFN stations via the lunar module. While the camera is mounted on the LRV, the TV signals will be transmitted directly to tracking stations as the astronauts explore the Moon.

Once the LRV has been parked near the lunar module, its batteries will have about 80 hours of operating life. This will allow ground controllers to position the camera for viewing the lunar module liftoff, post lift-off geology, and other scenes.

ENVIRONMENTAL IMPACT OF APOLLO/SATURN V MISSION

Studies of NASA space mission operations have concluded that Apollo does not significantly effect the human environment in the areas of air, water, noise or nuclear radiation. During the launch of the Apollo/Saturn V space vehicle, products exhausted from Saturn first stage engines in all cases are within an ample margin of safety. At lower altitudes, where toxicity is of concern, the carbon monoxide is oxidized to carbon dioxide upon exposure at its high temperature to the surrounding air. The quantities released are two or more orders of magnitude below the recognized levels for concern in regard to significant modification of the environment. The second and third stage main propulsion systems generate only water and a small amount of hydrogen. Solid propellant ullage and retro rocket products are released and rapidly dispersed in the upper atmosphere at altitudes above 70 kilometers (43.5 miles). This material will effectively never reach sea level and, consequently, poses no toxicity hazard.

Should an abort after launch be necessary, some RP-1 fuel (kerosene) could reach the ocean. However, toxicity of RP-1 is slight and impact on marine life and waterfowl are considered negligible due to its dispersive characteristics. Calculations of dumping an aborted S-IC stage into the ocean showed that spreading and evaporating of the fuel occurred in one to four hours.

There are only two times during a nominal Apollo mission when above normal overall sound pressure levels are encountered. These two times are during vehicle boost from the launch pad and the sonic boom experienced when the spacecraft enters the Earth's atmosphere. Sonic boom is not a significant nuisance since it occurs over the mid-Pacific ocean.

NASA and the Department of Defense have made a comprehensive study of noise levels and other hazards to be encountered for launching vehicles of the Saturn V magnitude. For uncontrolled areas the overall sound pressure levels are well below those which cause damage or discomfort. Saturn launches have had no deleterious effects on wildlife which has actually increased in the NASA-protected areas of Merritt Island.

A source of potential radiation hazard but highly unlikely, is the fuel capsule of the radioisotope thermoelectric generator supplied by the Atomic Energy Commission which provides electric power for Apollo lunar surface experiments. The fuel cask is designed to contain the nuclear fuel during normal operations and in the event of aborts so that the possibility of radiation contamination is negligible. Extensive safety analyses and tests have been conducted which demonstrated that the fuel would be safely contained under almost all credible accident conditions.

PROGRAM MANAGEMENT

The Apollo Program is the responsibility of the Office of Manned Space Flight (OMSF), National Aeronautics and Space Administration, Washington, D. C. Dale D. Myers is Associate Administrator for Manned Space Flight.

NASA Manned Spacecraft Center (MSC), Houston, is responsible for development of the Apollo spacecraft, flight crew training, and flight control. Dr. Christopher C. Kraft, Jr. is Center Director.

NASA Marshall Space Flight Center (MSFC), Huntsville, Ala., is responsible for development of the Saturn launch vehicles. Dr. Eberhard F. M. Rees is Center Director.

NASA John F. Kennedy Space Center (KSC), Fla., is responsible for Apollo/Saturn launch operations. Dr. Kurt H. Debus is Center Director.

The NASA Office of Tracking and Data Acquisition (OTDA) directs the program of tracking and data flow on Apollo. Gerald M. Truszynski is Associate Administrator for Tracking and Data Acquisition.

NASA Goddard Space Flight Center (GSFC), Greenbelt, Md., manages the Manned Space Flight Network and Communications Network. Dr. John F. Clark is Center Director.

The Department of Defense is supporting NASA during launch, tracking, and recovery operations. The Air Force Eastern Test Range is responsible for range activities during launch and down-range tracking. Recovery operations include the use of recovery ships and Navy and Air Force aircraft.

Apollo/Saturn Officials

NASA Headquarters

Dr. Rocco A. Petrone	Apollo Program Director, OMSF
Chester M. Lee (Capt., USN, Ret.)	Apollo Mission Director, OMSF
John K. Holcomb (Capt., USN, Ret.)	Director of Apollo Operations, OMSF
William T. O'Bryant (Capt., USN, Ret.)	Director of Apollo Lunar Exploration, OMSF

Kennedy Space Center

Miles J. Ross	Deputy Center Director
Walter J. Kapryan	Director of Launch Operations
Raymond L. Clark	Director of Technical Support
Robert C. Hock	Apollo/Skylab Program Manager
Dr. Robert H. Gray	Deputy Director, Launch Operations
Dr. Hans F. Gruene	Director, Launch Vehicle Operations
John J. Williams	Director, Spacecraft Operations
Paul C. Donnelly	Associate Director Launch Operations
Isom A. Rigell	Deputy Director for Engineering

Manned Spacecraft Center

Sigurd A. Sjoberg	Deputy Center Director, and Acting Director, Flight Operations
Brig. General James A. McDivitt (USAF)	Manager, Apollo Spacecraft Program
Donald K. Slayton	Director, Flight Crew Operations
Pete Frank	Flight Director
Phil Shaffer	Flight Director
Gerald D. Griffin	Flight Director
Eugene F. Kranz	Flight Director
Donald Puddy	Flight Director
Richard S. Johnston	Director, Medical Research and Operations (Acting)

Marshall Space Flight Center

Dr. Eberhard Rees	Director
Dr. Willian R. Lucas	Deputy Center Director, Technical
Richard W. Cook	Deputy Center Director, Management
James T. Shepherd	Director, Program Management
Herman F. Kurtz	Manager, Mission Operations Office
Richard G. Smith	Manager, Saturn Program Office
John C. Rains	Manager, S-IC Stage Project, Saturn Program Office
William F. LaHatte	Manager, S-II S-IVB Stages Project, Saturn Program Office
Frederich Duerr	Manager, Instrument Unit/GSE Project, Saturn Program Office
William D. Brown	Manager, Engine Program Office
James M. Sisson	Manager, LRV Project, Saturn Program Office

Goddard Space Flight Center

Ozro M. Covington	Director, Networks
William P. Varson	Chief, Network Computing & Analysis Division
H. William Wood	Chief, Network Operations Division

| Robert Owen | Chief, Network Engineering Division |
| L. R. Stelter | Chief, NASA Communications Division |

Department of Defense

Maj. Gen. David M. Jones, USAF	DOD Manager for Manned Space Flight Support Operations
Col. Alan R. Vette, USAF	Deputy DOD Manager for Manned Space Flight Support Operations, and Director, DOD Manned Space Flight Support Office
Rear Adm. Henry S. Morgan, Jr., USN	Commander, Task Force 130, Pacific Recovery Area
Rear Adm. Roy G. Anderson, USN	Commander Task Force 140, Atlantic Recovery Area
Capt. E. A. Boyd, USN	Commanding officer, USS Ticonderoga, CVS-14 Primary Recovery Ship
Brig. Gen. Frank K. Everest, Jr., USAF	Commander Aerospace Rescue and Recovery Service

CONVERSION TABLE

	Multiply	BY	To Obtain
Distance:	inches	2.54	centimeters
	feet	0.3048	meters
	meters	3.281	feet
	kilometers	3281	feet
	kilometers	0.6214	statute miles
	statute miles	1.609	kilometers
	nautical miles	1.852	kilometers
	nautical miles	1.1508	statute miles
	statute miles	0.8689	nautical miles
	statute miles	1760	yards
Velocity:	feet/sec	0.3048	meters/sec
	meters/sec	3.281	feet/sec
	meters/sec	2.237	statute mph
	feet/sec	0.6818	statute miles/hr
	feet/sec	0.5925	nautical miles/hr
	statute miles/hr	1.609	km/hr
	nautical miles/hr (knots)	1.852	km/hr
	km/hr	0.6214	statute miles/hr
Liquid measure, weight:	gallons	3.785	liters
	liters	0.2642	gallons
	pounds	0.4536	kilograms
	kilograms	2.205	pounds
	metric ton	1000	kilograms
	short ton	907.2	kilograms
Volume:	cubic feet	0.02832	cubic meters
Pressure:	pounds/sq. inch	70.31	grams/sq. cm
Thrust:	pounds	4.448	newtons
	newtons	0.225	pounds
Temperature:	Centigrade	1.8; add 32	Fahrenheit

Report No. M-933-72-16

MISSION OPERATION REPORT

APOLLO 16 MISSION

OFFICE OF MANNED SPACE FLIGHT

Prelaunch Mission Operation Report No. M-933-72-16

MEMORANDUM
3 April 1972
TO: A/Administrator
FROM: MA/Apollo Program Director

SUBJECT: Apollo 16 Mission (AS-511)

We plan to launch Apollo 16 from Pad A of Launch Complex 39 at the Kennedy Space Center no earlier than 16 April 1972. This will be the fifth manned lunar landing and the second of the Apollo "J" series missions which carry the Lunar Roving Vehicle for surface mobility, added Lunar Module consumables for a longer surface stay time, and the Scientific Instrument Module for extensive lunar orbital science investigations.

Primary objectives of this mission are selenological inspection, survey, and sampling of materials and surface features in a preselected area of the Descartes region of the moon; emplacement and activation of surface experiments; and the conduct of in-flight experiments and photographic tasks. In addition to the standard photographic documentation of operational and scientific activities, television coverage is planned for selected periods in the spacecraft and on the lunar surface. The lunar surface TV coverage will include remote controlled viewing of astronaut activities at each major science station on the three EVA traverses.

The 12-day mission will be terminated with the Command Module landing in the mid-Pacific Ocean about 150 NM north of Christmas Island.

Rocco A. Petrone

APPROVAL:
Dale D. Myers;
Associate Administrator
Manned Space Flight

NOTICE: This document may be exempt from public disclosure under the Freedom of Information Act (5 U.S.C. 552). Requests for its release to persons outside the U.S. Government should be handled under the provisions of NASA Policy Directive 1382.2.

MISSION OPERATION REPORTS are published expressly for the use of NASA Senior Management, as required by the Administrator in NASAManagement Instruction HQMI 8610. 1, effective 30 April 1971. The purpose of these reports is to provide NASA Senior Management with timely, complete, and definitive information on flight mission plans, and to establish official Mission Objectives which provide the basis for assessment of mission accomplishment.

Prelaunch reports are prepared and issued for each flight project just prior to launch. Following launch, updating (Post Launch) reports for each mission are issued to keep General Management currently informed of definitive mission results as provided in NASA Management Instruction HQMI 8610.1.

Primary distribution of these reports is intended for personnel having program/project management responsibilities which sometimes results in a highly technical orientation. The Office of Public Affairs publishes a comprehensive series of reports on NASA flight missions which are available for dissemination to the Press.

APOLLO MISSION OPERATION REPORTS are published in two volumes: the MISSION OPERATION REPORT (MOR); and the MISSION OPERATION REPORT, APOLLO SUPPLEMENT. This format was designed to provide a mission-oriented document in the MOR, with supporting equipment and facility description in the MOR, APOLLO SUPPLEMENT. The MOR, APOLLO SUPPLEMENT is a program-oriented reference document with a broad technical description of the space vehicle and associated equipment, the launch complex, and mission control and support facilities.

Published and Distributed by PROGRAM and SPECIAL REPORTS DIVISION (XP) EXECUTIVE SECRETARIAT - NASA HEADQUARTERS

SUMMARY OF   APOLLO/SATURN FLIGHTS

Mission	Launch Date	Launch Vehicle	Payload	Description
AS-201	2/26/66	SA-201	CSM-009	Launch vehicle and CSM development. Test of CSM subsystems and of the space vehicle. Demonstration of reentry adequacy of the CM at earth orbital conditions.
AS-203	7/5/66	SA-203	LH$_2$ in S-IVB	Launch vehicle development. Demonstration of control of LH$_2$ by continuous venting in orbit.
AS-202	8/25/66	SA-202	CSM-011	Launch vehicle and CSM development. Test of CSM subsystems and of the structural integrity and compatibility of the space vehicle. Demonstration of propulsion and entry control by G&N system. Demonstration of entry at 28,500 fps.
Apollo 4	11/9/67	SA-501	CSM-017 LTA-10R	Launch vehicle and spacecraft development. Demonstration of Saturn V Launch Vehicle performance and of CM entry at lunar return velocity.
Apollo 5	1/22/68	SA-204	LM-1 SLA-7	LM development. Verified operation of LM sub systems: ascent and descent propulsion systems (including restart) and structures. Evaluation of LM staging. Evaluation of S-IVB/IU orbital performance.
Apollo 6	4/4/68	SA-502	CM-020 SM-014 LTA-2R SLA-9	Launch vehicle and spacecraft development. Demonstration of Saturn V Launch Vehicle performance.
Apollo 7	10/11/68	SA-205	CM-101 SM-101 SLA-5	Manned CSM operations. Duration 10 days 20 hours.
Apollo 8	12/21/68	SA-503	CM-103 SM-103 LTA-B SLA-11	Lunar orbital mission. Ten lunar orbits. Mission duration 6 days 3 hours. Manned CSM operations.
Apollo 9	3/3/69	SA-504	CM-104 SM-104 LM-3 SLA-12	Earth orbital mission. Manned CSM/LM operations. Duration 10 days 1 hour.
Apollo 10	5/18/69	SA-505	CM-106 SM-106 LM-4 SLA-13	Lunar orbital mission. Manned CSM/LM operations. Evaluation of LM performance in cislunar and lunar environment, following lunar landing profile. Mission duration 8 days.
Apollo 11	7/16/69	SA-506	CM-107 SM-107 LM-5 SLA-14 EASED	First manned lunar landing mission. Lunar surface stay time 21.6 hours. One dual EVA (5 man hours). Mission duration 8 days 3.3 hours.
Apollo 12	11/14/69	SA-507	CM-108 SM-108 LM-6 SLA-15 ALSEP	Second manned lunar landing mission. Demonstration of point landing capability. Deployment of ALSEP I. Surveyor III investigation. Lunar surface stay time 31.5 hours. Two dual EVAs (15.5 man hours). Mission duration 10 days 4.6 hours.
Apollo 13	4/11/70	SA-508	CM- 109 SM-109 LM-7 SLA-16 ALSEP	Planned third lunar landing. Mission aborted at approximately 56 hours due to loss of SM cryogenic oxygen and consequent loss of capability to generate electrical power and water. Mission duration 5 days 22.9 hours.
Apollo 14	1/31/71	SA-509	CM-110 SM-110 LM-8 SLA-17 ALSEP	Third manned lunar landing mission. Selenological inspection, survey and sampling of materials of Fra Mauro Formation. Deployment of ALSEP. Lunar surface stay time 33.5 hours. Two dual EVAs (18.8 man hours). Mission duration 9 days.

| Apollo 15 | 7/26/71 | SA-510 | CM-112 SM-112 LM-10 SLA-19 LRV-1 ALSEP Subsatellite | Fourth manned lunar landing mission. Selenological inspection, survey and sampling of materials of the Hadley-Apennine Formation. Deployment of ALSEP. Increased lunar stay time to 66.9 hours. First use of Lunar Roving Vehicle and direct TV and voice communications to earth during EVAs. Total distance traversed on lunar surface 27.9 km. Three dual EVAs (37.1 man hours). Mission duration 12 days 7.2 hours. |

NASA OMSF MISSION OBJECTIVES FOR APOLLO 16 PRIMARY OBJECTIVES

* Perform selenological inspection, survey, and sampling of materials and surface features in a preselected area of the Descartes region.
* Emplace and activate surface experiments.
* Conduct in-flight experiments and photographic tasks.

Rocco A. Petrone
Apollo Program Director

Dale D. Myers
Associate Administrator Manned Space Flight
Date: 29 March 1972   Date: 31 March 1972

MISSION OPERATIONS

The following paragraphs contain a brief description of the nominal launch, flight, recovery, and post-recovery operations. For the second and third months launch opportunities, which may involve a T-24 or T+24 hour launch, there will be a revised plan. Overall mission profile is shown in Figure 1.

LAUNCH WINDOWS

The mission planning considerations for the launch phase of a lunar mission are, to a major extent, related to launch windows. Launch windows are defined for two different time periods: a "daily window" has a duration of a few hours during a given 24-hour period; a "monthly window" consists of a day or days which meet the mission operational constraints during a given month or lunar cycle.

Launch windows are based on flight azimuth limits of $72°$ to $100°$ (earth-fixed heading of the launch vehicle at end of the roll program), on booster and spacecraft performance, on insertion tracking, and on lighting constraints for the lunar landing sites. The Apollo 16 launch windows and associated lunar landing sun elevation angles are presented in Table 1.

TABLE 1 LAUNCH WINDOWS

	LAUNCH DATE WINDOWS*		SUN ELEVATION
	OPEN	CLOSE	ANGLE
16 April 1972	1254	1643	$11.9°$
14 May 1972	1217	1601	6.80
15 May 1972	1230	1609	$6.8°$
16 May 1972	1238	1613	$18.6°$
13 June 1972	1050	1423	$13.0°$
14 June 1972	1057	1426	$13.0°$
15 June 1972		(Still Under Review)	

* April times are Eastern Standard Time; all others are Eastern Daylight Time

LAUNCH THROUGH TRANSLUNAR INJECTION

The space vehicle will be launched from Pad A of launch complex 39 at the Kennedy Space Center. The boost into a 90-NM earth parking orbit (EPO) will be accomplished by sequential burns and staging of the S-IC and S-II launch vehicle stages and a partial burn of the S-IVB stage. The S-IVB/instrument unit (IU) and spacecraft will coast in a circular EPO for approximately 1.5 revolutions while preparing for the first opportunity S-IVB translunar injection (TLI) burn, or 2.5 revolutions if the second opportunity TLI burn is required. Both injection opportunities are to occur over the Pacific Ocean.

The S-IVB TLI burn will place the S-IVB/IU and spacecraft on a translunar trajectory targeted such that transearth return to an acceptable entry corridor can be achieved with the use of the reaction control system (RCS) during at least 5 hours (7 hours 39 minutes ground elapsed time (GET)) after TLI cutoff. For this mission the RCS capability will actually exist from 51-57 hours GET for the command service module/ lunar module (CSM/LM) combination and from 61-63 hours GET for the CSM only. TLI targeting will permit an acceptable earth return to be achieved using the service propulsion system (SPS) or LM descent propulsion system (DPS) until at least pericynthian plus 2 hours, if lunar orbit insertion (LOI) is not performed.

TRANSLUNAR COAST THROUGH LUNAR ORBIT INSERTION

Within 2 hours after injection the CSM will separate from the S-IVB/IU and spacecraft-LM adapter (SLA) and will transpose, dock with the LM, and eject the LM/CSM from the S-IVB/IU. Subsequently, the S-IVB/IU will perform an evasive maneuver to alter its circumlunar coast trajectory clear of the spacecraft trajectory.

APOLLO 16
FLIGHT PROFILE

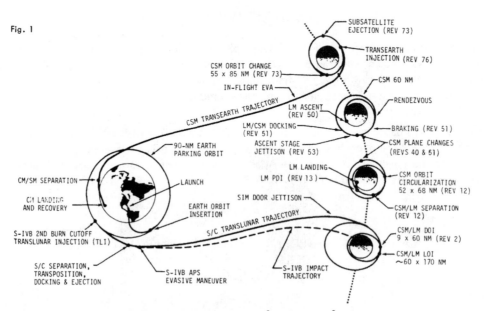

The spent S-IVB/IU will be impacted on the lunar surface at 2° 18'S and 31° 42'W providing a stimulus for the Apollo 12, 14, and 15 emplaced seismology experiments. The necessary delta velocity (Delta-V) required to alter the S-IVB/IU circumlunar trajectory to the desired impact trajectory will be derived from dumping of residual liquid oxygen (LOX) and burn(s) of the S-IVB/auxiliary propulsion system (APS) and ullage motors. The final maneuver will occur within about 10 hours of liftoff. The IU will have an S-band transponder for trajectory tracking. A frequency bias will be incorporated to insure against interference between the S-IVB/IU and LM communications during translunar coast.

Spacecraft passive thermal control will be initiated after the first midcourse correction (MCC) opportunity and will be maintained throughout the translunar-coast phase unless interrupted by subsequent MCCs and/or navigational activities. The scientific instrument module (SIM) bay door will be jettisoned shortly after the MCC-4 point, about 4.5 hours before LOI.

Multiple-operation covers over the SIM bay experiments and cameras will provide thermal and contamination protection whenever they are not in use.

A retrograde SPS burn will be used for LOI of the docked spacecraft into a 60 x 170-NM orbit, where they will remain for approximately two revolutions.

DESCENT ORBIT INSERTION THROUGH LANDING

The descent orbit insertion (DOI) maneuver, a SPS second retrograde burn, will place the CSM/LM combination into a 60 x 9-NM orbit.

A "soft" undocking will be made during the 12th revolution, using the docking probe capture latches to reduce the imparted Delta-V. Spacecraft separation will be executed by the SM RCS, providing a Delta-V of approximately 1 foot per second (fps) radially downward toward the center of the moon. The CSM will circularize its orbit to 60 NM near the end of the 12th revolution.

During the 13th revolution the LM DPS will be used for powered descent, which will begin approximately at pericynthian. These events are shown in Figure 2. A lurain profile model will be available in the LM guidance computer (LGC) program to minimize unnecessary LM pitching or thrusting maneuvers. A descent path of $25°$ will be used during the terminal portion of powered descent (from high gate) to enhance landing site visibility. The automatic vertical descent portion of the landing phase will start at an altitude of about 200 feet at a rate of 5 fps, and will be terminated at touchdown on the lunar surface.

DOI, SEPARATION, & CSM CIRCULARIZATION

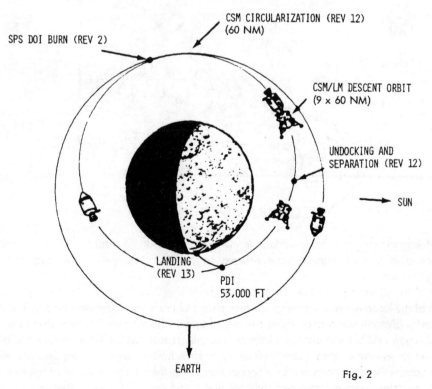

CSM CIRCULARIZATION (REV 12)
(60 NM)

SPS DOI BURN (REV 2)

CSM/LM DESCENT ORBIT
(9 x 60 NM)

UNDOCKING AND
SEPARATION (REV 12)

SUN

LANDING
(REV 13)

PDI
53,000 FT

EARTH

Fig. 2

LANDING SITE (DESCARTES REGION)

Descartes is designated as the landing site for the Apollo 16 Mission. The Descartes landing site lies in the central lunar highlands several hundred kilometers west of Mare Nectaris, and in hilly, grooved, and furrowed lurain which is morphologically similar to many terrestrial areas of volcanism. The Descartes area is also the site of extensive development of highland plains-forming material, a geologic unit of widespread occurrence in the lunar highlands.

There are three major units in the area: the first is a constructional unit which continues from the area of the site southward to the crater Descartes itself; the second is a highland unit controlled by the fracture system of Mare Imbrium; and the third unit is relatively flat, smooth, highland plains formation where the landing will be made. Knowledge of the composition, age, and extent of magmatic differentiation in a highland volcanic complex will be particularly important in understanding lunar volcanism and its contribution to the lunar highlands. Comparison of these highland materials with the mare samples of Apollo 11, 12, and 15 and the upland materials collected on Apollo 14 and 15 will provide a basis for conclusions relative to the gross compositional variations and early evolution of the lunar crust.

The planned landing point coordinates are 9°00'01"S, 15°30'59"E (Figure 3).

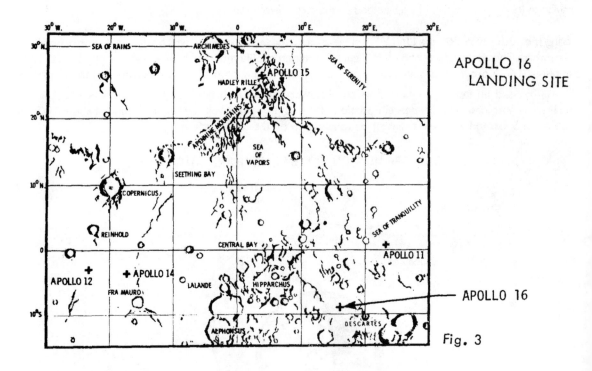

APOLLO 16
LANDING SITE

APOLLO 16

Fig. 3

LUNAR SURFACE OPERATIONS

The nominal stay time on the lunar surface is planned for about 73 hours, with the overall objective of optimizing effective surface science time relative to hardware margins, crew duty cycles, and other operational constraints.

Photographs of the lunar surface will be taken through the LM cabin window after landing. The nominal extra-vehicular activity (EVA) is planned for three periods of up to 7 hours each. The duration of each EVA period will be based upon real time assessment of the remaining consumables. As in Apollo 15 this mission will employ the lunar roving vehicle (LRV) which will carry both astronauts, experiment equipment, and independent communications systems for direct contact with the earth when out of the line-of-sight of the LM relay system. Voice communication will be continuous and color TV coverage will be provided at each

DESCARTES LRV TRAVERSES

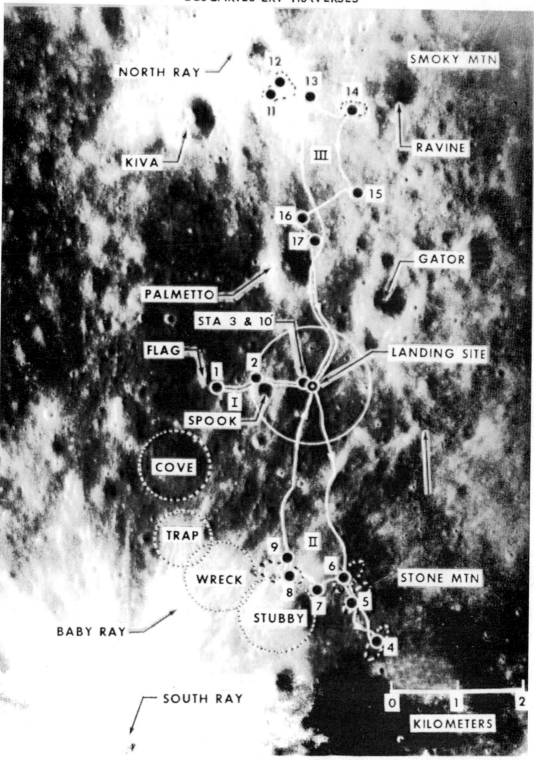

Fig. 4

major science stop (Figure 4) where the crew will align the high gain antenna. The ground controllers will then assume control of the TV through the ground controlled television assembly (GCTA) mounted on the LRV. A TV panorama is planned at each major science stop, followed by coverage of the astronauts' scientific activities.

The radius of crew operations will be constrained by the LRV capability to return the crew to the LM in the event of a portable life support system (PLSS) failure or by the PLSS walkback capability in the event of an LRV failure, whichever is the most limiting at any point in the EVA.

If a walking traverse must be performed, the radius of operations will be constrained by the buddy secondary life support system (BSLSS) capability to return the crew to the LM in the event of a PLSS failure.

NEAR LM LUNAR SURFACE ACTIVITY

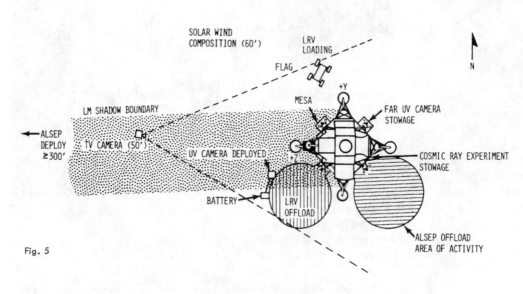

Fig. 5

EVA PERIODS

The activities to be performed during each EVA period are described below. Rest periods are scheduled prior to the second and third EVAs and prior to LM liftoff. Traverses performed after arming of the active seismic experiment (ASE) have been planned so as to avoid the line of fire of the ASE mortars.

The lunar communications relay unit (LCRU) and the GCTA will be used in conjunction with LRV operations. Television coverage will be provided by the GCTA during each major science stop when using the LRV.

The three traverses planned for Apollo 16 are designed for flexibility in selection of science stops as indicated by the enclosed areas shown along traverses II and III (Figure 4).

First EVA Period

The first EVA will include: LM inspection, LRV deployment and checkout, deployment of the far UV camera/spectroscope and cosmic ray deflector experiments, and deployment and activation of the Apollo lunar surface experiments package (ALSEP). Television will be deployed as soon as possible in this period for observation of crew activities near the LM (Figure 5). ALSEP deployment will be approximately 300 feet west of the LM (Figure 6). After ALSEP activation the crew will perform a geology traverse (see Figure 4).

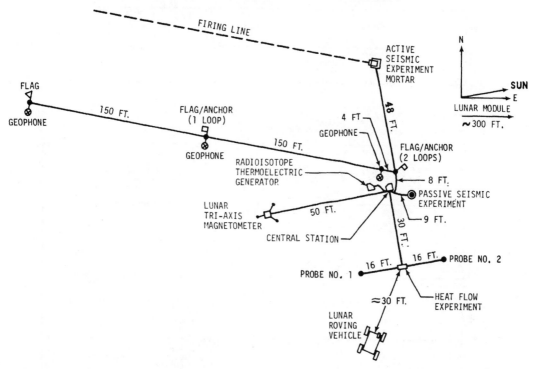

Fig. 6 **APOLLO 16 ALSEP DEPLOYMENT**

The data acquisition camera and Hasselblad cameras, using color film, will be used during the EVA to record lunar surface operations. High resolution photographic survey of surface features will be accomplished with the Hasselblad camera equipped with the 500mm lens. Lunar samples collected will be verbally and photographically documented. Sample return must be assured; therefore, a contingency sample of lunar soil will be collected in the event of a contingency during the EVA, but only if no other soil sample has been collected and is available for return to earth. The planned timeline for all EVA-1 activities is presented in Figure 7.

Second and Third EVA Periods

Traverses in the second and third EVA periods (Figures 8 and 9) are planned to maximize the scientific return in support of the primary objectives.

LRV sorties will be planned for flexibility in selecting stops and conducting experiments. Consumables usage will be monitored at Mission Control Center (MCC) to assist in real time traverse planning.

The major portion of the lunar geology investigation (S-059), portable magnetometer (S-198), and the soil mechanics experiment (S-200) will be conducted during the second and third EVAs and will include voice and photographic documentation of sample material as it is collected and descriptions of lurain features.

If time does not permit filling the sample containers with documented samples, the crew may fill the containers with samples selected for scientific interest.

The LRV will be positioned at the end of the EVA-3 traverse to enable GCTA-monitored ascent and other TV observations of scientific interest.

Fig. 7 # APOLLO 16 EVA-1 TIMELINE

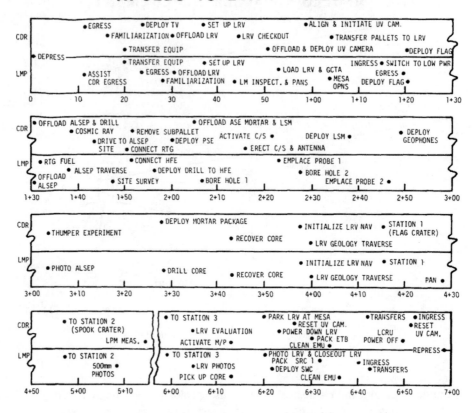

Fig. 8 # APOLLO 16 EVA-2 TIMELINE

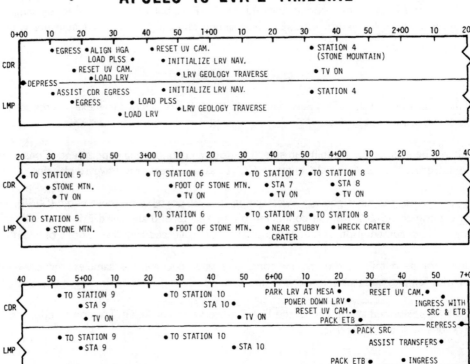

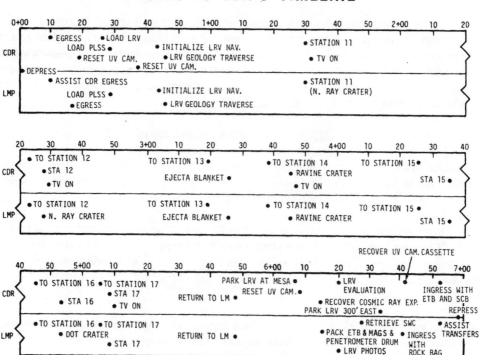

Fig. 9   **APOLLO 16 EVA-3 TIMELINE**

LUNAR ORBIT OPERATIONS

The Apollo 16 Mission is the second with the modified Block II CSM configuration.

An increase in cryogenic storage provides increased mission duration for the performance of both an extended lunar surface stay time and a lunar orbit science period. The SIM in the SM provides for the mounting of scientific experiments and for their operation in flight.

After the SIM door is jettisoned by pyrotechnic charges and until completion of lunar orbital science tasks, selected RCS thrusters will be inhibited or experiment protective covers will be closed to minimize contamination of experiment sensors during necessary RCS burns. Attitude changes for thermal control and experiment alignment with the lunar surface and deep space (and away from direct sunlight) will be made with the active RCS thrusters. Orbital science activities have been planned at appropriate times throughout the lunar phase of the mission and consist of the operation of five cameras (35mm Nikon, 16mm data acquisition, 70mm Hasselblad, 24-inch panoramic and a 3-inch mapping), a color TV camera, a laser altimeter, a gamma ray spectrometer, X-ray fluorescence equipment, alpha particle spectrometer equipment and mass spectrometer equipment.

Pre-Rendezvous Lunar Orbit Science

Orbital science operations will be conducted during the 60 x 9-NM orbits after DOI, while in the docked configuration. Orbital science operations will be stopped for the separation and circularization maneuvers performed during the 12th revolution, then restarted after CSM circularization.

In the event of a T-24 launch, the additional day in the 60 x 9-NM orbit prior to lunar landing will also be used for orbital science.

The experiments timeline has been developed in conjunction with the surface timeline to provide, as nearly as possible, 16-hour work days and concurrent 8-hour CSM and LM crew sleep periods. Experiment activation cycles are designed to have minimum impact on crew work-rest cycles.

Conduct of orbital experiments and photographic tasks have been planned in consideration of: mass spectrometer and gamma-ray spectrometer boom extend/retract requirements; outgassing, stand-by, and warm-up periods; experiments fields-of-view limitations; and Manned Space Flight Network (MSFN) data collection requirements. Water and urine dumps and fuel cell purges have been planned to avoid conflict with operation cycles. Prior to LM liftoff, the CSM will perform a plane change maneuver to provide the desired coplanar orbit at the time of the LM rendezvous.

LM Ascent, Rendezvous and Jettison

After completion of lunar surface activities and ascent preparations, the LM ascent stage propulsion system (APS) and RCS will be used to launch and rendezvous with the CSM. The direct ascent rendezvous technique initiated on Apollo 14 and subsequently used on Apollo 15 will be performed.

The LM ascent stage liftoff window duration is about 30 seconds and is constrained to keep the perilune above 8 NM. The ascent stage will be inserted into a 48 x 9-NM orbit so that an APS terminal phase initiation (TPI) burn can be performed approximately 47 minutes after insertion. The final braking maneuver will occur about 41 minutes later. The total time from ascent stage liftoff to the final braking maneuver will be about 95 minutes.

Docking will be accomplished by the CSM with RCS maneuvers. Once docked, the two LM crewmen will transfer to the CSM with lunar sample material, exposed films, designated equipment.

The LM ascent stage will be jettisoned and subsequently deorbited to impact on the lunar surface to provide a known stimulus for the emplaced seismic experiment. The impact will be targeted for 9°28'S and 14°58'E, about 23 km from the Apollo 16 ALSEP.

Post-Rendezvous Lunar Orbit Science

A period of orbital science activities will be conducted following LM jettison. If SPS Delta-V reserves permit, a plane change maneuver will be performed to increase the orbital inclination by at least 3°. An orbit shaping burn may be performed to insure at least 1 year of orbital lifetime for the subsatellite.

The subsatellite will be launched from the SIM in a predetermined orbit.

TRANSEARTH INJECTION THROUGH LANDING

After completion of the post-rendezvous CSM orbital activities, the SPS will perform a posigrade burn to inject the CSM onto the transearth trajectory. The nominal return time will not exceed 110 hours and the return inclination will not exceed 70° with relation to the earth's equator.

During the transearth phase there will be continuous communications coverage from the time the spacecraft appears from behind the moon until shortly prior to entry. MCCs will be made, if required. A 6-hour period, including pre- and post-EVA activities, will be planned to perform an in-flight EVA to retrieve film cassettes from the SIM bay in the SM and to conduct experiment M-191. TV and photographic tasks will be conducted during transearth coast. The CM will separate from the SM 15 minutes before the entry interface. Earth touchdown will be in the mid-Pacific and will nominally occur approximately 12.1 days after launch. Targeted landing coordinates are 5°0'N, 158°40'W.

POST-LANDING OPERATIONS

Flight Crew Recovery

Following splashdown, the recovery helicopter will drop swimmers and life rafts near the CM. The swimmers will install the flotation collar on the CM, attach the life raft, and pass fresh flight suits in through the hatch for the flight crew to don before leaving the CM. The crew will be transferred from the spacecraft to the recovery ship via life raft and helicopter and will return to Houston, Texas, for debriefing.

Quarantine procedures were eliminated prior to Apollo 15; therefore, the mobile quarantine facility will not be used. However, biological isolation garments will be available for use in the event of unexplained crew illness.

CM and Data Retrieval Operations

As a result of the partially collapsed main parachute experienced during the Apollo 15 landing, an attempt will be made to recover the earth landing system main parachutes on this mission.

In addition, the CM RCS propellants will not be vented during the Apollo 16 landing in order to preclude possible damage to the parachutes. After flight crew pickup by helicopter, the CM will be retrieved and placed on a dolly aboard the recovery ship, USS TICONDEROGA. The CM RCS helium pressure will be vented and the CM will be stowed near the ship's elevator to insure adequate ventilation.

Lunar samples, film, flight logs, etc., will be retrieved for shipment to the Lunar Receiving Laboratory (LRL). The spacecraft will be offloaded from the ship and transported to an area where deactivation of the propellant system will be accomplished. The CM will then be returned to contractor facilities.

ALTERNATE MISSIONS

If an anomaly occurs after liftoff that would prevent the space vehicle from following its nominal flight plan, an abort or an alternate mission will be initiated. An abort will provide for acceptable flight crew and CM recovery in the Atlantic or Pacific Ocean.

An alternate mission is a modified flight plan that results from a launch vehicle, spacecraft, or support equipment anomaly that precludes accomplishment of the primary mission objectives. The purpose of the alternate mission is to provide the flight crew and flight controllers with a plan by which the greatest benefit can be gained from the flight using the remaining systems capabilities.

The two general categories of alternate missions that can be performed during the Apollo 16 Mission are (1) earth orbital and (2) lunar orbital. Both of these categories have several variations which depend upon the nature of the anomaly leading to the alternate mission and the resulting systems status of the LM and CSM. An attempt will be made to launch the subsatellite in all the identified alternate missions. A brief description of these alternate missions is contained in the following paragraphs.

Earth Orbit

In case of no TLI burn, a mission of approximately 6-1/3 days will be conducted to obtain maximum benefit from the scientific equipment aboard. Subsequent to transfer of necessary equipment to the CM, the LM will be deorbited into the Pacific. A photography orbit of 240 x 114 NM will be established with the apogee over the United States to insure optimum SIM bay camera operation. The X-ray fluorescence spectrometer will be operated to investigate galactic X-ray sources.

The gamma ray spectrometer will be used to obtain data on the earth's gamma ray albedo and for gamma ray astronomy. The subsatellite will be jettisoned in the highest apogee orbit to insure the longest available lifetime for gathering science data from the particle detectors. Remaining SIM bay experiments will be operated on a non-interference basis to gather engineering data. Film cassettes will be retrieved by EVA on the last day of the mission.

Lunar Orbit

Lunar orbit missions of the following types will be planned if spacecraft systems will enable accomplishment of orbital science objectives in the event a lunar landing is not possible. If the SIM bay cameras are used, film cassettes will be retrieved by EVA during transearth coast. An attempt will be made to optimize orbital ground tracks in order to minimize real time flight planning activities.

CSM/LM (Operable DPS)

The translunar trajectory will be maintained to be within the DPS capability of an acceptable earth return until LOI plus 2 hours in the event LOI is not performed. If it is determined during translunar coast that a lunar landing mission cannot be performed, either the SPS or the LM DPS may be used to perform the LOI-1 maneuver to put the CSM/LM into an appropriate orbit. In the event the SIM bay door is not jettisoned prior to LOI, the LOI-1 maneuver will be performed with the SPS. The LOI-2 maneuver will place the CSM/LM in a 60-NM orbit. The orbital inclination is to be the maximum that is practicable considering total mission AV requirements. Orbital science and photographic tasks will be performed for up to approximately 6 days in lunar orbit.

If while in lunar orbit (following a nominal translunar coast and LOI trajectory) it is determined that a lunar landing mission cannot be performed, and the DPS is still available, the DPS will be used to perform a plane change to obtain maximum practicable orbital inclination. Orbital science and photographic tasks will be performed for up to approximately 6 days in lunar orbit.

An SPS capability to perform TEI on any revolution will be maintained.

CSM Alone

In the event the LM is not available following a nominal TLI burn, an SPS MCC-1 maneuver will place the CSM on a trajectory such that an acceptable return to earth can be achieved within the CM RCS capability. LOI will not be performed if the SIM bay door cannot be jettisoned. Orbital science and photographic tasks will be performed in a maximum practicable orbital inclination with the CSM remaining in a 60-NM orbit.

The duration in lunar orbit will be up to approximately 6 days.

If, following a nominal LOI maneuver it is determined that the DPS is inoperable, the LM will be jettisoned and an SPS circularization maneuver will be performed to obtain a maximum practicable orbital inclination. The CSM will generally remain in a 60-NM orbit. Orbital science and photographic tasks will be performed for up to approximately 6 days in the lunar orbit.

CSM Alone (From Landing Abort)

In the event the lunar landing is aborted, an orbital science mission will be accomplished by the CSM alone after rendezvous, docking, and LM jettison. The total lunar orbit time will be approximately 6 days.

EXPERIMENTS, DETAILED OBJECTIVES, IN-FLIGHT DEMONSTRATIONS, AND OPERATIONAL TESTS

The technical investigations to be performed on the Apollo 16 Mission are classified experiments, detailed objectives, in-flight demonstrations, or operational tests:

Experiment -   A technical investigation that supports science in general or provides engineering, technological, medical or other data and experience for application to Apollo lunar exploration or other programs and is recommended by the Manned Space Flight Experiments Board (MSFEB) and assigned by the Associate Administrator for Manned Space Flight to the Apollo Program for flight.

Detailed Objective - A scientific, engineering, medical or operational investigation that provides important data and experience for use in development of hardware and/or procedures for application to Apollo missions. Orbital photographic tasks, though reviewed by the MSFEB, are not assigned as formal experiments and will be processed as CM and SM detailed objectives.

In-flight Demonstration - A technical demonstration of the capability of an apparatus and/or process to illustrate or utilize the unique conditions of space flight environment. In-flight demonstration will be performed only on a noninterference basis with all other mission and mission-related activities. Utilization performance, or completion of these demonstrations will in no way relate to mission success.

Operational Test - A technical investigation that provides for the acquisition of technical data or evaluates operational techniques, equipment, or facilities but is not required by the objectives of the Apollo flight mission. An operational test does not affect the nominal mission timeline, adds no payload weight, and does not jeopardize the accomplishment of primary objectives, experiments, or detailed objectives.

EXPERIMENTS

The Apollo 16 Mission includes the following experiments:

Lunar Surface Experiments

Lunar surface experiments are deployed and activated or conducted by the Commander and the Lunar Module Pilot during EVA periods. Those experiments which are part of the ALSEP are so noted.

Lunar Passive Seismology (S-031) (ALSEP)

The passive seismic experiment is designed to monitor lunar seismic activity and to detect meteoroid impacts, free oscillations of the moon, surface tilt (tidal deformations), and changes in the vertical component of gravitational acceleration. The experiment sensor assembly is made up of three orthogonal, long-period seismometers and one vertical, short-period seismometer. The instrument and the near-lunar surface are covered by a thermal shroud.

Lunar Active Seismology (S-033) (ALSEP)

The active seismic experiment is designed to generate and monitor artificially stimulated seismic waves (3 Hz - 250 Hz) in the lunar surface and near subsurface. Naturally occurring seismic waves in the same frequency range will also be monitored on the experiment's emplaced geophones when they are active. Seismic waves will be produced by an astronaut-operated thumper device containing explosive initiators, as well as by an earth-commanded mortar package containing rocket-launched high explosive grenades. The grenades are designed to impact at various ranges from the geophones.

Lunar Tri-axis Magnetometer (S-034) (ALSEP)

The lunar surface magnetometer experiment is designed to measure the magnetic field on the lunar surface to differentiate any source producing the induced lunar magnetic field, to measure the permanent magnetic moment, and to determine the moon's bulk magnetic permeability during traverse of the neutral sheet in the geomagnetic tail.

The experiment has three sensors, each mounted at the end of a 3-foot long arm, which are first oriented parallel to obtain the field gradient and thereafter orthogonally to obtain total field measurements.

Lunar Heat Flow (S-037) (ALSEP)

The heat flow experiment is designed to determine the net lunar heat flux and the values of thermal parameters in the first 2.5 meters of the moon's crust.

The experiment has two sensor probes placed in bore holes drilled with the Apollo lunar surface drill (ALSD).

Lunar Geology Investigation (S-059)

The lunar geology experiment is designed to provide data for use in the interpretation of the geological history of the moon in the vicinity of the landing site. The investigation will be carried out during the planned lunar surface traverses and will utilize astronaut descriptions, camera systems, hand tools, core tubes, the ALSD, and sample containers. The battery-powered ALSD will be used to obtain core samples to a maximum depth of about 3 meters.

There are two major aspects of the experiment:

Documented Samples - Rock and soil samples representing different morphological and petrologic features will be described, photographed, and collected in individual pre-numbered bags for return to earth. This includes comprehensive samples of coarse fragments and fine lunar soil to be collected in pre-selected areas. Documented samples are the highest priority tasks in the experiment because they support many sample principal investigators in addition to lunar geology.

Geological Description and Special Samples - Descriptions and photographs of the field relationships of all accessible types of lunar features will be obtained.

Special samples, such as core tube samples, will be collected and documented for return to earth.

Solar Wind Composition (S-080)

The solar wind composition experiment is designed to measure the isotopic composition of noble gases in the solar wind, at the lunar surface, by entrapment of particles on an aluminum and platinum foil sheet.

A staff and yard arrangement is used to deploy the foil and maintain its plane perpendicular to the sun's rays. After return to earth, a spectrometric analysis of the particles entrapped in the foil allows quantitative determination of the helium, neon, argon, krypton, and xenon composition of the solar wind.

Cosmic Ray Detector (Sheets) (S-152)

The cosmic ray experiment is designed to measure, at the lunar surface, the flux, energy, spectrum, and the isotopic and charge distribution of solar cosmic rays heavier than helium, especially the abundant elements from carbon to iron, in the energy range up to 100 Mev/nucleon.

The instrument package consists of four types of detector material mounted on a panel: lexan polycarbonate plastics, aluminum foil, feldspar and pyroxene crystals, and mica sheets. This panel, which is bolted to the side of the LM, is exposed by the crew during the lunar stay time. At the end of the lunar surface mission the panel is returned to earth for detailed analysis.

Portable Magnetometer (S-198)

The portable magnetometer experiment is designed to measure the magnetic field at several points along a lunar surface traverse. The instrument is a fluxgate magnetometer having two ranges: +50 gamma and +100 gamma, with a resolution of 1 gamma. The tripod mounted instrument is connected by a 50-foot cable to a hand-held meter. Values along three orthogonal axes will be read by the astronaut and transmitted to earth over the voice communications link.

Soil Mechanics Experiment (S-200)

The soil mechanics experiment is designed to obtain data on the mechanical properties of the lunar soil from the surface to depths of tens of centimeters.

Data are derived from LM landing, flight crew observations and debriefings, examination of photographs, analysis of lunar samples, and astronaut activities using the Apollo hand tools. Experiment hardware includes an astronaut-operated self-recording penetrometer.

Far UV Camera/Spectroscope (S-201)

The far UV experiment is designed to measure the amount and excitation of hydrogen in nearby and distant regions of the universe (lunar surface, geocorona, solar wind, interstellar wind, and galaxy clusters) by obtaining imagery and spectroscopic data in the 500 to 15554 range.

Of particular interest is the Lyman-alpha line at 12164. The experiment will add to understanding the earth's magnetosphere, check the density of interplanetary and interstellar hydrogen clouds, and provide evidence of intergalactic hydrogen, as well as provide information on the suitability of a lunar-based astrophysical observatory.

The instrument is a tripod-mounted electronographic Schmidt camera with a lithium fluoride corrector plate, an objective grating, and Lyman-alpha interference filter. Film will be returned to earth for processing and analysis.

In-Flight Experiments

In-flight experiments may be conducted during all phases of the mission. They are performed within the CM, from the SIM located in sector 1 of the SM, and by a subsatellite launched in lunar orbit.

Gamma-Ray Spectrometer (S-160) (SIM)

The gamma-ray spectrometer experiment is designed to determine the lunar surface concentration of naturally occuring radioactive elements and rock-forming elements. This will be accomplished by the measurement of the lunar surface natural and induced gamma radiation while in orbit and by the monitoring of galactic gamma-ray flux during transearth coast.

The spactrometer detects gamma rays and discriminates against charged particles in the energy spectrum from 0. 1 to 10 Mev.

The instrument is encased in a cylindrical thermal shield which is deployed on a boom from the SIM for experiment operation.

X-Ray Fluorescence (S-161) (SIM)

The X-ray spectrometer experiment is designed to determine the concentration of major rock-forming elements in the lunar surface. This is accomplished by monitoring the fluorescent X-ray flux produced by the interaction of solar X-rays with surface material and the lunar surface X-ray albedo. The X-ray spectrometer, which is integrally packaged with the alpha-particle spectrometer, uses three sealed proportional counter detectors with different absorption filters. The direct solar X-ray flux is detected by the solar monitor, which is located 180° from the SIM in the SM sector IV. An X-ray background count is performed on the lunar darkside. Selected galactic sources are sampled during transearth coast.

Alpha-Particle Spectrometer (S-162) (SIM)

The alpha-particle experiment is designed to locate sources and to establish gross radon evolution rates, which are functions of the natural and isotopic radioactive material concentrations in the lunar surface. This will be accomplished by measuring the lunar surface alpha-particle emissions in the energy spectrum from 4.7 to 9.3 Mev.

The instrument employs 10 surface barrier detectors.

The spectrometer is mounted in an integral package with the X-ray spectrometer.

S-Band Transponder (S-164) (CSM/LM)

The S-band transponder experiment is designed to detect variations in the lunar gravity field caused by mass concentrations and deficiencies and to establish gravitational profiles of the ground tracks of the spacecraft.

The experiment data are obtained by analysis of the S-band Doppler tracking data for the CSM and LM in lunar orbit. Minute perturbations of the spacecraft motion are correlated to mass anomalies in the lunar structure.

Mass Spectrometer (S-165) (SIM)

The mass spectrometer experiment is designed to obtain data on the composition and distribution of the lunar atmosphere constituents in the mass range from 12 to 66 amu. The experiment will also be operated during transearth coast to obtain background data on spacecraft contamination.

The instrument employs ionization of constituent molecules and subsequent collection and identification by mass unit analysis. The spectrometer is deployed on a boom from the SIM during experiment operation.

Bistatic Radar (S-170) (CSM)

The bistatic radar experiment is designed to obtain data on the lunar bulk electrical properties, surface roughness, and regolith depth to 10-20 meters. This experiment will determine the lunar surface Brewster angle, which is a function of the bulk dielectric constant of the lunar material.

The experiment data are obtained by analysis of bistatic radar echos reflected from the lunar surface and subsurface, in correlation with direct downlink signals. The S-band and VHF communications systems, including the VHF omni and S-band high gain or omni antennas, are utilized for this experiment.

UV Photograph - Earth and Moon (S-177) (CM)

This experiment is designed to photograph the moon and the earth in one visual and three UV regions of the spectrum. The earth photographs will define correlations between UV radiation and known planetary conditions. These analyses will form analogs for use with UV photography of other planets.

The lunar photographs will provide additional data on lunar surface color boundaries and fluorescent materials.

Photographs will be taken from the CM with a 70mm Hasselblad camera equipped with four interchangeable filters with different spectral response. Photographs will be taken in earth orbit, translunar coast, lunar orbit, and transearth coast.

Subsatellite

The subsatellite is a hexagonal prism which uses a solar cell power system, an S-band communications system, and a storage memory data system. A solar sensor is provided for attitude determination. The subsatellite is launched from the SIM into lunar orbit and is spin-stabilized by three deployable, weighted arms. The following three experiments are performed by the subsatellite:

S-Band Transponder (S-164) (subsatellite) - Similar to the S-band transponder experiment conducted with the CSM and LM, this experiment will detect variations in the lunar gravity field by analysis of S-band signals. The Doppler effect variations caused by minute perturbations of the subsatellite's orbital motions are indicative of the magnitudes and locations of mass concentrations in the moon.

Particle Shadows/Boundary Layer (S-173) (Subsatellite) - This experiment is designed to monitor the electron and proton flux in three modes: interplanetary, magnetotail, and the boundary layer between the moon and the solar wind.

The particle experiment uses five curved plate particle detectors and two solid state telescopes to measure solar wind plasma (electrons in two ranges, 0-14 kev and 20-320 kev, and protons 0.05-2.0 Mev).

Magnetometer (S-174) (Subsatellite) - The subsatellite magnetometer experiment is designed to determine the magnitude and direction of the interplanetary and earth magnetic fields in the lunar region.

The biaxial magnetometer is located on one of the three subsatellite deployable arms. This instrument is capable of measuring magnetic field intensities in two ranges, 0 + 25 gammas and 0 + 100 gammas.

Gegenschein from Lunar Orbit (S-178) (CM)

The gegenschein experiment is designed to photograph the Moulton point region, an analytically defined null gravity point of the earth-sun line behind the earth. These photographs will provide data on the relationship of the Moulton point and the gegenschein (an extended light source located along the earth-sun line behind the earth). These photographs may provide evidence as to whether the gegenschein is attributable to scattered sunlight from trapped dust particles at the Moulton point.

Microbial Response in Space Environment (M-191)

The microbial response experiment is designed to determine the type and degree of alteration produced in selected biological systems when exposed to various types of radiation in a space environment. A self-contained microbial environment exposure device (MEED) will maintain representative types of microorganisms for exposure to space radiation for a specified period of time near the end of the transearth coast EVA. The MEED will be returned to earth for laboratory analysis of the exposed microorganisms.

Other Experiments

Additional experiments assigned to the Apollo 16 Mission which are completely passive are discussed in this section only. Completely passive connotes no crew activities are required during the mission to perform these experiments.

Apollo Window Meteoroid (S-176) (CM)

The objective of the Apollo window meteoroid experiment is to obtain data on the cislunar meteoroid flux of mass range 10^{-12} grams. The returned CM windows will be analyzed for meteoroid impacts by comparison with a preflight photomicroscopic window map.

Bone Mineral Measurement (M-078)

The bone mineral experiment is designed to determine the occurrence and degree of bone mineral changes in the Apollo crewmen, which might result from exposure to the weightless condition; and whether exposure to short periods of 1/6 g alters these changes. At selected pre- and post-flight times, the bone mineral content of the three Apollo crewmen will be determined using X-ray absorption techniques. The radius and ulna (bones of the forearm) and os calcis (heel) are the bones selected for bone mineral content measurements.

Biostack (M-211)

The biostack experiment is designed to study the interaction of biologic systems with the heavy particles of galactic cosmic radiation. Dormant biological systems will be sandwiched or stacked alternately between different physical detectors of heavy particle tracks. Post-flight analyses will correlate individual incident particles with the biological effects.

DETAILED OBJECTIVES

Following is a brief descritpion of each of the launch vehicle and spacecraft detailed objectives planned for this mission.

Launch Vehicle Detailed Objectives

Impact the expended S-IVB/IU in a preselected zone on the lunar surface under nominal flight profile conditions to stimulate the ALSEP passive seismometers.

Post-flight determination of actual S-IVB/IU point of impact within 5 km, and time of impact within 1 second.

Spacecraft Detailed Objectives

Evaluate LRV operational characteristics in the lunar environment.

Obtain SM high resolution panoramic and high quality metric lunar surface photographs and altitude data from lunar orbit to aid in the overall exploration of the moon.

Obtain CM photographs of lunar surface features of scientific interest and of low brightness astronomical and terrestrial sources.

Record visual observations of farside and nearside lunar surface features and processes to complement photographs and other remote-sensed data.

Obtain more definitive information on the characteristics and causes of visual light flashes.

Obtain data concerning exterior contamination induced by and associated with manned spacecraft.

Demonstrate that the improved gas/water separator can deliver gas-free water.

Evaluate the use of the improved fecal collection bag.

Determine the function of the Skylab food packages.

Evaluate the differences, correlation and relative consistency between ground-based and lunar surface task dexterity and locomotion performance.

Obtain data to support an understanding of the degree of body fluid and electrolyte disturbance during weightlessness.

Obtain data from subsatellite tracking for investigating new navigation techniques.

Obtain data via LM voice and data relay to provide a better understanding of the capability to transmit voice and PLSS data from extravehicular communications-1 (EVC-1) to MSFN via the LM in case of a LCRU failure during LRV traverses.

IN-FLIGHT DEMONSTRATION

The in-flight demonstration described below will be performed by the crew on a non-interference basis, during translunar coast.

Electrophoretic Separation

This demonstration will show the feasibility of separating mixtures of biological molecules by electrophoresis in a liquid medium. A comparison will be made of the separation resolution obtained in simple electrophoresis cells under weightless conditions and on earth.

This demonstration will not be flown for the May and June T-24 hour launch opportunities. In the event there is a scrub from a T-24 launch date and a rapid turnaround to a T-0 or T+24 hour launch opportunity, the electrophoretic separation demonstration will be installed.

OPERATIONAL TEST

The following significant operational test will be performed in conjunction with the Apollo 16 Mission.

Acoustic Measurement

The noise levels of the Apollo 16 space vehicle during launch and the CM during entry into the atmosphere will be measured in the Atlantic launch abort area and the Pacific recovery area, respectively. The data will be used to assist in developing high-altitude, high-Mach number, accelerated flight sonic boom prediction techniques. The Manned Spacecraft Center (MSC) will conduct planning, scheduling, test performance, and reporting of the test results.

Personnel and equipment supporting this test will be located aboard secondary recovery ships, the primary recovery ship, and at Fanning Island uprange from the prime recovery site.

MISSION CONFIGURATION AND DIFFERENCES

MISSION HARDWARE AND SOFTWARE CONFIGURATION

The Saturn V Launch Vehicle and the Apollo Spacecraft for the Apollo 16 Mission will be operational configurations.

CONFIGURATION	DESIGNATION NUMBERS
Space Vehicle	AS-511
Launch Vehicle	SA-511
First Stage	S-IC-11
Second Stage	S-II-11
Third Stage	S-IV B-511
Instrument Unit	S-IU-511
Spacecraft-LM Adapter	S LA-2Q
Lunar Module	LM-11
Lunar Roving Vehicle	L RV-2

Service Module	SM-113
Command Module	CM-113
On-board Programs	
Colossus 3	Command Module
Luminary 1F	Lunar Module
Apollo 16 Experiments Package	ALSEP
Launch Complex	LC-39A

CONFIGURATION DIFFERENCES

The following summarizes the significant configuration differences associated with the AS-511 Space Vehicle and the Apollo 16 Mission:

Spacecraft

Command/Service Module

Replaced 42-second timer with 61-second timer in the RCS control box.	Extend Mode 1-A abort sequence to T+61 seconds to reduce possible hazard of land landing with pressurized propellant tanks.
Strengthened meter glass.	Installed transparent Teflon shields to strengthen meter glass and to retain glass particles in case of breakage.
Installed Inconel parachute links in place of nickel plated links.	Reduced probability of parachute riser link failures due to flaws in links.
Replaced selected early series switches with 400 series switches	Reduced the possibility of switch failure by inspection and replacement as required.

Crew Systems

Modified swage fitting in suit assembly.	Redesigned swage fitting at front termination of crotch and thigh pulling cables to provide greater freedom of movement and reliability.

Lunar Module

Descent stage batteries modified.	Installation of Teflon separators between cells and battery case to prevent adhesion and cell case cracking. Also, increased thickness of plate tabs. Increased electrical capacity.
Installed battery coolant bypass to maximize capacity.	Addition of Glycol shutoff valve to increase battery temperature, if required, to maximize electrical capacity
Strengthened meter glass.	Added an exterior glass doubler to the range/range rate meter window to reduce stress. Added tape and particle shield as required to other meters.

SLA

Changed ordnance adhesive in pyro train.

Replaced potting compound (DC30-121) with GE-577 RTV to avoid a lead acetate reaction.

LRV

Improved seat belts on LRV.

New stiffer seatbelts were installed to eliminate adjustment and latching problems.

Launch Vehicle

S-IC

Added four retro-rocket motors.

Added four retro-rocket motors (previously deleted for Apollo 15) to improve S-IC/ S-II separation characteristics.

S-II

Modified S-II structure.

Increased factor of safety from 1.3 to 1.4. Improved POGO stability.

Modified S-II engines start/cutoff circuitry.

Eliminated single point relay failure modes in start/cutoff circuitry.

S-IVB

Installed 2-ply fuel and LOX feedline bellows.

Vendor change from stainless steel products duct to 2-ply solar duct.

IU

Modified LVDC to distinguish between failures of lower or upper engines.

LVDC modification provides indication of which engine has failed and initiates proper abort guidance program.

Redesigned command decoder.

Added solder joint stress relief to eliminate solder joint cracks for improved reliability.

Support Equipment

Ground/Electrical
Added redundant IU umbilical paths and logic changes for critical functions.

Reduced the possibility of an undesired S-IC/F-1 engine shutdown or failure to shutdown when desired.

Added redundant hardware command line through S-IC umbilical for each F-1 engine start control valve.

Minimized the possibility of an engine failing to start or run due to an open circuit through the umbilical.

TV AND PHOTOGRAPHIC EQUIPMENT

Standard and special-purpose cameras, lenses, and film will be carried to support the objectives, experiments, and operational requirements. Table 2 lists the TV and camera equipments and shows their stowage locations.

Table 2

Nomenclature	CSM at Launch	LM at Launch	CM to LM	LM to CM	CM at Entry
TV, Color, Zoom Lens (Monitor with CM System)	1	1			1
Camera, Data Acquisition, 16mm	1	1			1
Lens - 10mm	1	1			1
- 18mm	1				1
- 75mm	1				1
Film Magazines	13				13
Camera, 35mm Nikon	1				1
Lens - 55mm	1				1
Cassette, 35mm	9				9
Camera, 16mm		1			
Battery Operated (Lunar Surface)					
Lens - 10mm		1			
Film Magazines	8		8	8	8
Camera, Hasselblad, 70mm	1				1
Electric					
Lens - 80mm	1				1
- 250mm	1				1
- 105mm UV (4 band-pass	1				1
Film Magazines filters)	7				7
Film Magazine, 70mm UV	1				1
Camera, Hasselblad					
Electric Data (Lunar Surface)		2			
Lens - 60mm		2			
Film Magazines	11		11	11	11
Polarizing Filter		1			
Camera, 24-in Panoramic (In SIM)	1				
Film Magazine (EVA Transfer)	1				1
Camera, Lunar Surface Electric		1			
Lens - 500mm		1			
Film Magazines	2		2	2	2
Camera, 3-in Mapping Stellar (SIM)	1				
Film Magazine Containing 5-in Mapping and 35mm Stellar Film (EVA Transfer)	1				1
Camera, Ultraviolet, Lunar Surface		1	1		
Film Magazine, UV, LS		1		1	1

FLIGHT CREW DATA

PRIME CREW (Figure 10)

Commander: John W. Young (Captain, USN)

Space Flight Experience: Captain Young was selected as an astronaut by NASA in September 1962. Captain Young was Command Module Pilot for the Apollo 10 Mission, which included all phases of a lunar mission except the final minutes of an actual lunar landing.

He also served as pilot for the Gemini 3 Mission and Commander of the Gemini 10 Mission. Captain Young has logged more than 267 hours in space.

Command Module Pilot: Thomas K. Mattingly, II (Lieutenant Commander, USN)

Space Flight Experience: Lieutenant Commander Mattingly is one of 19 astronauts selected by NASA in April 1966. He served as a member of the support crews for the Apollo 8 and 11 Missions. He was selected as the Command Module Pilot for Apollo 13.

However, he had been exposed to German measles and was replaced by John L. Swigert, Jr. 72 hours before Apollo 13 liftoff.

Lieutenant Commander Mattingly has begin on active duty since 1960.

Lunar Module Pilot: Charles M. Duke, Jr. (Lieutenant Colonel, USAF)

Space Flight Experience: Lieutenant Colonel Duke was selected as an astronaut by NASA in April 1966. He served as backup Lunar Module Pilot for the Apollo 15 Mission.

Lieutenant Colonel Duke has been on active duty since graduating from the U.S. Naval Academy in 1957.

Fig. 10   APOLLO 16 PRIME CREW

BACKUP CREW

Commander: Fred W. Haise, Jr. (Mr.)

Space Flight Experience: Mr. Haise was selected as an astronaut by NASA in April 1966.

He served as a member of the backup crews for Apollo 8 and 11 Missions.

He was the Lunar Module Pilot for the Apollo 13 Lunar Landing Mission which was modified in flight to a lunar fly-by mission due to SM cryogenic oxygen system anomalies.

He has logged 142 hours 54 minutes in space.

Command Module Pilot: Stuart A. Roosa (Lieutenant Colonel, USAF)

Space Flight Experience: Lieutenant Colonel Roosa was assigned to the astronaut crew by NASA in April 1966. He was a member of the astronaut support crew for the Apollo 9 flight and Command Module Pilot for the Apollo 14 Lunar Landing Mission.

He has spent 216 hours 42 minutes in space.

Lunar Module Pilot: Edgar D. Mitchell (Captain, USN)

Space Flight Experience: Captain Mitchell was in the astronaut group selected in April 1966. He served as a member of the astronaut support crew for Apollo 9, backup Lunar Module Pilot for Apollo 10, and Lunar Module Pilot for Apollo 14.

Captain Mitchell has logged more than 216 hours in space.

MISSION MANAGEMENT RESPONSIBILITY

Title	Name	Organization
Director, Apollo Program	Dr. Rocco A. Petrone	OMSF
Mission Director	Capt. Chester M. Lee, USN (Ret)	OMSF
Saturn Program Manager	Mr. Richard G. Smith	MS FC
Apollo Spacecraft	Brig. Gen. James A. McDivitt	MSC Program Manager
Apollo Program Manager,	Mr. Robert C. Hock	KSC KSC
Director of Launch Operations	Mr. Walter J. Kapryan	KSC
Director of Flight Operations	Mr. Sigurd A. Sjoberg	MSC
Launch Operations Manager	Mr. Paul C. Donnelly	KSC
Flight Directors	Mr. M. P. Frank	MSC
Mr. Eugene F. Kranz	MSC	
Mr. Gerald Griffin	MSC	

ABBREVIATIONS AND ACRONYMS

AGS	Abort Guidance System	LSM	Lunar Surface Magnetometer
ALSEP	Apollo Lunar Surface Experiments Package	LV	Launch Vehicle
AOS	Acquisition of Signal	MCC	Midcourse Correction
APS	Ascent Propulsion System (LM)	MCC	Mission Control Center
APS	Auxiliary Propulsion System (S-I V B)	MESA	Modularized Equipment Stowage Assembly
ARIA	Apollo Range Instrumentation Aircraft	MH_z	Megahertz
AS	Apollo/Saturn	MOCR	Mission Operations Control Room
BIG	Biological Isolation Garment	MOR	Mission Operations Report
BSLSS	Buddy Secondary Life Support System	MPL	Mid-Pacific Line
CCATS	Communications, Command, and Telemetry System	MSC	Manned Spacecraft Center
		MS FC	Marshall Space Flight Center
CCGE	Cold Cathode Gauge Experiment	MSFEB	Manned Space Flight Evaluation Board
CDR	Commander	MSFN	Manned Space Flight Network
CPLEE	Charged Particle Lunar Environment Experiment	NASCOM	NASA Communications Network
		NM	Nautical Mile
CM	Command Module	OMSF	Office of Manned Space Flight
CMP	Command Module Pilot	OPS	Oxygen Purge System
CSI	Concentric Sequence Initiation	ORDEAL	Orbital Rate Display Earth and Lunar
CSM	Command/Service Module	PCM	Pulse Code Modulation
DAC	Data Acquisition Camera	PDI	Powered Descent Initiation
DDAS	Digital Data Acquisition System	PGA	Pressure Garment Assembly
DOD	Department of Defense	PGNCS	Primary Guidance, Navigation, and Control System (LM)
DOI	Descent Orbit Insertion		
DPS	Descent Propulsion System	PLSS	Portable Life Support System
DSKY	Display and Keyboard Assembly	PSE	Passive Seismic Experiment
ECS	Environmental Control System	PTC	Passive Thermal Control
EI	Entry Interface	QUAD	Quadrant
EMU	Extravehicular Mobility Unit	RCS	Reaction Control System
EPO	Earth Parking Orbit	RR	Rendezvous Radar
EST	Eastern Standard Time	RLS	Radius Landing Site
ETB	Equipment Transfer Bag	RTCC	Real-Time Computer Complex
EVA	Extravehicular Activity	RTG	Radioisotope Thermoelectric Generator
FM	Frequency Modulation	S/C	Spacecraft
fps	Feet Per Second	SEA	Sun Elevation Angle
FDAI	Flight Director Attitude Indicator	SEVA	Stand-up EVA
FTP	Fixed Throttle Position	S-IC	Saturh V First Stage
GCTA	Ground Commanded Television	S-II	Saturn V Second Stage
GET	Ground Elapsed Time	S-IVB	Saturn V Third Stage
GNCS	Guidance, Navigation, and Control System (CSM)	SIDE	Suprathermal Ion Detector Experiment
		SIM	Scientific Instrument Module
GSFC	Goddard Space Flight Center	SLA	Spacecraft-LM Adapter
HBR	High Bit Rate	SM	Service Module
HFE	Heat Flow Experiment	S PS	Service Propulsion System
HTC	Hand Tool Carrier	SRC	Sample Return Container
IMU	Inertial Measurement Unit	SSB	Single Side Bard
IU	Instrument Unit	SSR	Staff Support Room
IVT	Intravehicular Transfer	SV	Space Vehicle
KSC	Kennedy Space Center	SWC	Solar Wind Composition Experiment
LBR	Low Bit Rate	TD&E	Transposition, Docking and LM Ejection
LCC	Launch Control Center	TEC	Transearth Coast
LCRU	Lunar Communications Relay Unit	TEI	Transearth Injection
LDMK	Landmark	TFI	Time From Ignition
LEC	Lunar Equipment Conveyor	TLC	Translunar Coast
LES	Launch Escape System	TLI	Translunar Injection
LET	Launch Escape Tower	TLM	Telemetry
LGC	LM Guidance Computer	TPF	Terminal Phase Finalization
LH2	Liquid Hydrogen	TPI	Terminal Phase Initiation
LiOH	Lithium Hydroxide	T-time	Countdown Time (referenced to liftoff time)
LM	lunar Module	TV	Television
LMP	Lunar Module Pilot	USB	Unified S-Band
LOI	Lunar Orbit Insertion	USN	United States Navy
LOS	Loss of Signal	USAF	United States Air Force
LOX	Liquid Oxygen	VAN	Vanguard
LPO	Lunar Parking Orbit	VHF	Very High Frequency
LR	Landing Radar	ZNIV	Differential Velocity
LRL	Lunar Receiving Laboratory		
LRRR	Loser Ranging Retro-Reflector		

GPO 930-168

Post Launch
Mission Operations Report No. M-933-72-16

April 28, 1972

TO: A/Administrator

FROM: MA/Apollo Program Director

SUBJECT: Apollo 16 Mission (AS-511) Post Mission Operation Report No. 1

The Apollo 16 Mission was successfully launched from the Kennedy Space Center on Sunday, 16 April 1972. The mission was completed successfully, with recovery on 27 April 1972, one day earlier than originally planned. An anomaly in the backup Thrust Vector Control (TVC) system caused a delay in initiation of powered descent. Analysis and duplication of the anomalous condition in ground simulator systems indicated that the cause was loss of rate damping in the backup yaw control system and that the resultant gimbal oscillation was self limiting. The backup TVC system was therefore considered operable and the decision was made to "GO" for powered descent three revolutions later than initially scheduled. To minimize the remaining SPS engine firings, lunar orbit plane change 2 and the subsatellite shaping burn were deleted. Subsequently, it was decided to shorten the mission one day. Initial review of the mission events indicates that all mission objectives were accomplished. However, the Apollo Lunar Surface Experiments Package (ALSEP) Heat Flow Experiment (HFE) was terminated after drilling of the first bore hole, due to the inadvertent separation of the HFE cable from the connector at the central station. Detailed analysis of all data is continuing and appropriate refined results of the mission will be reported in the Manned Space Flight Centers' technical reports.

Attached is the Mission Director's Summary Report for Apollo 16 which is submitted as Post Launch Mission Operations Report No. 1. Also attached are the NASA OMSF Primary Objectives for Apollo 16. The Apollo 16 Mission has achieved all the assigned primary objectives and I judge it to be a success.

Approval: Rocco A. Petrone

Dale D. Myers
Associate Administrator for Manned Space Flight
Attachments

NASA OMSF MISSION OBJECTIVES FOR APOLLO 16

PRIMARY OBJECTIVES

Perform selenological inspection, survey, and sampling of materials and surface features in a preselected area of the Descartes region.

Emplace and activate surface experiments.

Conduct in-flight experiments and photographic tasks.

Rocco A. Petrone
Apollo Program Director
Date: 29 March 1972

Dale D. Myers
Associate Administrator Manned Space Flight
Date: 31 March 1972

ASSESSMENT OF THE APOLLO 16 MISSION

Based upon a review of the assessed performance of Apollo 16, launched 16 April 1972 and completed 27 April 1972, this mission is adjudged a success in accordance with the objectives stated above.

Rocco A. Petrone
Apollo Program Director
Date:2 May 1972

Dale D. Myers
Associate Administrator Manned Space Flight
Date:5 May 1972

NATIONAL AERONAUTICS AND SPACE ADMINISTRATION
WASHINGTON, D.C. 20546

REPLY TO ATTN OF: MAO

April 27, 1972

TO: Distribution

FROM: MA/Apollo Mission Director

SUBJECT: Mission Director's Summary Report, Apollo 16

INTRODUCTION

The Apollo 16 Mission was planned as a lunar landing mission to accomplish selenological inspection, survey, and sampling of materials and surface features in a preselected area of the Descartes region of the moon; emplace and activate surface experiments; and conduct in-flight experiments and photographic tasks.

Flight crew members were Commander (CDR) Captain John W. Young (USN), Command Module Pilot (CMP) Lieutenant Commander Thomas K. Mattingly (USN), and Lunar Module Pilot (LMP) Lieutenant Colonel Charles M. Duke, Jr. (USAF). Significant detailed information is contained in Tables 1 through 14. Initial review indicates that all primary mission objectives were accomplished (reference Table 1). Table 2 lists the Apollo 16 achievements.

PRELAUNCH

The Apollo 16 prelaunch countdown was accomplished with no unscheduled holds; however, at T-5 hrs 51 min there was an abnormal null shift for 2 seconds in the spare yaw rate gyro channel in the Instrument Unit (IU). A failure mode analysis was performed and it was concluded that the shift, should it occur, would not have an adverse effect on the mission. Launch day weather conditions were clear, visibility 10 miles, winds 13 knots, and scattered cloud cover 3,000 feet.

LAUNCH, EARTH PARKING ORBIT, AND TRANSLUNAR INJECTION

The Apollo 16 space vehicle was successfully launched from Kennedy Space Center, Florida, on time at 12:54 p.m. EST, April 16, 1972.

The S-IVB/IU/LM/CSM was inserted into an earth orbit of 95 x 90 nautical miles (NM) at 11:56 GET (Min:Sec).

During earth orbit, the IU temperature control system gaseous nitrogen (GN_2) bottle pressure started to leak at 10 psi/min. It was predicted that the GN_2 would be lost at approximately 5 hours GET. The IU was given another 2 hours before components would reach temperature redlines and start to fail. The S-IVB auxiliary propulsion system (APS) module No. 2 helium regulator also malfunctioned and caused continuous overboard venting. During the first revolution, a leak was also noted in the APS No. 1 helium supply. These problems did not affect the translunar injection (TLI) burn which occurred at 2:33:34 GET (hrs:min:sec). TLI was nominal.

TRANSLUNAR COAST

The Command Service Module (CSM) separated from the S-IVB/IU/LM at 3:05:01 GET (hr:min:sec), transposed, and then docked with the Lunar Module (LM). Color TV was transmitted for approximately 18 minutes during transposition and docking.

Due to expected early APS helium depletion, the liquid oxygen (LOX) dump from the S-IVB was retargeted closer to the desired lunar impact in order to reduce the length of the APS-1 burn required.

The first S-IVB APS burn, aimed for impact of the lunar surface at 2.3°S and 31.7°W was near nominal. Because of the APS Module No. 1 helium depletion and potential trajectory disturbances from stage venting, the APS-2 maneuver was not performed, and lunar impact operations were terminated.

Tracking of the S-IVB was lost at 27:09:07 GET due to the loss of signal from the IU command and communications system.

During the CSM/LM docking, light colored particles were noticed coming from the LM area. These particles were unexplained. At 7:18 GET, the crew reported a stream of particles emitting from the LM in the vicinity of aluminum close-out panel 51 which covers the mylar insulation over reaction control system (RCS) system A. This panel is located below the docking target on the +Z face of the ascent stage. To determine systems status, the crew entered the LM at 8:17 GET and powered up. All systems were normal. There had been no appreciable change in consumables since close-out on the pad, and the LM was powered down at 8:52 GET. CM TV was turned on at 8:45 GET in order to give the Mission Control Center (MCC) a view of the particle emission.

In order to point the high gain antenna (HGA), panel 51 was rotated out of sunlight and a marked decrease was then noted in the quantity of particles. On the TV picture, the source of the particles appeared to be a growth of grass-like particles at the base of the panel. TV was turned off at 9:06 GET.

Results of the investigation determined that the particles were shredded thermal paint.

It was further determined that the degraded thermal protection due to the paint shredding would have no effect on subsequent LM operations.

The spacecraft trajectory was such that midcourse correction (MCC-1) was not required.

The electrophoresis in-flight demonstration commenced on schedule at 25:05 GET. The demonstration appeared to be successful, based on the verbal description from the crew.

Ultraviolet (UV) photography of the earth from approximately 58,000 and 117,000 nm was accomplished as planned.

MCC-2 maneuver was performed on time at 30:39 GET and the burn was normal.

The service propulsion system (SPS) burn of 1.8 seconds produced a velocity change of 12.5 feet per second (fps). The MCC-2 maneuver was also used to track an SPS tank pressure transducer anomaly which occurred earlier in the mission. The transducer responded to the change in tank pressure, indicating that the transducer reference cavity was leaking. The transducer presented no problem during subsequent SPS maneuvers.

However, procedures were uplinked to the crew to account for the transducer reading.

At approximately 38:19:02 GET the command module computer (CMC) received an indication an Inertial Measurement Unit (IMU) gimbal lock had occurred. The computer correctly downmoded the IMU to "coarse align" mode and set the appropriate alarms. Due to the large number of LM panel particles floating near the spacecraft and blocking the CMP's vision of the stars, realignment of the platform was accomplished using the sun and moon.

It was suspected that the gimbal lock indication was an electrical transient caused by actuation of the thrust vector control (TVC) enable relay when exiting the IMU alignment program (P-52). An erasable software program was uplinked to the crew and entered in the CMC which caused the CMC to ignore gimbal lock indication during critical periods.

Visual light flash phenomenon was started about 2 hours late at 49:10 GET. Numerous light flashes were reported by the crew prior to terminating the experiment at 50:16 GET. The crew also reported the flashes left no after-glow, were instantaneous, and were white.

MCC-3 was not performed since the CSM/LM combination was nearly on the planned trajectory.

UV photography of the earth at approximately 177,000 nm was completed as planned.

During suiting operations prior to scheduled intravehicular transfer (IVT) to the LM, the CDR commented that it was very difficult to zip up the LMP suit. Also, the LMP expressed concern that his suit felt short and requested that he be allowed to let out his leg lacings. Lengthening the legs is possible during flight, but the adjustment was not recommended since it was felt that under pressurized conditions the length would be correct.

No corrective action was recommended for the difficulty with the zipper as this is characteristic of tight fitting suits.

The second IVT/LM housekeeping commenced about 54:30 GET and completed at 55:11 GET. All LM systems checks were nominal.

The Skylab food test was conducted about 6 hours later than planned, mainly for crew convenience.

The spacecraft entered the moon's sphere of influence at 59:19 GET and at approximately 33,000 nm from the moon.

MCC-4 was not performed since the spacecraft trajectory was near nominal.

The Scientific Instrumentation Module (SIM) door was successfully jettisoned at 69:59:00 GET.

LUNAR ORBIT INSERTION AND S-IVB IMPACT

Lunar Orbit Insertion (LOI) was performed at 74:28:27 GET. The 374.3-second maneuver produced a velocity change (pV) of -2802 fps and inserted the CSM/LM into a lunar orbit of 170.3 x 58.1 nm.

The resultant orbit was very close to the prelaunch predicted orbit of 170.6 x 58.5 nm.

S-IVB impact occurred at 75:08:00 GET about 37 minutes later than the prelaunch prediction.

The event was recorded by the Apollo 12, 14, and 15 Apollo Lunar Surface Experiment Package (ALSEP) sites. The best estimate of the impact point location is 1°50'N, 23° 18'W.

The estimate was based on the last data received at S-IVB loss of signal at 27:09:07 GET.

DESCENT ORBIT

The Descent Orbit Initiation (DOI) maneuver of -209.5 fps for 24.1 seconds resulted in an orbit of 58.5 x 10.9 nm. The prelaunch planned orbit parameters were 58.6 x 10.9 nm.

IVT/LM activation occurred at 93:34 GET, about 11 minutes early. The LM was powered up and all systems were nominal. The LMP did not report any difficulty with his suit before or after the IVT.

A pressure rise was noted at the LM reaction control system (RCS) system A helium pressure regulator.

In order to prevent a rupture of the burst disc and ensure sufficient ullage entrapped for expulsion of the propellant, 53.8 pounds of LM RCS fuel and oxidizer were transferred to the LM ascent propulsion system (APS) tanks to reduce the pressure and gain ullage in the RCS tanks. At 93:03 GET the burst disc ruptured. Subsequently the helium source pressure decreased at a rate which varied from about 4 psi/hr initially to about 1.25 psi/hr.

UNDOCKING, POWERED DESCENT, AND LANDING

The CSM and LM performed the undocking and separation maneuver on schedule at 96:13:31 GET.

The CSM was scheduled to perform the circularization maneuver on the 13th lunar revolution at 97:41:44 GET; however, during gimbal checks, a problem was discovered in the SPS yaw gimbal drive servo loop. As a result, the Powered Descent Initiate (PDI) maneuver was delayed. While the flight controllers were evaluating the servo loop problem, the LM and CSM maneuvered into a stationkeeping situation and prepared to either dock or continue the mission. During troubleshooting of the gimbal drive servo loop, checkout of the backup TVC indicated no rate feedback was obtained and the gimbal position indicator showed yaw oscillations. Analysis of this problem resulted in the conclusion that it would not preclude the lunar landing and that backup capabilities still existed for a safe mission. Based on this conclusion, the spacecraft were given a GO for PDI.

The CSM RCS performed a second separation maneuver at 102:30:00 GET. The SPS was then fired for 4.6 seconds and produced a velocity change of 81.6 fps.

The circularization maneuver at 103:21:42 GET placed the CSM into a resultant orbit of 68.0 x 53.1 nm.

PDI with the descent propulsion system (DPS) was performed at 104:17:25 GET on the 16th revolution. Landing occurred in the Descartes area at 104:29:35 GET. The estimate of the landing coordinates based on LM guidance and interpretation of crew visual reports are 8°59'S and 15°30'E, about 230 meters northwest of the planned target point. Since the LM had stayed in lunar orbit longer than planned, the LM was powered down to conserve electrical power. Due to the 6-hour delay in landing caused by the SPS gimbal drive servo loop problem, extravehicular activity (EVA) I was rescheduled with a sleep period occurring before the EVA. At 118:06:31 the GET clock was advanced 11 minutes 48 seconds.

LUNAR SURFACE

EVA-1 commenced at 119:05:11 GET. Almost all of the scheduled events were accomplished as planned except additional time was used at station 1, some time was deleted at station 2, additional time was required during LM closeout, and the second heat flow experiment (HFE) bore hole was cancelled during Apollo lunar surface experiment (ALSEP) deploy after the LMP drilled the first HFE bore hole.

The cancellation was due to the CDR inadvertently tripping over the HFE cable causing the cable to separate at the connector. The remaining ALSEP components were deployed successfully and functioned nominally at central station activation.

After deployment of the lunar roving vehicle (LRV), the volt and ampere readings for battery # 2 read off-scale low and the rear steering was inoperative.

Approximately 40 minutes later, subsequent to LRV loading operations, all meters and the rear steering were operating properly. There was no explanation for the off-scale meters or the inoperative steering.

The LRV performed nominally throughout the remainder of the EVA; terminated at 126:16:22 GET. Total EVA time was 7 hours 11 minutes 11 seconds.

EVA-2 started at 142:51:15 GET.

All stations were explored except station 7 which was deleted prior to crew egress to provide more time at station 10. At 146:48 GET while ascending a ridge and traversing very rocky terrain, full throttle was applied to the LRV, but there was no response at the rear wheels. The LRV continued to move, although the front wheels were "digging in." At 147:14 GET at station 8, a rear-drive troubleshooting procedure was commenced. During this procedure, a mismatch of power mode switching was identified as the cause of the problem. After a change in switch configuration, the LRV was reported to be working properly.

Between EVA stations 9 and 10, at GET 148:45, the LRV range, bearing and distance (all derived from odometer inputs) were reported to be inoperative. However, navigation heading was reported to be working. Later in the EVA the crew reset the power switches and the navigation system began operating nominally.

After the crew arrived at station 10 (LM and ALSEP area), the EVA was extended about 20 minutes. The extension was allowed since the crew's portable life support systems (PLSS) consumables usage was lower than predicted. At approximately 149:19 GET the LMP examined the damaged HFE. Visual inspection of the HFE revealed the cable was separated at the connector.

Results of troubleshooting a model of the damaged HFE connector and circuit board indicated that a fix could be accomplished with fairly high probability of success.

However, the fix was not attempted because the time required in both EVA 2 and 3 and while in the LM could escalate and impact the 3rd EVA. Additionally, there was some risk to the remainder of the ALSEP components involved in the fix.

After completing surface activities, the crew began ingress. During ingress, approximately a 2-inch portion of the CDR's PLSS antenna was broken off. Subsequently, a 15-18 db drop in signal strength was observed. Since the CDR's backpack radio relays the LMP's information to the LM and the lunar communications relay unit (LCRU) for transmission to ground stations, the broken antenna would have limited communications throughout EVA-3 to a maximum range of 1.9 to 2.1 km from the LM or LCRU if the oxygen purge system (OPS) were not interchanged. A decision was made later to have the CDR use the LMP's OPS which supports the PLSS antenna.

Following completion of tasks, EVA-2 was terminated at 150:14:41 GET. The total EVA time was 7 hours 23 minutes 26 seconds.

EVA-3 commenced approximately 30 minutes early at 165:43:15 GET.

The additional 30 minutes were spent in the vicinity of North Ray crater. Two additional stations were explored. All planned activities were accomplished during the EVA; however, some difficulty was experienced in configuring the cosmic ray detector for stowage and return to earth. Television coverage was excellent throughout the EVA. LM closeout at 171:23:29 GET was slightly later than planned. Total EVA-3 time was 5 hours 40 minutes 14 seconds.

A total of 27.0 km were traversed during the three EVA's (see Figure 1).

The combined EVA's were 20 hours 14 minutes 54 seconds, and is the longest total lunar EVA time recorded.

The CSM lunar orbit plane change (LOPC) 1 was performed on time. The SPS burn of 7 seconds produced a change in velocity of 124.0 fps and placed the CSM in a 64.6 x 55 nm orbit.

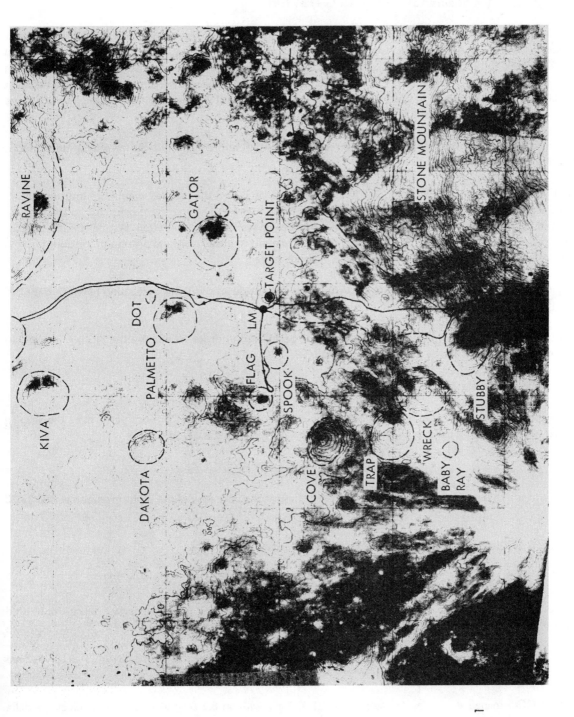

Fig. 1

ASCENT, RENDEZVOUS, AND DOCKING

Ascent stage liftoff occurred as planned at 175:43:35 GET.

Television of the liftoff and ascent was excellent. The ascent stage was inserted into a 40.2 x 7.9 nm orbit. At insertion, the range between the two spacecraft was about 33,000 feet too close and a 10 fps tweak burn was initiated at 175:54:05 GET to attain the proper separation. The terminal phase initiation (TPI) burn of 3.1 seconds was executed on time at 176:37:52 GET with a nominal velocity change of 78 fps and a resultant orbit of 64.2 x 40.1 nm. After CSM/LM docking, the LM crew transferred all the samples, the film, and some nonscientific equipment to the CSM.

After the crew rest period, the LM crew reentered the LM and transferred the remaining nonscientific equipment to the CSM.

LM jettison occurred at 195:12:00 GET about 2 minutes later than the planned time. After jettison the LM lost attitude and began tumbling at a rate of about 3° per second. This may have been due to the attitude and translation controller assembly (ATCA) primary guidance and navigation system circuit breaker inadvertently being left open. The LM ascent stage is currently in lunar orbit, and present predictions are that the ascent stage will impact the lunar surface in about 1 year.

At 195:15:00 GET the CSM executed a burn of 2 fps to provide an additional separation between the CSM and LM.

The LOPC-2 and the shaping maneuvers were not performed in order to not fire the SPS any more than absolutely necessary due to the degraded backup SPS TVC.

Subsatellite launching occurred at 196:13:55 GET.

Its expected lifetime is about 6 to 9 months. The shorter lifetime results from not performing the shaping burn which was to be used to optimize the orbit for a 1-year expected lifetime.

TRANSEARTH INJECTION AND COAST

The transearth injection (TEI) maneuver was performed at the beginning of the 65th revolution at 200:33:20 GET.

The SPS 162.4-second burn produced a change in velocity of 3,370.9 fps.

A GET clock update of 24 hours 34 minutes 12 seconds was made at approximately 202:25 GET.

Good quality television pictures inside the CM and from the LRV on the lunar surface were transmitted between 227:53 and 228:08 GET and 228:25 and 229:08 GET, respectively.

The spacecraft entered the earth's sphere of influence at 237:28 GET traveling at 3,751 fps and 187,827 nm from the earth.

The MCC-5 maneuver was performed at 239:21:02 GET with an 8.2-second RCS burn producing a delta velocity of 3.4 fps.

The CMP EVA commenced at 243:24 GET. The CMP retrieved the panoramic and mapping camera film in two trips to the SIM bay. The CMP observed the SIM bay to determine the condition of the instruments (see in-flight science). The microbial environment exposure device (MEED) was deployed and exposed for approximately 10 minutes. After returning the MEED to the CM, the CMP experienced some difficulty in closing the cover.

The EVA lasted approximately 1 hour 24 minutes.

A scheduled TV press conference started at 268:13 GET and terminated at 268:31 GET. During the conference the crew gave a brief description of the backside of the moon. An item of particular interest was the crew's description of the crater Guyot which appeared to be full of material.

The material seemed to have overflowed and spilled down the side of the crater. The crew compared their observations with similar geological formations in Hawaii.

MCC-6 was not performed since the spacecraft trajectory was near nominal.

The MCC-7 maneuver at 287:23:24 GET with the RCS firing of 3.5 seconds produced a change in velocity of 1.4 fps.

ENTRY AND LANDING

The CM separated from the SM at about 290:08 GET, 15 minutes before entry interface (EI) at 400,000 feet. Drogue and main parachutes deployed normally; landing occurred in the mid-Pacific Ocean at approximately 156°11 .4'W longitude and 00°43.2'S latitude. The CM landed in a stable two position, about 2.7 nm from the prime recovery ship, USS Ticonderoga, and about .3 nm from the planned landing point.

Weather in the prime recovery area was as follows: visibility 10 miles, wind 110° at 10 knots, scattered cloud cover 2,000 feet, and wave height of 4 feet.

ASTRONAUT RECOVERY OPERATIONS

Following CM landing, the recovery helicopter dropped swimmers who installed the flotation collar and attached the life raft. Fresh flight suits were passed through the hatch for the flight crew. The post ventilation fan was turned off, the CM was powered down, the crew egressed, and the CM hatch was secured.

The helicopter recovered the astronauts and transferred them to the recovery ship. After landing on the recovery ship, the astronauts proceeded to the Biomed area for a series of examinations.

Following the examinations the astronauts departed the USS Ticonderoga the next day, were flown to Hickam Air Force Base, Hawaii, and then to Ellington Air Force Base, Texas.

COMMAND MODULE RETRIEVAL OPERATIONS

After astronaut pickup by the helicopter, the CM was retrieved and placed on a dolly aboard the recovery ship. Since the CM had propellants onboard, it was stowed near the elevator shaft to insure adequate ventilation. All lunar samples, data, equipment will be removed from the CM and subsequently returned to Ellington Air Force Base, Texas.

The CM will be offloaded at San Diego, California, where deactivation of the CM propellant system will take place.

SYSTEMS PERFORMANCE

The Saturn V stages performed nominally, with some discrepancies that did not impact the mission.

The spacecraft systems performance was also near nominal throughout the mission with only one notable exception. During the lunar orbit precircularization burn checkout, at approximately 97:40 GET, checkout of the backup TVC indicated no rate feedback was obtained and the SPS engine gimbal position indicator showed yaw oscillations.

This delayed the lunar landing some 6 hours until 104:29:35 GET, caused some revision of lunar surface activities, and a 1 day earlier end-of-mission.

Other problems were: the HFE cable became entangled in the CDR's legs causing it to break at the connector to the central station, the LM steerable antenna yaw drive was inoperative at 94:35 GET, at 95:03 GET the LM RCS system A helium regulator failed.

A number of "glitches" occurred on the coupling data unit (CDU) circuitry in that the caution and warning (C&W) light and alarms were activated. The exact cause of the anomaly is unknown but it could have been contamination, loose wiring, or some similar condition since the crew could clear the "glitch" by tapping the panel above the display and keyboard assembly (DSKY). It did not appear that the failure was in either the CDU, the inertial measurement unit (IMU) or pulse integrated pendulous accelerometers (PIPA). All indications are that the C&W light indication was faulty and the guidance and navigation (G&N) system would function properly. The glitches did not occur during CM flight.

The stabilization and control system (SCS) was available for backup during the final phase of the mission but was not needed.

All anomalies were rapidly analyzed and either resolved or workaround procedures developed to permit the mission to continue. All anomalies are listed in Tables 9 through 14.

FLIGHT CREW PERFORMANCE

The Apollo 16 flight crew performance was excellent throughout the mission.

All information and data in this report are preliminary and subject to revision by the normal Manned Spaceflight Centers' technical reports.

SURFACE SCIENCE

The first Apollo 16 surface science event was the impact of the S-IVB stage at 75:08:00 GET.

The best estimate of the impact point location is 1°50'N and 23°18'W. This impact is approximately 155 km north of the Apollo 12 site, 250 km northwest of the Apollo 14 site, and 1100 km southwest of the Apollo 15 site. The seismometers at all three sites recorded the impact.

The distance of the impact from the seismometers will facilitate analysis of subsurface structure to depths of 30 to 200 km.

The Apollo 14 cold cathode gauge experiment (CCGE) recorded a brief event approximately 11.5 minutes after predicted S-IVB impact time.

A possible gas cloud passage was also recorded by the CCGE at Hadley/Apennine. The charged particle lunar environment experiment (CPLEE) recorded a 1-minute burst of low energy protons about 5 minutes after predicted impact time.

Suprathermal ion detectors at Apollo 12, 14, and 15 sites did not record a positive indication of the impact.

The LM touched down on the Descartes plateau at 104:29:35 GET. The crew reported observing expected landmarks during descent. From visual descriptions, the preliminary coordinates of the landing site were 8°59'S and 15°30'E which is about 230 meters northwest of the nominal landing site. The landing site was in a subdued crater lying on a ray from South Ray crater.

The surface is densely cratered with 30-40% of the surface covered with blocks up to 1/2 meter in diameter.

The rocks and boulders were bright and appeared to be primarily breccias. Detailed verbal descriptions were made of both distant and near field objects.

Lineations similar to those observed by the Apollo 15 crew were observed. Stone Mountain appeared to have many terraces. Trafficability for the LRV was considered to be good and an acceptable deployment site for the ALSEP was felt to be to the west of the LM.

The crew commenced the 1st EVA at 119:05:11 GET. The Far UV camera was deployed near the LM and the ALSEP was transported 110 meters WSW for deployment.

Deployment of the ALSEP was nominal except for the HFE. The CDR inadvertently tripped over the HFE cable, separating the cable at the connector after the first hole was bored. The second HFE bore hole was cancelled. The remaining ALSEP components deployed nominally. The deep core (8-3/4 feet) was drilled successfully and removed from the regolith by the LMP using the core extractor.

Scheduled traverse events were accomplished as planned with the crew having no difficulty in reaching Flag and Spook craters, although there was some difficulty in the identification of particular craters. Traverse and LRV evaluation test speeds were as high as 11 km per hour. The ground traversed and sampled was characterized by unconsolidated deposits, with or without abundant blocks.

Neither bedrock outcrops nor ridge-like thinly mantled bedrock were recognized.

Documented and grab samples were obtained. The most common rock type encountered was a breccia with dominantly light-colored clasts in light to dark-colored matrices.

The surface layer of dark soil appears to be underlain at a depth of about 1-3 cm by white soil.

A successful lunar portable magnetometer (LPM) site measurement was made at Station 2 (Spook crater). The local magnetic field measured was 180 gamma, approximately straight down. EVA-1 terminated at 126:16:22 GET for a total EVA time of 7 hours 11 minutes 11 seconds.

The crew commenced the 2nd EVA at 142:51:15 GET. All preplanned stations were explored except Station 7, which was deleted from the traverse plan prior to crew egress. The crew proceeded south to Station 4. The surface at Station 4 was largely rocky with rocks having a white clast and glass coating. Documented samples, double core, penetrometer readings, and soil samples were collected. The subsurface soil was found to be white. The crew then proceeded north to Station 5 on a 50-meter wide bench. Blocks were present, but no ledges or bedrock. Blocks were up to 5 cm and almost all angular.

No white soil was found.

Documented and rake samples were collected. A successful LPM measurement was made. The local magnetic field measured was 125 gamma, nearly vertically up, in contrast to the measurement at Station 2 which was vertically down.

Station 6 NNW of Station 5 was on a bench on the lower flank of Stone Mountain.

The area was found to have many large, hard rocks. The rocks appeared to be breccias containing needle-like crystals. The samples were definitely different from previous samples. The soil was gray at the surface and to the depth of soil sampled. The crew traveled to the northwest to Station 8.

Fields of blocks averaging 10-15 cm, but ranging up to 2 meters were found on the surface. The surface soil was predominantly black, but some white breccia and white crystalline rock fragments were obtained.

A double core sample in an area of glass beads as well as rock fillet samples were collected. Documented, grab, and rake samples were obtained. The crew then traveled north to Station 9, an old subdued crater. The crew obtained special soil samples. These samples should provide material from the upper 100-500 micron depth which were probably not exposed to contamination of the LM descent. Rock samples included a breccia with white clasts and a breccia with blue crystals.

The crew overturned a boulder and collected samples of the exposed surface. A drive tube sample was also obtained. At Station 10 in the vicinity of the ALSEP, documented samples, a double core, and penetrometer measurements were obtained. Before the crew ingressed the LM, the Far UV camera was retargeted. EVA-2 terminated at 150:14:41 GET after a total EVA time of 7 hours, 23 minutes 26 seconds.

EVA-3 commenced at 165:43:15 GET. After loading the LRV and retargeting the Far UV camera, the crew drove directly to North Ray crater. They described this part of the Cayley formation as more subdued and less densely cratered than that covered the previous 2 days. Blocks became less common north of the LM, more common on the flanks of Palmetto crater, and the rim of North Ray crater had many large blocks.

Two main types of rocks were sampled at North Ray: (1) white breccias and (2) dense, dark breccias typified by "House Rock," a 10 x 10 x 30-meter block. The area between Stations 11 and 12 was extensively sampled.

Rake, soil samples, documented samples, and grab samples were collected. A shatter cone was also collected from "House Rock." 500mm photos of the interior of the crater were taken, but the crew could not see the bottom.

The crew then drove to Station 13, approximately 1/2 km southeast of Station 11.

In addition to sampling and photography, a soil sample was collected from under a 5 meter boulder. This sample is believed to have been in permanent shadow since the rock came to rest on the lunar surface. A LPM reading was taken showing a local field of 313 gammas in a southwesterly direction.

From Station 13 the crew returned to the LM and Station 10; recording, at times, speeds as high as 17 km/hr. At Station 10, a double core, rake/soil sample and other samples were collected. A final LPM measurement was also made after parking the LRV east of the LM, both with and without a locally collected igneous rock placed on the LPM.

EVA-3 terminated at 171:23:29 GET, after a total EVA time of 5 hours 40 minutes 14 seconds.

A brief summary of some of the estimated statistics of the three EVAs follows:

Total EVA man-hours	40 hours 30 minutes
Total stop time at sampling	9 hours 7 minutes stations
Total traverse distance	27.1 km
Sample return	213 pounds

Apollo Lunar Surface Experiments Package

The ALSEP was deployed during the early part of EVA-1. Total deployment time was 2 hours 5 minutes. All experiments were successfully deployed and turned on, with the exception of the HFE, The radioisotope thermoelectric generator (RTG) is now supplying 70.94 watts.

The downlink signal strength remains constant at -139 dbm. All central station functions are normal.

Heat Flow Experiment

During deployment of the mortar package the CDR caught his foot in a loop of the HFE cable. He was unaware of the interference and continued to walk away from the central station, severing the cable at the central station connector.

Although the first HFE hole was drilled and a probe emplaced, the experiment could not be activated since the cable was severed.

Passive Seismic Experiment

The passive seismic experiment (PSE) was uncaged and leveled a few hours after activation. Seismic and tidal data channels have functioned normally. The instrument recorded LRV motion and crew activities at all times during the EVAs except on the inbound leg of EVA-3 when the active seismic experiment (ASE) was on, precluding the return of other data.

Crew activity to the maximum range of 4.4 km was recorded and LRV range could be estimated to +1/2 km. The instrument also recorded equipment jettison, LM venting, and its own settling.

The PSE sensor temperature is not stabilizing to the nominal value which is expected to preclude measuring tidal data during lunar daytime.

It is expected to exceed the measurement range shortly before lunar noon. An operating procedure has been devised which should minimize the effect of this anomaly during future lunations.

Lunar Surface Magnetometer

The lunar surface magnetometer (LSM) has been operating normally since turn-on. A field of 230 gamma was observed.

Since that time the instrument has recorded the passage of the moon from the solar wind region through the transition region and into the earth's geomagnetic tail (Figure 2).

Two flip calibrations have been performed. A site survey will be performed when the moon is deep in the geomagnetic tail. The LSM is operating with its digital filter in and will remain so for one lunation.

Active Seismic Experiment

The ASE geophones observed the scheduled 19 thumper firings.

It was also used to observe the last kilometer of the return of the LRV on EVA-3 to verify its use during Apollo 17 operations. It also was activated during ascent stage liftoff.

The mortar package roll sensor has shown off-scale high since turn-on.

The pitch sensor is normal (-2.3°). Crew observation verified that the mortar is in fact nearly level, indicating failure of the roll sensor.

The ASE will be operating in the listening mode once a week for 30 minutes. Firing of the mortars is currently planned in about 3 months.

Far Ultra-Violet Camera/Spectroscope

The Far UV camera/spectroscope was targeted on the scheduled 11 targets. Film advance was observed to be normal.

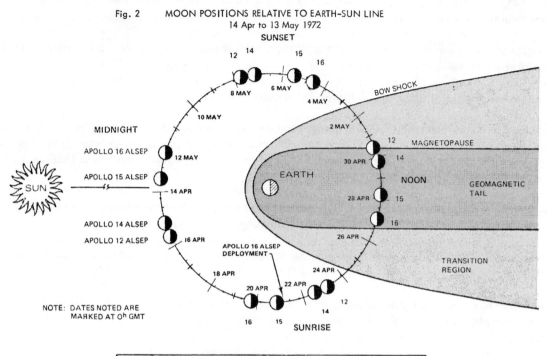

Fig. 2   MOON POSITIONS RELATIVE TO EARTH-SUN LINE
14 Apr to 13 May 1972

APOLLO	DAY/HOUR, GMT				
(ALSEP)	MIDNIGHT	SUNRISE	NOON	SUNSET	MIDNIGHT
16		19 APR/2032	(1ST) 27 APR/0546	4 MAY/1504	11 MAY/2348
15		20 APR/1950	(10TH) 28 APR/0506	5 MAY/1422	12 MAY/2301
14	15 APR/0430	22 APR/1323	(16TH) 29 APR/2244	7 MAY/0753	
12	15 APR/1607	23 APR/0132	(31ST) 30 APR/1024	7 MAY/1827	

The camera and battery were repositioned twice to maintain proper temperature. The film cassette was recovered at the end of EVA-3.

Solar Wind Composition

The solar wind composition (SWC) was deployed for 45 hours 5 minutes. A slight tear made during retrieval will not affect scientific results.

Lunar Portable Magnetometer

Five measurements were made with the LPM; at Spook Crater, Station 5, Station 13, LM site and the LM site with an igneous rock placed on the LPM. Readings ranged from 180 gammas down to 125 gammas up. The rock caused a change of 5 gammas, well within the expected range.

Cosmic Ray Detector

During activation by the crew, a pull lanyard broke, possibly precluding activation of the neutron flux detector, radon emission detector, and solar wind detector. Visual verification by the crew was inconclusive. At the end of EVA-1, temperature indicators showed the cosmic ray detector (CRD) to have exceeded specified values. As planned, it was placed in the LM footpad, in the shadow of the leg.

Indicators prior to CRD stowage indicated that the temperatures were not excessive at the Lexan plates which form the bulk of the CRD Experiment.

IN-FLIGHT SCIENCE

X-ray and gamma-ray astronomical data and alpha particle cislunar and instrument background data were acquired during transearth coast (TEC). The X-ray sources Scorpius X-1 was observed for a total of 6 hours during three observation periods, Cygnus X-1 was observed for 5 hours during two observation periods, the Super Galactic Plane was scanned for 10 hours, and the Galactic Anti-Center Point was observed for 1 hour. Gamma-ray astronomical data was also acquired with the boom extended (the CSM in the passive thermal control (PTC) mode) and scanning the Super Galactic Auxiliary Plane, the Ecliptic Auxiliary Plane, and the Super Galactic Plane in addition to observations of Scorpius X-1 and Cygnus X-1.

The X-ray, gamma-ray, and alpha-particle spectrometers operated nominally throughout the lunar orbit phase of the mission. A preliminary X-ray fluorescence map has been plotted and shows regions of high aluminum concentration over the highlands, including the Descartes region, with low concentrations of aluminum over the maria. A map of the gamma-ray activity in the range 0.54-2.75 Mev indicates that the only significant region of high radioactivity is the western maria, from 101° to 60°W and falls off rapidly outside this band. The highest level of radioactivity (10 ppm Thorium) is at 15°W near the Fra Mauro formation, and the lowest over the backside highlands. On the basis of the gamma-ray spectrometer data the radioactivity of the returned samples is predicted to be 1-2 ppm Thorium. No intense regions of radon emission were observed by the alpha-particle spectrometer. However, a significant variation of 210p. was observed over the lunar surface, indicating that radon emissions have occurred in the last tens of years.

Mapping camera operation was nominal throughout the mission, except for longer than nominal extension/retraction times. This has been attributed to the glare shield for the stellar camera being jammed against the handrail paralleling the SIM bay (as reported by the CMP during the inflight EVA). Therefore, glare may be present on the stellar frames and complicate the star field data reduction. Approximately 2900 mapping camera frames were acquired in lunar orbit and approximately 60 of the post-TEI photographs will be useful scientifically. Approximately 5.6% of the lunar surface was covered by the vertical mapping camera photography with about 4.0% being areas not covered by Apollo 15. The mapping camera obliques covered approximately 20% of the surface with about 7% being of areas not similarly photographed on Apollo 15.

The laser altimeter was operated on all vertical mapping camera passes in lunar orbit. A total of 2395 laser altimeter measurements were made with an overall reliability of 63.5%. The percentage of valid measurements gradually deteoriated during the mission, decreasing to about 50% by Rev 60 and decreasing abruptly to about 5% on Rev 63, the last photographic pass.

Except for the Rev 63 data, the number of valid altitude measurements is sufficient to provide the necessary cartographic control for the photography.

The mechanical operation of the panoramic camera was nominal throughout the mission. An undervoltage in the power supplied to the camera necessitated termination of the photopass on Rev 3 after only four frames were taken.

Operation was nominal on the remaining passes. However, the automatic exposure control indicated consistently low light levels, resulting in overexposure of the film for all areas away from the terminator. It is expected that this overexposure will be largely compensated during film processing. Approximately 1,425 panoramic frames were acquired in lunar orbit and about 175 frames post-TEI. The total area photographed by the panoramic camera was about 10% less than would have been obtained on the nominal mission.

The mass spectrometer operated successfully in lunar orbit acquiring about 85 hours of data during water dumps, fuel cell purges, and during nominal experiment operating periods; however, the attempt to retract the mass spectrometer boom just prior to TEI was unsuccessful.

Subsequent attempts at troubleshooting the retraction problem were also unsuccessful and the boom was jettisoned prior to the SPS burn for TEI. The data acquired in lunar orbit should be adequate to overcome the major problem of determination of constituents of the lunar atmosphere. The main purpose of mass spectrometer operation during TEC was to obtain additional data which would help unravel the more puzzling aspects of the contamination problem.

Because data was obtained in TEC with the Apollo 15 mass spectrometer, the present loss of data is not of overwhelming importance.

The dual frequency bistatic radar experiment (BRE) was performed on Rev 40. The S-band signal received at Goldstone was quite strong and properly polarized, and signal lock was maintained throughout the experiment.

The S-band data are of good quality and all pre- and post-calibrations were good. The reflected VHF signal received at the Stanford 85-foot antenna was of poor quality and the usefulness of the data is questionable.

The VHF only bistatic radar test was initiated on Rev 42. As in the case of the dual frequency BRE, the reflected VHF signal received at Goldstone was of poor quality. The CSM omniantenna was switched off and then back on and the VHF signal problem cleared up. Three and a half frontside passes of good quality VHF bistatic radar data were acquired. The VHF only data can be used in lieu of the poor quality VHF data obtained during the dual frequency portion of the experiment on Rev 40.

S-band transponder (Doppler tracking) data from the CSM, LM, and subsatellite have been processed over short time intervals. The quality of the data is excellent and many new features can be resolved. The amplitudes are not as large as recorded from Apollo 15 since the trajectory path was not over any mascon areas.

However, craters and highland features >40 km were evaluated. A preliminary gravity profile acquired on Rev 10 (CSM/LM) from 60°E to 50°W indicates a positive anomaly over Fecunditatis, a positive anomaly over the highlands east of Descartes with a small negative anomaly over the Descartes region, and positive anomalies over the rims of Albategnius and Ptolemaeus, with Ptolemaeus itself a negative anomaly.

The Apollo 16 subsatellite was launched at 196:13:55 GET into a 66.6 x 52.8 nm orbit with a longitude of the descending node on the initial rev of 117° W. Power was applied to the experiments on Rev 26 at 241:18 GET.

Preliminary analysis of the data indicates that the magnetometer and solid state telescopes are working normally. The high voltage for the electrostatic analyzers was turned on at 271:11 GET after allowing sufficient time for outgassing. The subsatellite lifetime is estimated between 195-222 days based on the Jet Propulsion Laboratory (JPL) 15-8 gravity model.

The Apollo 15 subsatellite was reactivated at 261:46 GET, Rev 3200 after being inactive for 1 month. No degradation was noted.

The low light level astronomy photographic tasks were performed and 75% of the planned tasks were completed in spite of early termination of the experiment. The CM photography of 22 of 26 planned lunar surface targets were also successfully completed. Visual observations of 10 surface areas were made by the CMP, using the unaided eye as well as the 10X binoculars flown for the first time on Apollo 16, and significant details not otherwise available were noted. The Skylab Contamination Study was conducted on the TEC phase of the mission.

Total SIM data acquired in lunar orbit:

Mass spectrometer	72.5 hr (-X) 12.5 hr (+X)
X-ray/alpha particle spectrometers	63 hr prime data 11.5 hr slightly degraded (LM attached) 13.8 hr non-SIM attitude
Gamma-ray spectrometer	56.3 hr prime data 7.7 hr slightly degraded (boom partially retracted) 11.5 hr seriously degraded (RTG attached) 13.8 hr non-SIM attitude (boom extended) 26.5 hr background during TEC
Mapping camera	5 term-to-term obliques 9.3 term-to-term vertical passes 6 darkside passes with concurrent laser altimeter operation
Panoramic camera	1425 frames of stereo photography
Bistatic radar	1 dual frequency pass 4 VHF only passes

Astronomical data acquired during TEC:

X-ray spectrometer	23 hr
Gamma-ray spectrometer	33 hr

TABLE I APOLLO 16 OBJECTIVES AND EXPERIMENTS

PRIMARY OBJECTIVES

The following were the NASA OMSF Apollo 16 Primary Objectives:

Perform selenological inspection, survey, and sampling of materials and surface features in a preselected area of the Descartes region.

Emplace and activate surface experiments.

Conduct in-flight experiments and photographic tasks.

APPROVED EXPERIMENTS

The following experiments were performed:

Apollo Lunar Surface Experiments Package (ALSEP)
S-031   Lunar Passive Seismology
S-033   Lunar Active Seismology
S-034   Lunar Tri-Axis Magnetometer
S-037   Lunar Heat Flow (incomplete)*

Lunar Surface

S-059	Lunar Geology Investigation
S-080	Solar Wind Composition
S-152	Cosmic Ray Detector (Sheets)
S-198	Portable Magnetometer
S-200	Soil Mechanics
S-201	Far UV Camera/Spectroscope

In-Flight

S-160	Gamma-ray Spectrometer (SIM)
S-161	X-ray Fluorescence (SIM)
S-162	Alpha-particle spectrometer (SIM)

* The Lunar Heat Flow second bore hole was not accomplished because of the inadvertent breakage of the HFE cable to the central station connector. This experiment is not active (see page 7 - Lunar Surface Section).

S-164	S-band Transponder (CSM/LM) (Subsatellite)
S-165	Mass Spectrometer (SIM)
S-170	Bistatic Radar (CSM)
S-173	Particle Shadows/Boundary Layer (Subsatellite)
S-174	Magnetometer (Subsatellite)
S-177	UV Photography - Earth and Moon (CM)
S-178	Gegenschein from Lunar Orbit (CM)
M-191	Microbial Response in Space Environment
-	Visual Observations from Lunar Orbit
-	Visual Light Flash Phenomena

Other

M-078	Bone Mineral Measurement
M-211	Biostack
S-176	Apollo Window Meteoroid

DETAILED OBJECTIVES

The following detailed objectives were assigned to and accomplished on the Apollo 16 Mission:

Subsatellite Tracking for New Navigation Techniques
LRV Evaluation
Gas/Water Separation Delivery of Gas-free
Water Fecal Collection Bag Evaluation
Skylab Food Package Function
SM Photographic Tasks
CM Photographic Tasks
Apollo Time and Motion Study
LM Voice and Data Relay Evaluation
Visual Observations from Lunar Orbit
Visual Light Flash Phenomena
Impact S-IVB on Lunar Surface
Postflight Determination of S-IVB Impact Point
Body Fluid and Electrolyte Disturbances Understanding
Skylab Contamination Study

SUMMARY

Fulfillment of the primary objectives qualifies Apollo 16 as a successful mission.

The experiments and detailed objectives which supported and expanded the scientific and technological return of this mission were successfully accomplished.

TABLE 2
APOLLO 16 ACHIEVEMENTS

Fifth Manned Lunar Landing
Largest Payload Placed in Lunar Orbit (76,100 pounds)
First Cosmic Ray Detector Deployed on Lunar Surface
First Use of Far UV Camera on Lunar Surface
Longest Lunar Surface Stay Time (71 hours 14 minutes)
Longest Lunar Surface EVA (20 hours 15 minutes)
First Landing In and Exploration of Lunar Highlands
Largest Amount of Lunar Samples Returned to Earth (approximately 213 pounds)

TABLE 3
APOLLO 16 POWERED FLIGHT SEQUENCE OF EVENTS END OF MISSION

EVENT

PRELAUNCH

	PLANNED (GET) HR:MIN:SEC	ACTUAL (GET) HR:MIN:SEC
Guidance Reference Release	-17.6	-17.6
Liftoff Signal (TB-1)	0	0
Pitch and Roll Start	11.8	11.8
Roll Complete	29.8	29.8
S-IC Center Engine Cutoff (TB-2)	2:17.4	2:17.2
Begin Tilt Arrest	2:38.7	2:38.5
S-IC Outboard Engine Cutoff (TB-3)	2:41.4	2:41.2
S-IC/S-II Separation	2:43.2	2:43.0
S-II Ignition (Command)	2:43.8	2:43.6
S-II Second Plane Separation	3:13.1	3:12.9
S-II Center Engine Cutoff	7:41.4	7:41.2
S-II Outboard Engine Cutoff (TB-4)	9:18.6	9:19.0
S-II/S-IVB Separation	9:19.6	9:20.0
S-IVB Ignition	9:19.7	9:20.1
S-IVB Cutoff (TB-5)	11:45.0	11:45.8
Insertion	11:54.8	11:55.6
Begin Restart Preps (TB-6)	2:23:57.2	2:23:58.0
Second S-IVB Ignition	2:33:35.2	2:33:36.0
Second S-IVB Cutoff (TB-7)	2:39:19.7	2:39:18.1
Translunar Injection	2:39:29.5	2:39:27.9

Prelaunch planned times are based on MSFC Launch Vehicle Operational Trajectory.

TABLE 4
APOLLO 16 MISSION SEQUENCE OF EVENTS END OF MISSION

Liftoff 12:54.6 April 16	00:00:00	00:00:00
Earth Parking Orbit Insertion	00:11:54	00:11:56
Second S-IVB Ignition	02:33:35	02:33:36
Translunar Injection	02:39:29	02:39:29
CSM/S-IVB Separation, SLA Panel Jettison	03:03:50	03:05:01
CSM/LM Docking	03:13:50	03:22:10
Spacecraft Ejection From S-IVB	03:58:50	03:59:20
S-IVB APS	04:23:01	04:18:35
Evasive Maneuver	11:38:50	Not Performed
Midcourse Correction-1	30:38:50	30:39:00
Midcourse Correction-2	52:28:39	Not Performed
Midcourse Correction-3	69:28:39	Not Performed
Midcourse Correction-4	69:58:39	69:59:00
SIM Door Jettison	74:28:39	74:28:27
Lunar Orbit Insertion (Ignition) S-IVB Impact	74:30:08	75:08:00
Descent Orbit Insertion (Ignition)	78:35:30	78:33:44
CSM/LM Undocking	96:13:31	96:13:31
CSM Separation	96:13:31	96:13:31
CSM Separation No. 2	Not Planned	102:30:00
CSM Circularization	97:41:44	103:21:42
Powered Descent Initiate	98:34:41	104:17:25
LM Lunar Landing	98:46:42	104:29:35*
Begin EVA-1 Cabin Depress	102:25:00	119:05:11*
Terminate EVA-1 Cabin Repress	109:25:00	126:16:22
Begin EVA-2 Cabin Depress	124:50:00	142:51:15
Terminate EVA-2 Cabin Repress	131:50:00	150:14:41
Begin EVA-3 Cabin Depress	148:25:00	165:43:15
CSM Plane Change (LOPC)	152:28:48	169:17:40
Terminate EVA-3 Cabin Repress	155:25:00	171:23:29
LM Liftoff	171:45:09	175:43:35
LM Tweak Burn	Not Planned	175:54:05
Terminal Phase Initiate Maneuver	172:39:23	176:37:52
LM/CSM Docking	173:50:00	177:53:13
LM Jettison	177:31:15	195:12:00
CSM Separation	177:36:15	195:15:00
Ascent Stage Deorbit	179:16:29	Not Performed
Ascent Stage Lunar Impact	179:39:29	Not Performed
LOPC-2	193:13:46	Not Performed
Shaping	216:49:12	Not Performed
Subsatellite Launch	218:02:08	196:13:55
Transearth Injection	222:20:33	200:33:20*
Midcourse Correction-5	239:23:03	239:21:02*
CMP EVA Depress	242:00:00	243:24:55
CMP EVA Repress	243:10:00	244:38:05
Midcourse Correction-6	268:22:45	Not Performed
Midcourse Correction-7	287:22:45	287:23:24
CM/LM Separation	290:07:45	290:08:31
Entry Interface (400,000 ft)	290:22:45	290:23:31
Landing	290:36:03	290:37:01**

*11 min 48 sec GET clock update at 118:06:31 GET and an additional 24 hr 34 min 12 sec GET clock update at approximately 202:25:00 GET. ** Mission duration was approximately 265 hours 51 minutes.

TABLE 5
APOLLO 16 TRANSLUNAR AND MANEUVER SUMMARY

Maneuver	GROUND ELAPSED TIME (GET) AT IGNITION (HR:MIN:SEC) Pre-Launch Plan	Real-Time Plan	Actual	BURN TIME (SECONDS) Pre-Launch Plan	Real-Time Plan	Actual	VELOCITY CHANGE (FEET PER SECOND-FPS) Pre-Launch Plan	Real-Time Plan	Actual	GET OF CLOSEST APPROACH HT (NM) CLOSEST APPROACH Pre-Launch Plan	Real-Time Plan	Actual
TLI (S-IVB)	02:33:15	02:33:34	02:33:34	335	342.7	342.7	10386	10389.6	10389.6	74:32:13.4 / 71.4	74:30:07 / 71.4	74:32:22 / 146.7
CSM Sep	03:03:50	03:04:20	03:05:01	3	3		0.5	0.3		74:32:13.4 / 71.4	— — —	— — —
CSM Dock	03:13:50	03:14:20	03:22:10	NA	-	-	NA	-	-	74:32:13.4 / 71.4	— — —	— — —
SM EJT	03:58:50	03:59:00	03:59:20	3	3	4.6	0.3	0.3	0.5	74:32:13.4 / 71.4	74:32:22 / 146.7	74:32:21 / 136.2
S-IVB Evasive	04:23:01	04:18:33	04:18:35	80.2	16.0	17.1	9.8	10.2	10.9	74:30:08 / 0	74:22:53 / 0	74:23:07 / 0
MCC-1 (SPS)	11:38:50	NP		0		NP	0	NP		74:32:13.4 / 71.4	— — —	— — —
MCC-2 (SPS)	30:38:50	30:39:00	30:39:00	0	1.9	1.8	0	12.6	12.5	74:32:13.4 / 71.4	74:32:12 / 71.5	74:32:07 / 71.7
MCC-3	52:28:39	NP		0		NP	0	NP		74:32:13.4 / 71.4	NP	
MCC-4	69:28:39	NP		0		NP	0	NP		74:32:13.4 / 71.4	NP	
SIM Door Jettison	69:58:39	69:59:00	69:59:00	NA			NA			NA		

NA – Not Applicable   NP – Not Performed

TABLE 6
APOLLO 16 LUNAR ORBIT SUMMARY END OF MISSION

Maneuver	GROUND ELAPSED TIME (GET) AT IGNITION (HR:MIN:SEC) Pre-Launch Plan	Real-Time Plan	Actual	BURN TIME (SECONDS) Pre-Launch Plan	Real-Time Plan	Actual	VELOCITY CHANGE (FEET PER SECOND-FPS) Pre-Launch Plan	Real-Time Plan	Actual	RESULTING APOLUNE/PERILUNE (NM) Pre-Launch Plan	Real-Time Plan	Actual
LOI	74:28:39	74:28:27	74:28:27	375	374.3	374.3	2807	2802	2802	170.6 / 58.5	170 / 58.3	170.3 / 58.1
S-IVB Impact	74:30:08	75:07:03	75:08:00	NA			NA			NA		
DOI	78:35:30	78:33:44	78:33:44	24.1	24.2	24.2	206	210.3	209.5	58.6 / 10.9	58.5 / 10.3	58.5 / 10.9
DOI Trim	NA			NA			NA			NA		
Undock	96:13:31	96:13:31	96:13:31	NA			NA			NA		
CSM Sep	96:13:31	96:13:31	96:13:31	1.4	6.8	6.8	1.0	1.0	1.0	60.5 / 8.9	59.2 / 10.2	59.2 / 10.4
CSM 2nd Sep		102:30:00	102:30:00	NA	6.8	6.8	NA	1.0	1.0	NA	59.7 / 11.2	59.7 / 11.2
CSM Circ	97:41:44	103:21:42	103:21:42	5.9	4.6	4.6	99.6	81.6	81.0	68.2 / 51.8	68.0 / 53.1	68.0 / 53.1
PDI	98:34:41	104:17:25	104:17:25	721.5	720.5	730	6696.3	6703.9	6703	0 / 0	0 / 0	0 / 0
Landing	98:46:42	104:29:25	104:29:35	NA			NA			NA		

NA – Not Applicable

TABLE 7
APOLLO 16 LUNAR ORBIT SUMMARY END OF MISSION

Maneuver	GROUND ELAPSED TIME (GET) AT IGNITION (HR:MIN:SEC) Pre-Launch Plan	Real-Time Plan	Actual	BURN TIME (SECONDS) Pre-Launch Plan	Real-Time Plan	Actual	VELOCITY CHANGE (FEET PER SECOND-FPS) Pre-Launch Plan	Real-Time Plan	Actual	RESULTING APOLUNE/PERILUNE (NM) Pre-Launch Plan	Real-Time Plan	Actual
CSM LOPC	152:28:48	169:17:40	169:17:40	9.1	7.1	7.0	158.7	124.7	124	62 / 57.3	64.5 / 55.3	65.6 / 55
ASCENT	171:45:09	175:43:35	175:43:35	434.3	436.8	3.1	6047.9	6054.9		45.4 / 9	41.1 / 8.9	40.2 / 7.9
TWEAK		175:54:05	175:54:05	-	-	-	-	10	10			
TPI*	172:39:23	176:37:52	176:37:52	2.5	3.1	3.1	50.3	77.7	78	61.9 / 44	63.9 / 40.1	64.2 / 40.1
Docking	173:50:00	177:48:26	177:53:13	NA			NA			NA		
LM JETT	177:31:15	195:10:00	195:12:00	NA			NA			NA	68.0 / 53.7	67.8 / 53.8
CSM SEP	177:36:15	195:15:00	195:15:00	13.2	6.8	6.8	2.0	2.0	2.0	61.7 / 59.5	66.4 / 52.7	66.3 / 52.6
ASC Deorb	179:16:29	203:08:09	NP	95.5	150.9	NP	229.6	364.0	NP	NA	68.0 / -130.9	NP
ASC Impact	179:39:29	203:25:16	NP	NA			NA			NA		
LOPC-2	193:13:46	NP	NP	15.8			282.5			62.9 / 57.9		
Shaping	216:49:12	NP	NP	2.2			38			85 / 55		
SAT JETT	218:02:08	196:13:55	196:13:55	NA			NA			85 / 55	66.6 / 52.8	66.6 / 52.8

*APS only, does not include the nominal 10-sec RCS ullage (21.8 fps)

NA - Not Applicable   NP - Not Performed

TABLE 8
APOLLO 16 TRANSEARTH MANEUVER SUMMARY

Maneuver	GROUND ELAPSED TIME (GET) AT IGNITION (HR:MIN:SEC) Pre-Launch Plan	Real-Time Plan	Actual	BURN TIME (SECONDS) Pre-Launch Plan	Real-Time Plan	Actual	VELOCITY CHANGE (FEET PER SECOND-FPS) Pre-Launch Plan	Real-Time Plan	Actual	GET ENTRY INTERFACE (EI) / VELOCITY (FPS) AT EI / FLIGHT PATH ANGLE AT EI Pre-Launch Plan	Real-Time Plan	Actual
TEI (SPS)	222:20:33	200:33:20	200:33:20	150.5	162.1	162.4	3212.2	3370.9	3370.0	290:22:45 / 36,175.8 / -6.5	265:48:44 / 36,196.3 / -6.51	265:48:40 / 36,196.9 / -7.44
MCC-5	239:23:03	239:20:55	239:21:02	0.0	8.3	8.3	0.0	3.4	3.4	290:22:45 / 36,175.8 / -6.5	290:23:36 / 36,196.4 / -6.5	290:23:34 / 36,196.4 / -6.5
MCC-6	268:22:45		NP	0.0		NP	0.0		NP	290:22:45 / 36,175.8 / -6.5	NP / NP / NP	
MCC-7	287:22:45	287:23:00	287:23:24	0.0	3.5	3.5	0.0	1.4	1.4	290:22:45 / 36,175.8 / -6.5	290:23:32 / 36,196.1 / -6.5	290:23:32 / 36,196.4 / -6.5
CM/SM SEP	290:07:45	290:09:31	290:08:31	NA			NA			NA / NA / NA		
ENTRY	290:22:45	290:23:31	290:23:31	NA			NA			290:22:45 / 36,175.8 / -6.5		290:23:31 / 36,196.2 / -6.55
SPLASH	290:36:03	290:36:55	290:37:01	NA			NA			NA / NA / NA		

NA - Not Applicable   NP - Not Performed

TABLE 9
APOLLO 16 CONSUMABLES SUMMARY* END OF MISSION

CONSUMABLE (TOTAL QUANTITY	FLIGHT LOAD	FLIGHT PLANNED REMAINING	ACTUAL REMAINING
CM RCS PROP (POUNDS)	233.2	173.9	181.7
SM RCS PROP (POUNDS)	1,342.4	580.8	570.0
SPS PROP (POUNDS	40,544.2	1,681.6	2,399.0
SM HYDROGEN (POUNDS)	79.0	20.9	24.0
SM OXYGEN (POUNDS)	961.0	356.4	424.0
LM RCS PROP (POUNDS)	631.2	0.0	345.0**
LM DPS PROP (POUNDS)	19,482.9	757.6	1,155.0
LM APS PROP (POUNDS)	5,288.3	244.7	426.0**
LM A/S OXYGEN (POUNDS)	4.8	4.2	4.6
LM D/S OXYGEN (POUNDS)	93.4	47.0	50.8***
LM A/S WATER (POUNDS)	85.0	52.4	46.2***
LM D/S WATER (POUNDS)	393.2	67.5	RESIDUALS 7.8
LM A/S BATTERIES (AMP-HOURS)	592.0	NOT AVAILABLE	
LM D/S BATTERIES (AMP-HOURS)	2,075.0	NOT AVAILABLE	
LRV BATTERIES (AMP-HOURS)	121.0/121.0	64.5/64.5	158 TOTAL****

* All values adjusted to be compatible with quantities carried in Real-Time Computer Complex (tanked values)

** 53.8 pounds of LM RCS propellant transferred inflight to LM APS tanks.

*** LM Jettison

**** 158 total based on real-time analysis of power consumed during EVAs.

TABLE 10
SA-511 LAUNCH VEHICLE DISCREPANCY SUMMARY

S-II/J-2 engine No. 4 helium tank decay at ESC.	Open
S-IVB APS-1 helium leak.	Open
S-IVB APS-2 helium leak.	Open
S-IVB Battery No. 2 depletion time.	Open
IU thermal conditioning system GN_2 leak.	Open
IU CCS lost at 27 hours.	Open
IU EMR Bits 13 and 14 at 6:46 to 7:01 hours.	Open
ST-124 cross range velocity shift at F +0.8 hours.	Open
IU EDS spare yaw rate gyro ramp for a 2-second period during prelaunch checks.	Open

TABLE 11
COMMAND/SERVICE MODULE 113 DISCREPANCY SUMMARY

ECS mixing valve fluctuation in auto mode.	Open
Onboard TV monitor horizontal lines.	Open
SPS oxidizer tank pressure reference lost.	Open
Hydrogen tank No. 3 heat leakage excessive.	Open
False gimbal lock at GET 38:18:55.	Open
Unable to uplink commands at GET 44:00.	Open
First Mapping Camera retract cycle time twice normal.	Open
Mass Spectrometer boom talkback barberpole.	Open
SPS secondary engine yaw actuator oscillations.	Open
Twenty percent of laser altimeter data erroneous.	Open
Pan camera film overexposure.	Open
LiOH cannisters sticking.	Open
Gamma ray cover	Open
approximately 30°.	Open
Event timer erratic at 244:20 GET.	Open
CM cabin fan made unusual noise at 202:17 GET.	Open
Master alarm at 195:06 GET.	Open
Battery vent manifold pressure went to 14 psi during battery charging.	Open
Mapping camera stellar lens glare shield crushed against EVA handrail.	Open
Intermittent loss of signal. HGA would not acquire at 260:00 GET.	Open
Inertial subsystem coupling data unit fail alarms.	Open

TABLE 12
LUNAR MODULE 11 DISCREPANCY SUMMARY

Strips of paint peeling off panels at 7:18 GET.	Closed
Steerable antenna locked in yaw at 94:35 GET.	Open
RCS System A helium regulator failure at 95:05 GET.	Open
Landing Radar indicated bad data at 96:37 GET.	Open
Apparent blockage in ECS suit loop.	Open
Panels 186, 187, 188, and 189 tore loose at ascent stage liftoff.	Open
LM ascent stage tumbling after jettison at 195:17 GET.	Open
Ascent engine chamber pressure measurement fluctuated during ascent at 175:32 GET.	Open
Rendezvous radar drifting prior to lunar liftoff.	Open
MESA deployment abnormal.	Open

TABLE 13
LUNAR ROVING VEHICLE DISCREPANCY SUMMARY

Rear wheel steering inoperative at 119:34 GET.	Open
Battery 2 read off scale low at 119:34., GET.	Open
Battery temperature meters read off scale low at 119:34 GET.	Open
Attitude indicator broken at 144:12 GET.	Open
Right rear fender knocked off.	Open
Battery 2 "ampere-hour remaining" meter failure during EVA-3.	Open
Battery 1 "temperature" meter failure during EVA-3.	Open
Batteries 1 and 2 failure to cool down between EVAs.	Open
Battery 1 amp-hours remaining meter readings decreased excessively.	Open

TABLE 14
APOLLO 16 CREW/EXPERIMENT DISCREPANCY SUMMARY

LMP reported PGA suit legs too short at 55:49 GET.	Open
ALSEP subpackage 2 fell off carrier bar at 120:40 GET.	Open
ALSEP Heat Flow Experiment cable broken by crew member.	Open
ALSEP No. 3 spike on mortar package pallet did not deploy.	Open
ALSEP Thumper arm and fire knob required 2 actuations to fire at position No. 2.	Closed
Cosmic ray detector panel 4 curtain raising cable broke.	Open
Purge valve pull pin came off CDR's suit during EVA.	Open
CDR broke 2 inches off OPS antenna during ingress after EVA-2.	Open
During Thumper deployment at 121:19 GET taut cables pulled central station out of alignment.	Open
Vertical staff gnomon separated from leg assembly at 146:22 GET.	Open
Dispensor of documented sample bags fell off several times during EVAs.	Open
Velcro came off sample bags.	Open
TV camera sun shade came loose at 107:40 GET.	Open
Mortar box pitch angle data erratic.	Open
Pan camera lens torque current erratic on Rev 47.	Open
Unexplained MESA temperature response during lunar stay.	Open
Gas separator inoperative.	Open
CDR and LMP suit wrist disconnects hard to lock before SIM EVA.	Open
LMP biomedical readouts bad.	Open
Suitloop pressure readout 0.5 psi high during CMP EVA at 243:50 GET.	Open
ALSEP Passive Seismic Experiment temperature rise at 261:00 GET.	Open
Vacuum cleaner became inoperative at 192:52 GET.	Open

CONFIDENTIAL

MSC-06805

APOLLO 16
TECHNICAL
CREW DEBRIEFING
(U)

MAY 5, 1972

PREPARED BY
TRAINING OFFICE
CREW TRAINING AND SIMULATION DIVISION

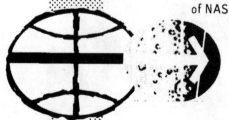

MANNED SPACECRAFT CENTER
HOUSTON, TEXAS

CONFIDENTIAL

1.0 SUITING AND INGRESS

YOUNG — I didn't notice any problems associated with suiting ingress. Cabin Close-out was nominal throughout. As everyone knows, it is very difficult to reach the dump handle to dump the cabin for the purge; but you can do it with your elbow. That's the way I did it. The other thing that made me nervous about ingress was the way Troy Stuart leaned over the abort handle. I know the pyros are not armed, but man, we should caution the suit technicians to stay away from that kind of stuff.

SLAYTON — Speaking of Troy reminds me that we do need to get a comment from Charlie here on the suiting; whether he had some difficulty, based on your first go-around.

YOUNG — Yes. On the day of launch, he said the legs of his were tight. And, everybody just sort of poo-pooed that.

SLAYTON — We can talk about that later.

YOUNG — Suit circuit check was good; ingress and cabin close-out were nominal. In fact, we were about 20 minutes ahead the whole time through the launch.

2.0 STATUS CHECKS AND COUNTDOWN

YOUNG — Ground communications and countdown were nominal; launch preparation was nominal; systems preparation was nominal. Crew Station Controls and Displays: I think there's a couple of things we can say here. One is that we knew that the H2 tank 1 pressure read-out was oscillating, didn't we?

MATTINGLY — Yes. They told us that. We had a caution and warning on it.

YOUNG — They told us that. But the other thing that we didn't know was that the SPS pressure - the fuel pressure was what - 7 psi low? Due to a transducer.

MATTINGLY — I've got all those numbers written down. Apparently the problem with the SPS pressure was something that they had known about and it was not something that we ever discussed. The problem that Charlie saw was that the Delta-P on the pad was greater than 20.

DUKE — That's right.

YOUNG — When we were sitting there, we were sitting with an abort condition for LOI. We had a greater-than-20 Delta-P laying there on the pad.

MATTINGLY — And they came back and said that's the way it is because there is a shift in there.

YOUNG — And then some other thing had failed and nobody told us anything about that. Some ground reference thing had failed and I guess it could have been our fault and they asked us if we wanted a final briefing on the actual systems, how they were operating, and I asked Dave Ballard if there was anything we ought to know about and he said not that he knew of. So I said, let's not do it then because we're kind of busy and I don't want to unnecessarily fill squares.

MATTINGLY — We talked to Don Arabian about it and he was sending us all the preflight problem tracking lists and I read every one of those reports.

YOUNG — It wasn't in there.

MATTINGLY —There was a transducer shift, I remember, and it was a telemetry transducer shift. That was the way it was described, and it was something that the MOCR had calibrated and that was all taken care of and whether this was the same problem or whether it was an additional problem, I don't know.

YOUNG — The point being, I think we could have had that problem all worked out before we ever left the ground with the mission rules and everything for LOI instead of going through a couple of days of LOI mission rules and coming up to LOI with an SPS light and that kind of thing. The bad feature about an SPS light, is that's not the only thing you can get an SPS warning light for.

72-H-440

MATTINGLY — That's right. It negates the value of the SPS pressure light because you've masked it by having it on all the time.

YOUNG — For future design, we ought to be able to reset lights if there's more than one variable going to them. For example, when you get a fuel cell light, there's five things going into a fuel cell. Charlie just walked into the room and he's going to say something about his suiting. His legs were too short. When we walked out of the suit room you said your legs were too short and I said, "Charlie."

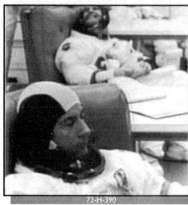

72-H-390

DUKE — Well, I was really kidding a little bit. They really felt a little tight, but I asked for them to be tight because Clyde said that they would stretch and sure enough they did. On the surface, they felt fine.

YOUNG — I'll tell you that fastener didn't stretch much.

DUKE —That was the zipper problem it turned out, not a suit fit problem. I had the same problem with John's on PDI day that he had with mine the day before. When you pull the restraint zipper together, it doesn't lie flat in the middle of the back. It gives a sort of series of W's, and when you try to pull it across there it takes three hands to really do it.

YOUNG — Charlie, yours just wasn't making on the first day. It was coming close. It really was nervous. I thought every EVA would be the last because I didn't know whether I'd make the restraint zipper every time.

72-H-391

DUKE —The fit was fine, though, after we got out on the surface and got pressurized.

YOUNG — Yes. The tighter you can get it, the better off you're going to be when you put that 3.5 psi in there.

DUKE — But sure enough, preflight, going out, they felt a little tight to me. But I'd asked for it, so I couldn't complain.

MATTINGLY — Had you changed it since the last time you wore it?

YOUNG — No. Well, the problem we got into there, and maybe we could have avoided it, is that, a couple of times preflight with the flight suits, Charlie and I zipped each other just to see how hard that would be. We ran about four or five suited exercises where we donned the suits in the evening and worked for a couple of hours inside the building where the QC guys could watch the suits, and the first couple of times we tried zipping them. Maybe it's the last time you want to do the zipping, after all the adjustments have been made.

DUKE — On every adjustment of the flight suit, when we got to the middle of the back, Troy had a tough time getting that restraint zipper, because of that W in the back.

YOUNG — I really recommend that there ought to be some way to adjust that thing down so that when the two guys are alone one guy can do it, because you need to be able to pull it together like that and you need to be able to stretch it out like that which takes four hands, only you got about two. You know what I mean? You need to be able to pull the thing together to get the teeth closed. You could probably do it with some kind of restraint that pulls the two pieces together, like a tie-down ring. I am not telling them how to fix it, but I thank it's a solvable problem. You really do want that thing tight because you're in a good suit when it is tight.

DUKE — Concerning the SPS fuel and oxidizer Delta-P; I couldn't believe it when I looked up there and it was out of limits. It was 7 psi low.

YOUNG — Was that what it was?

DUKE — Yes. It was about 165 or 163, and it normally runs around 175, because I've always seen it right in the green band.

YOUNG — How about the oxidizer?

DUKE — The oxidizer was in limits. At that time, it was okay. It looked like it was riding a little high, but it was still in the green band.

MATTINGLY — Didn't you have a greater-than-20 Delta-P?

DUKE — Yes. Right then, we had a greater-than-20 Delta-P which the rules say you don't burn. So we asked them about it and that was the first we heard that we had a bias shift on the fuel side.

YOUNG — It's one of those communications problems, I'm sure, but we ought to solve it.

DUKE — But they apparently knew about it.

MATTINGLY — I don't think the ground, the MOCR guys, knew about it.

DUKE — No, they didn't.

MATTINGLY — They were caught by surprise. They were doing a lot of scurrying there also just like we were.
YOUNG — Apparently so. Okay. Distinction of Sounds in Launch Vehicle Sequence, Countdown to Lift-off. I didn't hear any sounds, did you?

MATTINGLY — I was surprised at how quiet everything was.

DUKE — I didn't hear any valves or any valves opening or anything like that.

YOUNG — No, I think it's nominal. I didn't hear any on Apollo 10, either.

MATTINGLY — Not a thing; I was really surprised.

YOUNG — S-IC Ignition.

3.0 POWERED FLIGHT

DUKE — Wow!

YOUNG — Wow is right. There goes a train that is leaving. Lift-off - you can tell lift-off because everything is moving.

DUKE — It is like an elevator slowly lifting off. But, at ignition, I had the lateral frequency of something or other. It just kept shaking at the same frequency throughout the whole S-IC burn. You felt yourself going faster and faster and faster. I had the feeling it was a runaway freight train on a crooked track, swaying from side to side. That was all the way through the first stage.

MATTINGLY — Are you saying that the frequency changed?

DUKE — No, the frequency stayed the same. But, my impression was the g's made it feel like we were going faster and faster and faster.

YOUNG — My impression was unlike a fixed-base simulator, you sensed the yaw, roll, and pitch programs; but, the rate changes are negligible.

DUKE — I didn't sense any of that.

MATTINGLY — I didn't sense lift-off except for the lights. I didn't sense the program. To me, it was just like fixed-base simulator with the vibration on top of it.

DUKE — I felt the slow acceleration.

YOUNG — Some guys sense lift-off and some guys don't. Cabin pressure - Did you notice that relieving on time?

DUKE — Yes, I heard it, as a matter of fact.

MATTINGLY — I thought the simulation of that was very realistic.

YOUNG — It really was. Dynamic pressure noise builds right up to max Q. That thing is making some kind of racket. Engine gimbal and retro motion - I didn't notice it. S-IC inboard and outboard ECO - At inboard cut-off, we got this minor damped unloadings and when the outboard engines cut-off, we got at least four times as much.

MATTINGLY — I was well braced for it. I'm sure glad I was. That really gets your attention!

The Prime Crew of Apollo 16 from left to right Command Module Pilot Thomas K. Mattingly, Commander John Young and Lunar Module Pilot Charles Duke. John Young trains to deploy the Lunar Roving Vehicle (bottom left).
Duke comes to grips with the 500 mm lens and Hasselblad camera (bottom right)

The Prime crew pose for a publicity shot during training with the Lunar Roving Vehicle (above)

Lift off of Apollo 16 12:54 p.m. EST, April 16, 1972. (below)

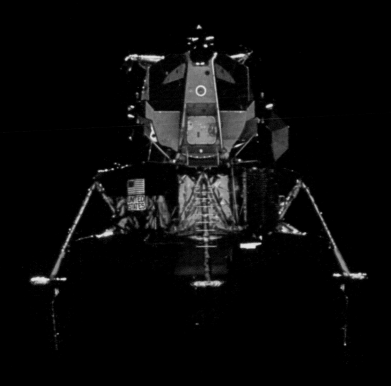

Two dramatic pictures of the Lunar Module "Orion" immediately after undocking with the CSM in Lunar Orbit.

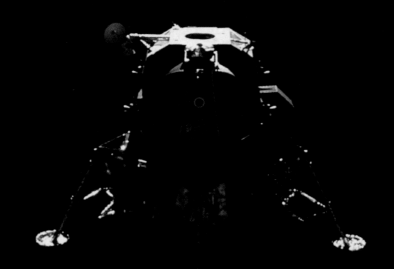

An exuberant John Young leaps into the air to salute the camera during the first EVA at Descartes. (above)
Charlie Duke takes his turn to salute in a more relaxed pose. (below) Duke bubbled with enthusiasm for the
duration of the three EVAs and, compared to his predecessors, talked incessantly to the ground.

The perfect clarity of the lunar vacuum is well illustrated in this shot taken about 300 feet from the Lunar Module. In the foreground is the Passive Seismic Experiment and at center left is the Active Seismic Experiment.

Charlie Duke works in the shadow of the LM. In the foreground is the Ultraviolet Telescope. In the background, the Lunar Rover.

Charlie Duke goes prospecting at the rim of Plum crater. Some of the craters were so steep Duke and Young couldn't get close enough to see the bottom. (above)

Once again the Boeing Lunar Roving Vehicle proved its worth by allowing the two explorers to move rapidly across kilometers of territory. (below)

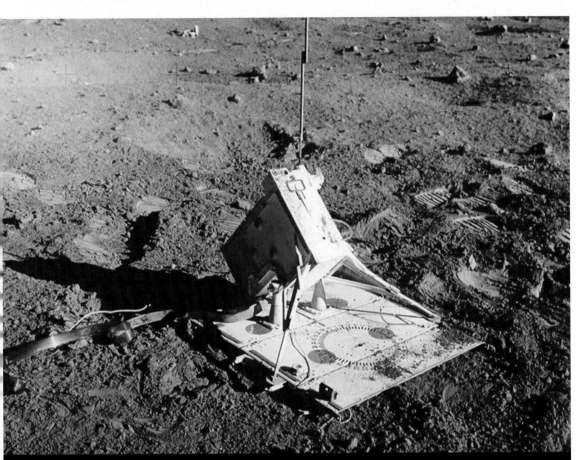

The Active Seismic Experiment included this Mortar package which was armed and fired after the crew were well clear. The science team on the ground activated four grenades. (above)

John Young works with his camera. Behind him can be seen the fourth deployed ALSEP

The Command/Service Module "Casper" in lunar orbit. From this vantage point CM pilot Ken Mattingly took hundreds of photographs of the lunar surface. (above)

A spectacular shot of home showing the California coast and the Baja peninsula. (below)
Splashdown (inset) 27th April 1972, one day early.

YOUNG — I mean to tell you it does. I was holding on to the bottom of the T-handle, at that point, because I sure didn't want to do the wrong thing.

DUKE — It's a good idea to brace yourself. And, I was surprised with the debris that I caught out of my left eye as it came by the hatch window from the staging.

YOUNG — Hey, that's another thing that you remarked on.

MATTINGLY — Yes. That amazing.

YOUNG — The debris was going right along with us.

MATTINGLY — It was passing us. I don't understand that.

DUKE — I think that was from retrofire.

MATTINGLY — No, sir. This was during the powered flight steady state. There were particles; I looked out John's window and particles were going past us in the same direction. I kept looking at that; there's no way. But, it did it. I don't remember it on the S-I; but, I remember it on the S-II and the S-IV.

SLAYTON — This wasn't during the staging sequence?

MATTINGLY — No, sir. This was steady state, powered flight well after staging; and, I don't know where they came from. I don't know what they were, but they were there.

YOUNG — S-II ignition was nominal. I got a feeling it took a little longer to get those lights out.

DUKE — I did, too.

YOUNG — For some reason we noticed the PU shift on the S-II, which was at a different time than nominal. Maybe 3 or 4 seconds early. I don't remember just when. It may be in the onboard tape.

MATTINGLY — I think it's on the tape because we commented on it. But, it was several seconds early.

YOUNG — Different. The launch escape tower and booster protective cover jettison. That was on time and you can see the whole works go off. I didn't see it on Apollo 10, but I sure saw it this time.

DUKE — It was great. Every pyro event was just beautiful. You could hear it, feel it, and see it.

YOUNG — S-II engine cut-off was on time and nominal.

DUKE — I thought the S-II was very smooth and very quiet. I had the sensation of very low acceleration or g's and no noise at all that I could tell. I felt like we were almost floating at that time.

YOUNG — Well, it was low g's. S-II/S-IVB separation was nominal. The S-IVB ignition was nominal. Communications - We didn't have a bit of problem with communications throughout the flight. Controls and displays told us just what we wanted to know. Crew comfort was okay.

DUKE — On control and displays, I might add a comment here that prelift-off, when we did that gimbal motor check, it was more apparent on the fuel cells than the battery buses. But, in flight, it was more apparent on the battery buses than in the fuel cells. When you started them at 6 minutes. I don't know why that is.

YOUNG — The reason is we were not on full internal power during the ground checks.

DUKE — That's probably the reason.

MATTINGLY — Oh, we were on full internal power.

YOUNG — I thought we were.

MATTINGLY — Sure we were. Once you bring the fuel cells up, we do not have any external power coming in. I'm sure that by the time we climbed in, we were running on fuel cells.

YOUNG — Maybe so.

MATTINGLY — Maybe it's typical. Maybe gravity has something to do with the responses.

YOUNG — It's no big deal.

DUKE — Crew comfort through powered flight. Good.

YOUNG — Man, I tell you, you just hope that mother keeps running because there isn't a lot you can do if it quits. Pogo comments - We noticed a buzz on the S-II.

MATTINGLY — S-II and S-IVB both.

YOUNG — And it stopped at 9:07, I remember that; 9:05 to 9:07, it stopped. I don't remember where it started. I thought it pulsed up and then died out and then pulsed up and died out. It had an amplitude variation that would go up like that and then come back off and then go up and come back off. I couldn't tell you what frequency it was, but I imagine it was well in excess of 40 cycles per second.

MATTINGLY — I think the tape will say what time it started.

YOUNG — Yes. I bet the accelerometers in the CSM picked that up also.

MATTINGLY — That might be.

DUKE — That was probably too high a frequency for the BMAGS.

SLAYTON — Was it the same frequency on both boosters?

DUKE — Yes.

YOUNG — So, it could be an air-stage problem or it could be something coming through the SLA.

DUKE — But I wasn't concerned about it, were you?

YOUNG — Yes, but there's nothing you can do about it. I mean, if it is going to damp or not, there's nothing you can do.

DUKE — I never had the feeling it was divergent. It was just a constant buzz.

MATTINGLY — On TLI, it seemed to be increasing in amplitude; although I thought the frequency was still the same.

YOUNG — Yes. That's what I thought all the way to ECO. But, there was a buzz on the S-IVB all the way to engine cut-off. And, it was a high frequency buzz.

YOUNG — Evaluation of Insertion Parameters; I forget what the numbers were, but they were good. Post-Insertion Systems Configuration Checks - I'd say that was nominal.

4.0 EARTH ORBIT AND SYSTEMS CHECKOUT

DUKE — I thought it went very smooth, smoother than in the simulator.

YOUNG — ORDEAL. No problem to unstow. Optics Cover Jettison (Debris). Want to say something about that Ken?

MATTINGLY — Once again, I heard the sextant cover go off, but I did not hear the telescope cover go. Like everyone else that has looked through there, I looked and didn't see a thing in the telescope and I was concerned about what to do next. In fact, we talked about it. What was the best thing to do? I didn't think the cover had gone. While we were playing around we came across a star pattern that was bright enough to see and it was an excellent telescope. There was no debris that you could see from it. COAS and horizon check. Right where it is supposed to be at 31.7 degrees or whatever. It was right on the line.

YOUNG — Unstowage, did you have any problems with that?

MATTINGLY — No, except it's so much easier in zero g than it is in one g. All those things would come out for a change.

YOUNG — Where it'd take you a week in one g to flip over and get the COAS you can do it in milliseconds, with a pressure suit on in zero g.

DUKE — Canaries had an antenna problem on the first time by and they dropped the uplink for awhile.

YOUNG — They dropped the uplink and we missed getting a ...

DUKE — Did we?

MATTINGLY — Yes sir.

YOUNG — It was the pyro arm one time that we missed because when they were coming up they dropped someone.

DUKE — That was the second time by.

MATTINGLY — I guess the first place we checked the BMAG GDC drift was during this Earth orbit period.

DUKE — Right.

MATTINGLY — They were running pretty high, I checked both BMAGs, I got BMAG package number 1, post-TLI and BMAG package number 2 pre-TLI and they were running like 6 degrees an hour. Just roughly, approximately, in all axes. And the reason that's significant to me is that they changed during the mission.

YOUNG — They were different they didn't maintain the same drift. If you're really serious about using that thing for backup, at the time you use it you ought to know what drift you can expect out of it. TLI Preparation, a piece of cake.

DUKE — Yes, we had - they changed a few numbers on us. On the yaw and the ORDEAL, and we were - there was some concern on the APS module, but it worked throughout.

YOUNG — We had all the procedures ready to go if we did have to fly it in roll. I am sure glad we didn't. Subjective Reaction to Weightlessness. It's really neat. Beats work.

DUKE — For the first rest period I had that fullness in the head that a lot of people have experienced. More of a pulsing in the temples, really than a fullness in the head.

YOUNG — I tried to outguess it by standing on my head for 5 minutes a night a couple of weeks before launch. Standing on your head is a heck of a lot harder. That's an overkill, but this is nice.

MATTINGLY — I really think going out and flying those airplanes helped us.

YOUNG — Oh, yes, I highly recommend that.

MATTINGLY — And I flew every day except one. I don't know if that helped, but I bet it didn't hurt.

YOUNG — Yes, rate of roll in particular, it's got to help. It tightens up your eardrums.

MATTINGLY — I think that was a good thing to do.

YOUNG — You ought to approach it with the idea that you're going in there and make yourself as uncomfortable as you can stand. Do rate of roll until you can't stand it anymore.

YOUNG — TLI burn was nominal, but I agree with Ken. I think that buzz increased throughout. It seemed to ramp up. Did you get that feeling?

MATTINGLY — I felt the frequency remained stable, but the amplitude was increasing.

5.0 TLI THROUGH S-IVB CLOSEOUT

YOUNG — S-IVB performance - What a champ. What was the first midcourse?

DUKE — Twelve feet per second.

MATTINGLY — That was after we skipped the midcourse 1, though. It had been pointed out in memos before, but everybody knows the P15 shutdown time calculation is not going to be valid until near the end of the burn because of the changing acceleration. It appeared to me that it was off by far more in powered flight than it had ever been in the simulator.

YOUNG — Hey, there was something wrong back in Earth orbit systems checkout. The problem we had there...

DUKE — The TEMP IN valve.

YOUNG — The TEMP IN valve, the auto position of the glycol evaporator TEMP IN valve was cycling MIN to MAX. I forgot all about that.

DUKE — We put the flow on manual and adjusted it on 381; and, we didn't touch it but a couple of times from then on.

YOUNG — I'm glad that fixed it. Because if we had to do that in lunar orbit, we would have been behind the power curve, to go down there and tweak that thing every time it came around.

MATTINGLY — That valve is exceptionally sensitive.

YOUNG — Yes. We've known that all along. On Apollo 7, that was the big thing, to go down there and do a manual adjustment of the TEMP IN valve to get it to do its thing. It is exceptionally sensitive.

MATTINGLY — It's even more sensitive than the adjustment of the DIRECT O2.

YOUNG — Okay, but that's a systems problem. I guess somebody's going to have to work that one. Maybe they won't do anything since it didn't bother us. Okay, TLI burn, nominal. S-IVB performance and ECO, nominal. Maneuver to separation attitude was nominal. During ORB RATE, we got some pictures. They're really something. You could see all of the United States. If the pictures come out, there will really be some pictures.

MATTINGLY — The Earth was right there in the window. And centered right in the middle of the Earth was the United States, without a cloud over it.

DUKE — All the way from the Great Lakes to Brownsville.

MATTINGLY — Just as if you had drawn it and set it up so you could take a picture of it.

DUKE — You couldn't, see the northeast, Maine and those areas.

YOUNG — That is something we should show at the press conference. That will be a good picture. The S-IVB maneuver to the TD and E attitude was nominal. We discussed the ramp buzz. It was too high a frequency to be characterized as a pogo. The amplitude was so low you couldn't characterize it as pogo either. It didn't seem like anything was in danger of coming apart. I was more worried about it quitting. Separation from the SLA was nominal. In fact, we saw one of the SLA panels floating away. How was the high gain antenna lockup, Charlie?

DUKE — Easy. It was just as advertised.

YOUNG — Formation flight? Why don't you talk about TD&E, Ken?

MATTINGLY —That's got to be the simplest maneuver performed in space flight. That was exactly like the simulator. When we pitched over, the crosshairs on the COAS were almost exactly centered on the target. It was just a matter of pushing it, sitting there and waiting for the two to come together. I made one lateral correction and one vertical correction. We didn't do another thing until contact. Whatever gas we used during TD&E we used after I hit in trying to get it recentered. They busted the 15 guys about forcing it in. I tried to center it up and that is a pretty expensive operation. It's very inefficient when you have your nose hooked to something you're trying to push. I was using the translation controller and I was really surprised. Either the friction on the probe head or something is a lot more than I expected. It was very, ineffective.

YOUNG — It seemed like you used about five times the fuel that you normally use to get in there.

MATTINGLY — I felt like I used 50 percent of the...

DUKE — It took a long time.

YOUNG — It took a long time and we never allowed any time for this. All the time we were closing in on it. When you pitched around, you were only about 60 feet out.

MATTINGLY — Or less. It was nominal.

YOUNG — It couldn't have been very far away. He was right there. I was three times that far away on 10. We went on in and docked. All the way in the particles were coming off the S-IVB and the LM. I had never seen anything like it. They were really streaming off. I didn't know whether they were S-IVB particles, LM particles, or what they were at the time.

DUKE — It looked to me as if they were coming right above the ascent propellant tank.

YOUNG — I'm sure they will be in the film.

DUKE — They were jetting off. They weren't just floating away.

YOUNG — That's the way it looked to me, too.

MATTINGLY — It sounds as if you saw streams of particles coming out and, I don't think that's what they were. They were just large clouds of material out there and they were coming away from the SLA. I couldn't see a source. I didn't think it was anything that you could pinpoint. It was just a lot of debris.

YOUNG — Thermal coating inspection - We didn't know it, but, that's what we were doing. We were watching the thermal coating float away. CSM handling characteristics during transposition and docking.

MATTINGLY — If you don't touch the controls, you're doing good.

YOUNG — Extraction.

MATTINGLY — That went as planned.

YOUNG — EMS behavior during TD&E.

DUKE — You had that big drift in there at that time.

MATTINGLY — It had a high drift. I didn't use it. I hadn't planned on using it, anyhow.

YOUNG — Sounds at SEP, RCS, retraction, and extraction.

MATTINGLY — Let me say something about the sounds of those engines. I think our impressions were different. I didn't hear any RCS sounds when I got off the S-IVB. I didn't hear any sounds during the turnaround; and, I didn't hear anything on closure until I got in real close. I would swear - I know it's not possible - but, I'd swear, I could hear the jets impinge on the LM before we docked. And you could certainly see it. Maybe I was visually seeing the skin of the LM kind of flutter and I knew that should make a noise. I heard the same noises every time we fired the engines after that. I don't know if there could be enough local atmosphere or whether you can get a reflected shock that you could hear. I don't know how it is, but, I know I could hear reflections off the LM before we docked.

YOUNG — I think that is possible, Ken, with the mass going out and coming back and bouncing off your vehicle. There are a lot of particles in there.

MATTINGLY — And, it was at this point that your TV monitor went out, wasn't it, John?

YOUNG — Yes. The TV monitor failed for the first time. This is no big thing, because it doesn't mean you're not getting a picture. It was intermittent and they should take a look at it. The attitude control you had to do once you got docked cost some gas. You said something about the ripple fire.

MATTINGLY — When we retracted I had the feeling that we had seen the two vehicles come together and I hadn't heard a latch. I was starting to get worried about what had happened. The motion looked to me as if it had stopped, and then the latches fired. It seemed as if it was very slow. Maybe they really hadn't come together. I had the feeling that those latches were very slow to fire.

YOUNG — The work in the tunnel was nominal. All the latches were made and everything was nominal except latch 10. The bungee was just a hair outboard. When you looked at it, it hadn't made - -

MATTINGLY — Subsequent to this (we didn't see it then) when we were going into the LM, I took a look at latch 10 and the bungee, the little cap that's on the top of the spring bungee, was recessed about a half to five-eighths of an inch. This, as I understand it, is typical of a latch that hadn't fired. When I looked at the latch I could see clear underneath the latch and it was not making contact with the tunnel ring at all.

YOUNG — This is not the kind of inspection you can make with the probe in.

MATTINGLY — You have to have the probe out to see that.

YOUNG — The separation and evasive maneuver was nominal. The S-IVB was just where they said it would be when we finally got around to looking at it. The S-band performance was nominal.
DUKE — You couldn't see anything but it slowly moved out of our field of view.

YOUNG — We didn't see the dump because it had gone by then.

DUKE — It did all it was supposed to do. The last time we saw it, it was doing what it was supposed to.

MATTINGLY — This was the last time we were on schedule.

YOUNG — Climbing out of that suit is really something.

MATTINGLY — We had a hard time getting the suits into the suit bag because we were trying to be careful of them. I don't know if taking our pockets off would have helped us or not.

YOUNG — Taking the pockets off is another 10 or 15 minutes per suit. The way they put those things on, they're on there to stay. The best way to get them off is to cut them off.

MATTINGLY — The only point I'm making is that I have stowed suits in the bag in stowage exercises. and it's not a big deal.

YOUNG — It's entirely different in flight and I think there are a couple of things that are different. One is the suits which we stowed here never had the pockets and all the paraphernalia straps on. The other thing is you have an old beat up suit that's a rag. You don't care if you step on it, kick it, or what you do to it but when you're on your way to the Moon, you're being ginger with those things.

MATTINGLY — The bag is just too small for the suits.

YOUNG — Here's something I feel pretty strong about. We haven't had any problems so far but I think that the crews that are going to have to take these suits off and stow them in a bag should be given a demonstration on how to properly roll up and stow an A-7LB full pressure suit by a suit technician close to launch so they don't forget what they've learned. Course, that's just another square to fill close to launch. If you damage that suit by stowing it in that bag, and I can see how you could do that because you wouldn't believe the kind of kicking, shoving, grunting, and heaving you have to do in zero g to get that suit in there, and if you damage it, there goes your mission because you can't fix it.

MATTINGLY — I think you should point out, we only stowed two suits in a three-suit bag.

YOUNG — Yes. We only put two suits in there. We got three suits in there on the way back and then found out we weren't supposed to have three suits in there. We were supposed to stow one of them under the couch. And, we only put two in there on the way out. We stowed Ken's suit underneath the left couch. But, I was always concerned that maybe we had folded a suit in the wrong direction and put some undue strain on the zipper. We had zipped the zipper up, but, we hadn't zipped the pressure sealing zipper up all the way. But, we zipped the restraint zipper all the way up and we put the neck ring's dust cover over it.

MATTINGLY — I think it's a shame to take a chance on the suits on your way to the Moon. I think you should make that bag bigger.

YOUNG — There's about this much room between the bag and the front hatch which is not used and I don't see why they don't make that big enough so that a guy is not ruining his EVA by stowing his pressure suit. Now, maybe the CSD people will come back and say you can't hurt those pressure suits. But if I know Ed Smiley, I reckon he feels that way about it, probably.

MATTINGLY — Well, the other thing is, when you fold these on the ground and you got one g helping you pull it up into a little ball. In flight, it is whatever you could get with your arms.

YOUNG — The other problem is that J-mission spacecraft, with those boxes under you, only one man can get in there to do the job at any one time. You can't get two men in the space to push the suit right. Well, it was of some concern to me in that we were treating suits properly when we stowed them that first time. We set a new world record for suit donning and doffing in zero gravity and 1/6-gravity seven times, something I would just as soon not have. We were behind the first day. We didn't have enough time in the timeline to doff the suits and stow them. We didn't have any time in there. With one man in the middle of suit doffing, another man is handicapped because he's helping. So, that leaves one man to mind the store.

MATTINGLY — We actually had time in the Flight Plan for doffing the suits. We just didn't have adequate time. We had an hour or less; and, we used almost 1 hour on the first suit.

YOUNG — That was one thing that put us so far behind going to bed that night. And, when we did that...

MATTINGLY — By the time you go to the bathroom, which there is no time for.

YOUNG — ...which took us by surprise. It is something we think we should eliminate preflight.

MATTINGLY — That put us another hour and a half behind, on top of that.

YOUNG — It was just the suit doffing that got to us.

MATTINGLY — I took a lot longer on the P23s.

DUKE — That took a lot of time. MIN impulse took time lining up. I remember writing that comment in the Flight Plan.

YOUNG — We were late starting P23s.

MATTINGLY — We were late starting because of the suit doffing.

DUKE — No, we were behind because we had to get in the LM.

YOUNG — Oh, yes. That's what really put us behind.

DUKE — That was not in the Flight Plan. We had to take the probe and drogue out and we had to go into the LM at about 13 hours.

YOUNG — I forgot all about that.

DUKE — Because of the thing spewing particles. It was during the P23s. Stop the P23s and let's get in the LM.

MATTINGLY — I never was able to use the telescope for anything, except to see the LM radar and the quad, from the time we picked the LM up until we got into lunar orbit. It was due to the tremendous number of particles that were floating around that, I guess, came from the LM. It was just like everyone talked about - if you do a sighting right after a water dump. We were continually populating the environment with these little things popping off. So the telescope - except for objects like the Earth and the Moon - is essentially useless. The sextant was beautiful. The auto optics put it in there. Every time we made our REFSMMAT change we used the same technique of going to SCS and recording the shaft and trunnion angles. And it's a good thing because the telescope was useless. The first time we did this, the auto optics did not place the stars in the sextant field of view after we had torqued it to the new REFSMMAT. We picked them up with no loss of time because we had the shaft and trunnion available and we could crank it in and press on.

6.0 TRANSLUNAR
COAST

YOUNG — I had never heard of that technique before, but man I think it ought to be mandatory if you're going to coarse align that platform translunar coast.

MATTINGLY — It makes you feel comfortable to know that you're not going to get lost.

YOUNG — Right. What it means is you would have to end up doing a P52 Earth/Moon probably and then going off from there. That would be a big waste of gas and time.

MATTINGLY — This technique worked well. I thought that the sextant simulation was just like the real sextant. The stars looked the same, they are just as obvious. It's really a refreshing thing.

YOUNG — That's really a beautiful technique. I never did that. It never occurred to me to do it. On Apollo 10, I was too sissy. I pulse torqued it all the time.

MATTINGLY — The only time I pulse torqued this mission was at ground command, and darned if it didn't go to gimbal lock. I had to drive away from it. We did get a chance to do something that people hadn't done before. That was the Sun/Earth alignment. We used the Sun filter and that works like a champ. I was really impressed. You put that Sun filter on the telescope and it's all blank until you find the Sun. Then it's a nice object and you can see the cross hairs on top of it and the Sun's illumination. The Earth was about two-thirds full then. I didn't find it very difficult to guess where the center of the Earth would be. I probably would have gotten a better angular difference had I used something small like the Moon.

YOUNG — You got a plus 07.

MATTINGLY — Which was certainly good enough to bring me up where auto optics…

YOUNG — That's as good as you can get in P57 almost.

MATTINGLY — From there on, we were back in business with no lost time to speak of. The optics calibration was a little bit different from the simulator in that it was easier to calibrate prior to P23. The system was very reproducible. In fact, I thought it was broken. I made a couple of marks - things that I knew were way off base to see if it was really reading the trunnion properly, and it was. It's a very sensitive measurement and it's easy to make. Right from the beginning whenever we would put the optics into zero, or move the shaft at a high rate, it really sounded bad. Sounded like something was clattering. It seemed to move in jerks and spurts - kind of like it was on a ratchet. It got worse as the mission progressed. It never seemed to affect the performance of the optics. But it was as noisy as I've ever heard on the simulator. It was kind of discomforting to hear that thing. If you moved it slowly, it was all right. But you couldn't get on high speed at all even at the beginning of the mission.

YOUNG — When you drove it in high speed it would go in a jerky motion. The line would go fast for a while, then it would slow down right in tune with this grinding noise that's in there. There is something wrong with those optics.

MATTINGLY — It was the actual drive of the shaft that was doing it. Initially the trunnion would overshoot like a quarter of a degree. By the time we reached lunar orbit the initial overshoot was generally the full sweep of the sextant. If you watched the star acquisition in the sextant you would see the star swing back. It would bounce. It would take three or four sweeps before it got down to keeping it within the cross hairs in the sextant. That seemed to get progressively worse throughout the mission. As soon as you went to MANUAL, it damped the motion immediately. The optics drift in MANUAL, CMC and MANUAL, RESOLVE and DIRECT. The drift is a great deal more in RESOLVE. There was no way I could find any combination that would not have drift in both the shaft and the trunnion. We got into a discussion about this preflight. They said if I went to MANUAL it would not move. But it does. I wrote down how much it drifts in the Flight Plan. The other thing on the optics is that the eyepieces back off. I think that is unforgivable.

YOUNG — They float off and go somewhere.

MATTINGLY — You can't afford to loose the eyepiece. This is the eyepiece that goes on the removable part.

YOUNG — They did that on Apollo 10. They said they were going to glue them on so it would never happen.

MATTINGLY — The jam nuts were so tight I couldn't get them loose. They weren't holding the optics in. The eyepiece screws on top of that and it's free floating. If you jam it down the little eyecup would be pointing someplace where your eye isn't. Charlie found the telescope eyepiece floating around one night. I taped them, but it doesn't work because the tape is not very sticky. The tape just isn't that adhesive to that metal. Later I had trouble with the focus on both the sextant and the telescope. Stu had asked me to look at the sextant image, because he said his was fuzzy. I found that I could make mine fuzzy by defocusing it. I had the feeling that I had cranked the focus down as far as it would go. I had run into the stop and it hadn't quite cleared up the image - like I needed to go another turn on the adjustment to get it focused. If you backed it off a turn it got very fuzzy and it sounded to me like what Stu had described. So I screwed it all the way down and it stayed in focus because I could jam it against the end. The telescope focus was a little bit better and it wasn't jammed against the nut. I had to tape it. Even underneath the tape it backed off - it's a pretty sensitive focus. In lunar orbit, I was unable to see star patterns in the telescope at night in the double umbra.

YOUNG — Black night.

MATTINGLY — You've got to be able to see them if they're there. They weren't there, and thought we had reproduced the Apollo 15 problem. After I noticed this a couple of times I went back and played with it. I played with the focus and apparently my problem was that it has unfocused itself. When I focused it there was apparently enough concentration of light then that could see stars. When they were out of focus, I didn't see anything. When I focused it up, the star pattern just popped out, and they were beautiful. It could easily happen. I had the thing taped and I thought it was in place. Just a small amount of movement will make it unfocus.

YOUNG — We never had any problem with that on Apollo 10. We could see all kinds of star patterns up there at night.

MATTINGLY — I just made it a practice after that to always check the focus, and then it was okay. You need the telescope right now, not after 30 minutes of looking for the eyepiece.

YOUNG — I don't want to tell you this but it's the truth. I went to sleep one night just before Ken did. While I was asleep the platform realigned itself into coarse align. Ken fixed it and I never woke up. I never knew anything about it until the next day. Couldn't believe it.

MATTINGLY — It was the last thing we had to do on the presleep checklist - take the voice mode off. John crawled back into the couch - floodlights were all on. I was going to look at something in the optics and I said, hey John, turn the voice mode off. There was no sound. I looked back and old John looked like a dead man. All the floodlights were on. About that time, the master alarm came on and the world turned upside down. We talked to the ground for an hour about it. We realigned the platform. John never batted an eyelash. I finally reached over and turned all the floodlights out wrote him a note and stuck it in front of him and went to bed.

YOUNG — I got up the next day and read the note. I thought, Ken has got the sickest sense of humor of any guy I've ever seen in my life. He told me the platform failed that night. He must have sat up all night thinking that up. Then I closed it up and thought - wait until he wakes up. Boy, what a sense of humor. He's worse than me. It was true, I couldn't believe it. (Laughter) That's funny.

MATTINGLY — I think you were tired. (Laughter)

YOUNG — I really slept good that first 3 or 4 hours. I really slept good.

DUKE — That's really the only systems problem we had on the way out.

YOUNG — Yes. That happened the second night.

DUKE — The ground said expect the SPS light during that first midcourse. Sure enough the SPS light came on due to the oxidizer side being high, 210.

YOUNG — Never went on again.

DUKE — During the burns, the pressures were where they said they would be. It was fine during the burns. Comm worked great the whole way.

YOUNG — Ground handling of comm at night was really good. I had a headset on many times and I never heard it. I think I was just asleep. They used to wake me up on Apollo 10. They never did this time. They stayed ahead of it the whole time this time.

DUKE — They said they lost comm with us once but they didn't wake me up.

YOUNG — Passive Thermal Control. Ken gets an attaboy for that. He's the only guy to ever set up passive thermal control correctly the first time, and every time. It was beautiful.

DUKE — Right on.

YOUNG — It really worked good. Let's say something about LM Ingress which sort of got us off the first days timeline. When Ken was down doing the navigation sights, I was looking out his window. I was sitting there watching. All of a sudden the stuff was sorta floating off. I saw it coming out from behind this place that we tried to point out to you on the TV, when we got the TV on it. I saw this stuff's coming out like something is shooting it out of there. I was as nervous as a cat. Didn't I show it to you guys?

DUKE — I saw it from my side.

YOUNG — You didn't see it though, Ken?

MATTINGLY — No.

YOUNG — Charlie saw it. It was directional. So I figured the only thing it could be was that something was making it squirt out of there. You've got the best gauge in the world when you can look at something and see it leaking. I wondered if there was some valve we could shut; now was the time. We went into the LM and powered up the TM, and they didn't see a thing. In fact, we turned on the RCS gauges and we didn't see anything..

DUKE — Nominal.

YOUNG — Boy, I don't know what it is. It could have been a thruster down in there, but I don't know why it would come off directional like that. Maybe we got some pictures of that.

DUKE — Later on during the flight, another panel started doing the same thing. It was shredding off paint, it was also directional, but it was not as much as that big panel.

YOUNG — When we finally got down on the Moon and looked around at the LM, it looked like a shaggy dog. We've got the pictures here, hair hanging out all over. Cislunar nav and navigational sightings cost us a hunk of time. It got Ken so far away from the cislunar navigation preliminary sightings that he really didn't see but one horizon.

MATTINGLY — There was never any question in my mind about what to mark on. There's only one thing you could mark on.

YOUNG — It's worth a little OJT and not waste any more time on it.

MATTINGLY — The auto maneuvers never placed the spacecraft at the substellar point. We took a long time, used a lot of gas on it, but it seemed to me there wasn't any reason to take a mark unless you got it in the right place. With that big stack up there, that can just take a lot of time.

DUKE — The procedure that you used for recycling through P23, is that a standard procedure that everybody uses?

SLAYTON — It's part of the flight data file.

YOUNG — That really saves you the gas, the ability to recycle through P23 and get right back in MARK. Ought to really be a gas saver. Midcourse Correction (External Delta-V, EMS Delta-V, Ullage Ignition, Accuracy) - all right on the money. The pressures were just what they told us they would be. We didn't notice the chamber pressure, like no one ever has noticed on the short burn. The Delta-V counters were right on. Did quite a bit of UV photography on the way to the moon.

SLAYTON — You guys really handled it good.

DUKE — It was a simple sequence.

YOUNG — I hope they get some good photos. There were some good pictures of Earth, that would be worth having. One time we were looking at the whole of Australia - all in the clear. High Gain Antenna Performance nominal the whole way.

MATTINGLY — I wrote down all the notes on the electrophoresis demonstration in the experiments checklist.

YOUNG — Daylight IMU Realign and Star check was nominal.

MATTINGLY — The electrophoresis demonstration didn't work like I anticipated it would. We allotted an hour for it. I think we ran it three times in about 30 minutes. As soon as I pulled the closing Mylar out of the way with the knob I got a spurt of stuff that came out and hit on the face of the glass on the box. Whether that violated the integrity of the experiment I have no idea. There were bubbles in the tubes. There was one bubble in each tube. They told us, if we had bubbles, tap them out to get rid of it. I tapped them, but I couldn't get rid of the darn things. I tried shaking the box, to put some acceleration on it and get the bubble to move. I just didn't have any luck.

MATTINGLY — Then, they told us after awhile, don't worry about it anymore.

YOUNG — They were big bubbles too; they were as big as the end of...

MATTINGLY — Well it'll be on the photographic record.

YOUNG — As big as the end of a pencil, or as big as the end of this match tip.

MATTINGLY — I'll buy that. It filled the tube, and it was just about a circular bubble. Hopefully the photographic record has all of the things we saw. The experiment was conducted according to the checklist. We were in PTC. They gave us something like 5 minutes to wait after we tapped it, and we didn't wait that long. We didn't because there was no motion. The bubbles were steady and weren't changing any. So, it seemed like a reasonable thing to proceed. When you've got auto optics you've got no problems. And when you don't have auto optics, you got lots of them.

YOUNG — ALFMED Experiment. Charlie's got very sensitive eyes. If he didn't get enough data, they won't ever get enough data. He was seeing three to my one.

DUKE — They looked just like the experimenter had described preflight during our little briefing with him. They looked exactly like that, from white dots to fuzzy lines to straight lines, to fuzzy cloud-like haze lightning. They seemed to come in batches. Then you wouldn't see any. Then you would see three.

YOUNG — Well, you can tell who's the hard-headedest in the whole bunch. I'm three times as hard-headed as you are and Ken's infinitely more hard-headed than I am.

DUKE — Hopefully it worked. The motors were driving right, things sounded right on the whole experiment. I think the experiment worked right. We sure gave them a lot of sightings.

MATTINGLY — The mechanical part of the experiment was done properly. I've played with the instrument before. It did all the right things. I'm sure the plates returned to zero when we got through. We took some photographs of the placement on Charlie's head from three different directions before and after. I think that we have it sufficiently documented, so that they can put something together with it.

YOUNG — I looks like they are out here in front of your eye. That's my subjective opinion, I can't believe that they could be back here in your head. Maybe they are.

MATTINGLY — I think it's all in your head. (Laughter)

YOUNG — No. They're real. I never noticed any difference in frequency that I could put my fingers on. It looked like one of the pictures of ions in a cloud chamber. It's weird. I went in there skeptical that I was going to see those things, because I didn't see them at all on Apollo 10. No, I never saw any on 10 that I remember.

SLAYTON — No one ever saw any until Apollo 11. Then all of a sudden, everybody, started to see them.

YOUNG — I can't close my eyes right now and see them. I thought it was psychological at first.

DUKE — Once one catches your attention, they're there.

YOUNG — I saw them on the lunar surface too.

DUKE — Yes.

MATTINGLY — Did you ever see them behind the Moon?

DUKE — Yes.

YOUNG — I don't remember.

DUKE — I remember that first night sleeping. I don't know where we were in orbit, but I saw them.

MATTINGLY — Did you see any when occulted from the Sun?

YOUNG — I don't remember that.

MATTINGLY — I'm still looking for my first one.

YOUNG — I think that's beautiful. I think that's great. There's a message right there but I don't know what it is. Charlie could see so many, and Ken didn't see any. Whatever kind of shielding Ken's got, we better get some more of it.

MATTINGLY — Send me back and see if I can see them again.

YOUNG — I knew you were going to say that. CM/LM Delta-P was nominal. I think we had a very low leak rate in the LM.

DUKE — Tight LM.

YOUNG — We pumped it up in accordance with the rules which changed when we were pumping it up there once. But it still worked out okay.

MATTINGLY — Ultra-violet photography went according to plan. For the record, we did use a cardboard shield.

YOUNG — I think we had a tight LM, because every night before we went to sleep, we pumped the cabin up to 5.7. That would give you a 0.6 or 0.7 LM/CM Delta-P. The next morning it'd be still the same CM/LM Delta-P. So, it wasn't leaking. Removal of Probe and Drogue - Ken finally got so good at it, he was doing it in milliseconds. The fourth time was PDI day.

MATTINGLY — Did you ever smell anything on the initial docking? I never smelled the old burnt metallic smell when I was up in the tunnel. I got a definite feel for that after rendezvous.

YOUNG — I got the odor when I was initially up there in the tunnel. It's a standard - don't ask me what it is - it's a standard smell.

MATTINGLY — I don't know, maybe it's the probe that we're smelling.

YOUNG — Maybe so.

MATTINGLY — Because it was a very definite odor, and I hadn't anticipated it after rendezvous.

YOUNG — Yes, maybe so. I don't think it's any big thing.

MATTINGLY — Passive Thermal Control, one more time just as good as the first time. Sounds (Service Module RCS) - I tried to tape them. So that somebody could hear them. I couldn't get it on the tape; it was just below the threshold. I checked it once specifically during a midcourse, or maybe it was during one of our rendezvous. I think the simulation at the Cape is called - our CMS reproduction of that...

YOUNG — You also get the physiological cue too that you don't get in the fixed base simulator.

MATTINGLY — You get a very definite sound when the engine is on and stays on. I noticed that during one of these translation maneuvers when I had a chance to listen for it, and sure enough there is a sound that goes with the translation. It's not as loud as the solenoids, but there sure is a sound that goes with it. But I tried to tape it, and I didn't get a thing.

YOUNG — Well, I don't know how it helps you in the real world, but just to know that it's there would help you if you ever had a stuck thruster. But in the simulator, it would sure be a helpful cue that you had a stuck thruster. You're really in trouble with stuck thrusters, in the CMS because you never know that the thing is firing until you've run out of gas almost.

MATTINGLY — But as far as motions in the CSM, I had the distinct impression that you'd have to have an awfully big rate before you would feel it. When the LM's on you know motion, because that thing heaves and bends, and creaks, and there's no question that something's happening. I think that jet monitor program is a very worthwhile thing. I sure was able to sleep a lot better knowing it was in there.

YOUNG — That's that EMP that tells you when the jets are firing when you sleep.

MATTINGLY — What it does, it measures when you exceed the dead band by a discrete amount. And I thought that was a super thing to have.

YOUNG — Resting, Comfort, Housekeeping, and Exercise.

SLAYTON — I think those things ought to be broken down in a little more detail.

MATTINGLY — Why don't we take them one at a time, because there are four separate subjects there. If you add hygiene to that we'll be here the rest of the day. Sleep, why don't we talk about that?

YOUNG — We were really behind the first day because of the extra ingress in the LM, and suit doff. What goes by the board, is your eat period, when you get into a situation like that you just forget to eat because you can't. If your LM is leaking, you are not going to stop to grab a bite to eat. Or if you suspect it's leaking. And the next day we were behind for some other reason. What was it?

MATTINGLY — That was the day we went to the John.

YOUNG — That's right, we went to the bathroom. All three guys went to the bathroom, and you can't do it simultaneously. The first time it's going to run you what - 40 minutes apiece?

DUKE — Yes.

YOUNG — To do it right. Once you get good at it, you can cut the time down considerably.

DUKE — My first night's sleep was miserable. But after that I was settled down and I slept like a baby the rest of the time.

YOUNG — I was really surprised, my first night's sleep was good, and so were all the rest of mine, and I can't account for it because I don't remember sleeping very good at all on Apollo 10.

MATTINGLY — The first night, you commented the same as I did, that you saw every hour on the clock.

YOUNG — Well, after the first four I saw every hour on the clock, but I'd wake up and go back to sleep.

MATTINGLY — I woke up every hour. And that was true of every night except two of them. Only 2 nights during the mission that I think I slept straight through. And I never felt tired, I never felt like I wanted to go to sleep - that was the problem. I'd lay there, and my eyeballs would be wide awake and my head would be thinking about things, and I guess I'm used to being physically a little tired too.

YOUNG — My rest period typically would run 4 to 5 hours, and sometimes in there I would wake up for a half an hour and make a head run and get a drink of water, and then I'd go back to sleep and sleep the rest of the night. And I don't know why, because we weren't doing any useful work to amount to anything, I didn't need that sleep.

DUKE — You can sit in a chair and not do anything and if they keep you awake for 15 or 16 hours you're going to be tired and you want to go to sleep, and it's the same thing in space flight.

YOUNG — You think so, huh?

DUKE — Yes. Yes, you're burning up calories just sitting in that chair.

YOUNG — There's a lot of mental exercise that goes into some of those things.

DUKE — I was tired. I was ready to go to bed every night.

MATTINGLY — You didn't have any trouble, you were able to dive under there and sack out. I just couldn't go to sleep, and had no desire to go to sleep.

YOUNG — I think it would help for us all to have the same place, like Charlie was under the left couch. I slept on the right couch, and Ken slept in the left, couch, and that was a pretty good standard place to be, because if anything came up we were right there at the EPS and the ECS and Ken had it controlled and Charlie had a place where he could get out of the way and sleep.

DUKE — It was nice sleeping under that couch. I appreciate you guys letting me have it, because I really enjoyed that.

MATTINGLY — I slept the last 2 nights without the sleeping bag. I felt like the only advantage to the sleeping bag was a way to stay warm. Other than that, if it was inconvenient to get it out, you had just as soon leave it stowed.

YOUNG — Yes, but I really needed it to stay warm there a couple of times.

MATTINGLY — We talked about it before, but I guess each of us has a different thermostat. But I felt like I was a little too warm until we got into the 60-mile orbit around the Moon. And up to then, I felt like every night I wished it was a little cooler. And after that I needed the sleeping bag and my jacket in order to stay warm enough to sleep. I tried all the things Stu had suggested about anchoring your head and doing different things that he thought would help. Charlie put his head in the sleeping bag.

DUKE — I did, and that worked great for me. I liked the bag because I felt a little pressure on the head from it, and it worked fine for me.

MATTINGLY — I tried all of that. None of those things seemed to have any effect on me.

DUKE — I didn't wear the coveralls. I stayed in my long Johns the whole time and was very comfortable in that. But at night it seemed to chill down low in the spacecraft down under the couch, and so I needed a sleeping bag.

MATTINGLY — One other thing on that sleeping - it sort of falls in with it. It's really controlled for this place, but I've always badmouthed that EL. - -

YOUNG — Hey, that's good stuff.

MATTINGLY — - - as being a waste of the Government's money. But I found that that instrument panel was just perfect. I could turn the lights down so they didn't disturb anything, and all I had to do is glance at it, and I could read all the things I needed to read.

YOUNG — Without waking everybody up.

MATTINGLY — Without waking anybody up. It was there at instant reference - no problem focusing on it. And I really thought that stuff was super. And I thought that was really a good thing to have.

YOUNG — Yes. But the thing that I think that a lot of people don't realize, that if anybody moves around in that cockpit, it wakes up the other two guys. There's a couple times there when I don't think I woke up Charlie, but every time I reached for the water gun, I know I woke up Ken. Every time he reached for it he woke me up.

DUKE — That's why we got in the habit of getting a drink and taking it to bed with us. I think that's a good idea, just to keep from reaching over there and riding around the cockpit. Seems like when you bang something you actually do it on purpose - it reverberates the whole cockpit. We were taking the urine bags to bed with us, too - in case you had to make a head run. One of the things we didn't do was wear our couch loops. Anytime any one of us would move, it would swing the couch. But if it hadn't been loose there was some distance in there where it would have banged. We had it loose so we could get in that 382 count (?). I think if we had put the strut out we might have minimized some of that, I don't know. Maybe this is the chance to mention we need a smaller trash bag. The Jett bags are just monstrous.

YOUNG — And when it gets filled, it's bigger than Ken.

DUKE — And there's no air rim bag, you need something about half that size that you can fill up really.

MATTINGLY — Housekeeping is probably one of the biggest things. It's the thing you don't practice enough.

YOUNG — It takes you a couple of days to get going at it.

MATTINGLY — You've got to get into the swing of it. You need to plan way ahead to do things very slowly and take a lot of extra time so you learn the little details of housekeeping that go on. When we finally learned how to eat, we had a system where one guy would fill the bags and one guy would cut them open and mash them up. We put the trash can right in front us so that when you got through the pills were right here together. You put the pill in, drop it in the trash and get rid of things as you came across them.

DUKE — That was pretty efficient. I think we were eating meals in 35 minutes or so and getting it all done.

MATTINGLY — But our first meals must have been an hour and a half. We were all over the place. Charlie pointed out the trash cans that's really true. We started out using the Jett bag because that's what we were going to throw it away in. That thing is just too big. It's a real nuisance. You pick up a piece of trash - something floats by you and you pick it up. You say, I'll throw it in the trash can, but you got to go over to this big monster and wrestle with it and after it's about half full you sort of reach a static state. Every time you open it and put something in, you're going to get two pieces out. You want to collect enough to make it worthwhile to open the bag. We never really felt like we used the bottoms of those things because they were so deep and hard to get into. But the idea of having a set of jaws on that thing that you could just pull open and drop something in was real convenient. And I think maybe something the size of that purse you guys had is a good intermediate trash bag. When I was by myself I was using the Volkswagen pouches in the LEB, I'd just fill up one of those until it was full. When it was full, I'd make one run to the trash can and dump it all in. And that saved me a lot of time.

YOUNG — What you're really saying is that we should have used that in the first place. I don't think with three guys in there if you use the Volkswagen pouches. You can't use them for anything else. Not only that, but every time a guy goes by he's going to hit them and the stuff will come tumbling out.

MATTINGLY — I don't think that's an adequate solution, but that kind of approach sure works a lot better.

DUKE — An extra TSB would be nice, because it's got the clamp on it. We could use it for an interim trash bag. It's deep enough.

MATTINGLY — The amount of trash is really surprising. We ignored one of the things that kind of caught us later. We could have kept the size of the trash bag smaller, if we had thought about squeezing the air and things out of the food packages when we got through with them. What we were doing is just taking a food bag, putting a pill in it, wrapping it up and throwing it in the trash. And everyone of these things with some air in them is bigger than they were packaged initially.

DUKE — I was squeezing mine down.

MATTINGLY — Well, towards the end I was working real hard at mashing all the air out of them and rolling them up, which made a lot smaller bundle.

YOUNG — A food package is 6 inches by 4 inches by 3 inches, vacuum packed. And after you eat it, it is twice that big.

MATTINGLY — By the time you throw in tissues and all the little miscellaneous things that you never think about, it's really quite a bit of trash. I jettisoned two jettison bags in the LM. One of them was as full as you could get it and the other one was probably two-thirds to three-fourths full. I have no idea what they weighed. Based on what we saw after we got back it probably was a pretty heavy burden.

YOUNG — Housekeeping. We didn't allow any time in the time line for the second day when we all had to make a head run. That got us behind.

DUKE — Every morning when you get up and every night when you go to bed they have a presleep and a postsleep checklist. But there's never any time allotted for the thing. There needs to be a period when you get up in the morning, you ought to have 30 minutes from wake-up until the first word you hear from the ground.

YOUNG — They'd say, "Good morning, here's your Flight Plan update." Here's your Flight Plan update - where am I - and you're off and running. You know that it took us on the order of 10 to 15 minutes to prepare that crew status report, write it all down every morning. We didn't figure on that either.

MATTINGLY — No one factor you can point your finger at and say, if I could eliminate this I'd buy a lot of time. It's just an accumulation of all these things. You need a chance to just go around and pick up your toys and put things away. There just wasn't any time anywhere in the Flight Plan to do that sort of thing.

YOUNG — Exercise?

DUKE — We did it.

YOUNG — There were a couple of periods we missed.

DUKE — We missed all of them the first day.

YOUNG — Yes.

DUKE — We wouldn't have missed the first day but we had the LM problems and the P23s.

YOUNG — There was another period we missed because we were engaged in post EVA or entry stowage.

MATTINGLY — Post-EVA stowage. I think I scrubbed one in lunar orbit, because it was that or go to the head. Really I didn't have any choice. In retrospect I don't know how to assess the value of exercise. The Exer-genie was the only practical thing I could think of. We gave a lot of thought preflight to how we could exercise.

YOUNG — I did a bunch of big 4s and it worked pretty good. But I could tell I wasn't doing the kind of work I was doing on the ground. I subjectively could feel that.

MATTINGLY — I felt better when I got through, I'll say that for it. I just had a better feeling, but I just can't believe that the amount of exercise I had justified 30 minutes. I really think I'd have been just as well off to just forget the whole thing.

DUKE — Well, I got my heartbeat up to 100 or so, but that's about the max you could do was a hundred.

YOUNG — They said I got mine up to 114 one time.

MATTINGLY — I question that, because I had just checked it. If you go out there and work up a sweat, really do exercise like you ought to, the ECS will not handle that kind of a load. The ECS is marginal. It's designed for three marshmallows laying there. It isn't designed for you to go out and do any exercise. The old Exer-genie gets hot. I guess it's not going to start any fires, but I just don't like reaching over that big heat sink.

YOUNG — It's the barrel that gets hot. That heat has got to go somewhere, I guess.

MATTINGLY — Charlie, did that thing cool off faster when you put it on the bulkhead? We should have measured that.

DUKE — We didn't measure the time, Ken.

MATTINGLY — But it was hot to touch and that barrel got pretty hot. The other thing I worried about was laying there and banging into things. Because you can't do any reasonable exercise and maintain your body position.

DUKE — You know what we should have done, I think - - put the couches in the 180 position.

MATTINGLY — You're always banging in to things when you do that. Kick the optics. I found working in the center couch is the best place, I could get in there pretty well.

YOUNG — But even then, if you strapped yourself in the center couch and did a big 4, you would probably be all right, because you are restrained.

MATTINGLY — No, still kick the optics.

YOUNG — Don't want to kick the optics.

MATTINGLY — Always have one eye open for what you're doing.

YOUNG — What I'm really sorry about is that we couldn't do them in an uninterrupted fashion. I'd start my exercise period and get the ISS light on the way back. It really slows your exercise down. They've got to handle the biomed harness a little better from the ground standpoint. On occasion, they wanted us to put it on and take it off waiting until after you got to that place in the Flight Plan before deciding what they wanted to do. Ken had just taken his off, because he was going to be off that night, but Charlie's wasn't reading right. So the ground calls up and says, okay, we're going to watch Ken's. He had just taken it off. To put it back on is 15 to 20 minutes of playing with these little sticky things in zero gravity. They just don't hang in there very good, they float all over the place. It seems to me the decision to watch somebody ought to be made long enough ahead in the Flight Plan so the guy can be expecting it, and doesn't cost us 15 minutes which we don't have.

MATTINGLY — Well, we really stayed on the schedule except for that one night. As far as I'm concerned I'd keep the things-on rather than go to all the Mickey Mouse of changing them.

YOUNG — The thing that bothers you about the biomed is those big flat patches. If you could leave those off.

DUKE — That's an individual thing, now. The sensors bothered me. I've got these little round blisters where the sensors were.

YOUNG — I do too.

DUKE — The big tape is itchy but the blisters came from the sensors. In my case, it was necessary for me to doff mine. My only complaint was that there was 2 minutes in the Flight Plan, LMP don biomed, CMP doff biomed. Well, you can't do it in 2 minutes. You need 20.

MATTINGLY — You can do it in 15 if everything works perfectly, but you're really making a mistake to count on less than 20.

YOUNG — Well, the first thing you have to do is float over and get the medical kit, which means everybody has to stop while you're doing it, or else they have to reach down and give it to you. It's not all in the same place in the medical kit. The biomedical stuff is in one side and you have to turn it over and get it out of the other side. Then you have to put the little sensors in the little boxes. Then you have to rip off the old and put on the new tapes and stick the thing on you. Doesn't sound like it takes 15 minutes, but that's what it was running us.

MATTINGLY — Using those tattoos as a way of knowing how to place things is a waste of time. I used the Pen-tel. I wouldn't consider doing it any other way. I never had time to look for that stuff.

YOUNG — The only way I could get them on the tattoo was if Charlie helped me put them on.

DUKE — I did it with a mirror and it worked fine for me.

YOUNG — The biomed donning and doffing schedule should be done well ahead, so that one guy hasn't taken his off when they still want to look at it. They just sort of arbitrarily came out and said Charlie's is not working, we'll watch Ken. Ken had already taken his off. We didn't want to argue with them, but we didn't have the 15 minutes available for that.

MATTINGLY — One time they called me during lunar orbit and said that I was getting a noisy signal. They wanted me to change it. I ignored it for one rev, because there wasn't any place in there to do it. They bugged me on the next rev. I had the distinct impression that if I ignored it one more rev that I was going to be talking to Chris or somebody. So I went ahead and did it, but it really frosted me. I didn't go to the Moon to change biomed sensors, and I think that sort of thing ought to be put in some kind of proper perspective. If there's a hardware malfunction or if you get a scratchy sensor or signal, there's got to be some judgment about when it's worth breaking into your time line to replace it.

YOUNG — Right in the presleep checklist there ought to be a check for the guy that's on biomed, a biomed sensor check. From then on if it craps out completely, they ought to leave you alone, because they can't wake you up to do it. One night I was supposed to be on biomed the whole night. When I got up the next morning they said we didn't read your biomed. I don't know why they didn't read it, but nobody else asked me to check it before I went to sleep that night. If they really want to see it they ought to have bugged us before we went to sleep.

DUKE — Sim door jett. We saw it.

MATTINGLY — You can hear it. The Apollo 15 guys said they never heard it, but I think that's because they had their helmet and gloves on.

YOUNG — The other two LM checkouts were nominal. Concerning PTC, you ought to stow the optics to get the light out of the cockpit. Every time you go around the light comes through the telescope and lights up the whole cockpit. I knew that from Apollo 10.

MATTINGLY — I'd like to give them an "atta boy" for those window shades.

YOUNG — They really work - work good - that's another first.

MATTINGLY — They did a good job on that.

YOUNG — The gas separator on the water gun leaked pretty bad.

MATTINGLY — There were two gas separators, the standard cigar separator that's been carried before, and we had the new one - the plastic the Lexan new type. I think we should have gone back and taken the gas separator off later in the flight, because we put it on when we were getting gas. That was the day before LOI or the day of LOI, I don't remember which. It was when we were getting ready for that Skylab meal when we first noticed we were getting gas in the regular food port. We put the cigar on and initially got a lot of gas. I think that was just purging the gas that was already in the cigar. After that it seemed to work pretty good. You always got more gas in the hot water than in the cold water. But once we started being a little slower about getting the water out the gas accumulation decreased, also when we started giving the water guns more time. The only real drawback to the cigar was that if you forgot to put the little plastic cap on the bottom of it, you were going to get water, and you could really get a lot of water out of it. All you do is put that cap on there and that seems to seal it. I never saw any water come out around that cap. But if you forgot the thing just for a few minutes, a big bowl of water could collect. I think we have some pictures of that to show what it looks like and where it goes. We'd put about three shots of water into it and nothing came out. I guess that's when Charlie noticed that the thing was cracked across the top. I guess it started to put water out of that crack. We never got any water out of the bottom. Water did not come out the gas separator part either. It all came out of the crack. When we took it off, you could find out where the water had been because it all compressed inside there. So, we just bundled that one up into the bag and put it back in stowage and never played with it again.

YOUNG — I thought the gas separator worked pretty good.

MATTINGLY — We never did have a lot of trouble with gas. I would spin bags and look at it and I was getting maybe 10 percent volume.

YOUNG — I think 10 to 15 percent of the stuff in the bags was gas.

MATTINGLY — Cold water is a little better.

YOUNG — And it varied. Some bags seemed to have less gas in them than others. Don't ask me to explain that.

MATTINGLY — I think there's something in the way. You pull them out and release it. It squirts an ounce in there.

YOUNG — I think you're getting gas in it.

MATTINGLY — I don't understand how it does but initially we were in too big a hurry. As soon as that thing would pop in, we'd pull it back out for another shot. In fact, Charlie was pushing it in there for a while. When we were doing that we had our gas problem. When we went back to taking our time and letting these things run their full course, we didn't. I think there may be something in the technique on how you use those things. Maybe you can induce some gas from the cockpit and around that little plunger if you push it too fast. After that we started getting uniform quantities of water. For a while there we were getting differences in quantity between the hot and the cold ounce. After we finally got on to it we were getting very similar quantities out of both guns. I think we were inducing a lot of that gas by pushing on those water panels.

YOUNG — These are system anomalies. - the broken chlorine ampoule, remember that one?

MATTINGLY — There was nothing that caught my eye in putting it in or I probably wouldn't have tried it. At the time I squeezed the ampoule down and stuffed it in, I didn't feel anything abnormal. When I pulled it out, it was obvious that it had broken. We had a lot of chlorine in the ...

YOUNG — Chlorine turns your hands white as everyone knows. I was concerned that I had gotten some chlorine into the system so I wanted to put some buffer in. This was my mishandling of the buffer. I didn't put it in completely. I didn't jam down on the buffer ampoule because I had just broken one. I was afraid to squeeze down on the ejector too hard the second time. I think Tony is right. I didn't have the plunger down tight and as soon as I opened the valve going into the port, it came back and it went out this release ports on the side. That's when we thought we had a septum nut that had come loose. In fact we really did not. We had a lot of water there after a while because we were mopping up at a pretty good rate.

MATTINGLY — We went back and put in another buffer ampoule and it all worked fine. The urine collection is no problem because we do that for...

DUKE — We just forgot to time it sometimes.

YOUNG — It's very difficult to time urine in the dark if you don't want to wake up your buddies.

MATTINGLY — You can time it, but what are you timing? Who knows?

YOUNG — Another thing is those Gemini bags have a non-repeatable back pressure on them and you are not going to urinate so fast to leak around the seal. You are going to slow it down and then if you urinate as fast as you want to and you have a stuck valve you have urine all over the cockpit.

YOUNG — So that time is invalid. It couldn't be right.

MATTINGLY — I wouldn't dare let that thing run free. You have to make sure that there are no leaks. It seems to me that the timing is just ridiculous.

YOUNG — I'm not sure that you can properly time your start and stop times with a urine bag like that.

MATTINGLY — I always had some question in my mind when it stopped.

YOUNG — Yes.

MATTINGLY — I couldn't be sure.

DUKE — I think all I got was the bag pressure from the valve.

MATTINGLY — I was just never sure.

DUKE — I had the same feeling. It was a guess with me.

YOUNG — We were unfortunate in that we had the medical requirements document in the Flight Plan. Suppose you got up in the middle of the night, everything is dark. The medical requirements document where you log the urine is in the Flight Plan and it is in the dark. You have to float over to it, break the Flight Plan loose from where you have it stowed, turn on some kind of light - which would wake everybody up - and log in the urine. If you don't, you're going to forget it. It seems to me you're compromising more than you're getting, from an operational standpoint.

MATTINGLY — That happened to us in the daytime. The Flight Plan was always busy. Somebody was always using it for something. It was very seldom that you could stop and either log your food, or urine, or fluid, or anything. That wasn't a convenient place. I think maybe we could have broken it out and put that in a separate place. That would have solved a lot of our problems. It seemed to me that just having to do something like that ... your effort. It sounds silly to say that writing down that I had a cup of coffee and two graham crackers took up much of my time but when you have enough things to do to keep busy all day that's just one thing that you can tell contributes nothing. It's just a real nuisance to go through it and I really can't explain why it is such a nuisance. But it sure was.

DUKE — What we finally ended up doing was in the mornings, if we had the time - in the postsleep checklist - to get that all done. If the ground wouldn't talk to you at this time that would be the place to bring all that up to date.

MATTINGLY — I couldn't always remember what I had the other day. Every time you get a squirt of water, you guess how many ounces you get with each drink.

DUKE — We haven't even discussed that part. I think the medical part of timing the urine was useless and timing the water was a pain in the rear end. We never did it. We just squirted it in our mouths and guessed how much we had.

MATTINGLY — And I'm sure we over guessed.

DUKE — It's just a pain to fill those bags. Once you open them, they leaked at the end. If you didn't want to drink it all, you had a leaky bag.

MATTINGLY — I just don't think that's a good way to log normal functions. If the guys that come up with these ideas would just walk around and try to do a day's work and not let their logging interfere with them, they'd understand.

YOUNG — They wouldn't drink a hell of a lot of coffee, I'll tell you that.

MATTINGLY — It just gets you to the place where it's too much trouble to do it. It's constraining. When you're doing something and all of a sudden you have to stop to urinate the next thing you don't want to have to do is reach for the Flight Plan. You may be operationally constrained at how much time you have to urinate any way.

DUKE — You're right. As it turned out none of us went more than three times a day any way.

MATTINGLY — Always at the wrong time.

DUKE — It was locked.

MATTINGLY — It was just amazing to me. It was almost a direct opposite of dieresis.

YOUNG — I had the distinct impression that dieresis was not water.

DUKE — That was not my problem at all.

MATTINGLY — That's right. I agree with that.

YOUNG — Maybe we learned something by noticing that.

MATTINGLY — But it sure wasn't coming out that way.

DUKE — It wasn't. Jim Irwin said he had just one all the time and the three of us, on most days, went twice a day.

YOUNG — That's right.

DUKE — Once when I got up and once before I went to bed.

YOUNG — I think it was going out through the skin, because the cockpit was really dry.

DUKE — Our blood plasma wasn't down any more than anybody else's. It was down but it was normal to everybody else. I don't think they're going to get much from the time ...

YOUNG — I don't think it's valid data.

DUKE — Now the 24 hours where we collected it all, at least they got a total volume measurement - that is we didn't dump it. We collected the whole bit.

YOUNG — We didn't dump it.

MATTINGLY — We did put them all in one overwrap cause we couldn't get three overwrap bags.

YOUNG — That's all right. I mean it doesn't mix. We talked about that before. While we're on urine, let me talk about the URA.

DUKE — I think we ought to throw that off of the spacecraft.

YOUNG — I never used it.

MATTINGLY — Well I tried it a couple of times because I kept thinking that there must be a better way to operate it. You just don't get enough vacuum to pull fluid in and you end up with big bubbles and puddles. It's a much bigger mess to clean up than using the bags. The big problem is you want to do everything you can to keep that stuff contained. Yours was working properly.

DUKE — The trouble is once you jettison the SIM bay, you can't use that thing anyway.

MATTINGLY — But my point was that for something like Skylab, it isn't going to work.

YOUNG — You have periods of time a day that you can dump the urine. The rest of the time, you have to keep it. If you just happened to hit that perfect moment you're in business but the rest of the time you can't.

MATTINGLY — I still contend that the way ours was operating, given a choice, if I had complete freedom to jettison at any time, I still would have used the bag because it was a lot cleaner. When you pull that cap off there's always a big bubble that runs around and you have to be real careful to clean it all up. It just seemed to me that it was a lot cleaner and neater just to use the bag and forget it. You never had that problem with the bag.

DUKE — I have no opinion. I didn't use it.

YOUNG — I didn't get a chance to use it either. I never used anything but the Gemini bag. I never once had the chance to dump directly into space because we were always too busy to take the time to do it and the rest of the time we had the door off and couldn't.

MATTINGLY — I tried it two or three times. We didn't have that problem on Apollo 10. It worked like a champ. You could go either in the bag or directly through the bag through the dump panel over the side. It worked great but we just never had that chance on the G-mission. It's a urine collection business that is what it is.

7.0 LOI, DOI, AND LUNAR MODULE CHECKOUT

MATTINGLY — The most significant thing about LOI, to start with, is the alteration of our procedures, based on EMP 509, that we put in because of the attempt to avoid any problems with transients. Coarse aligning of our platform is part of the LOI. I think, considering the fact that you only get one chance in a mission to do LOI, that we did the right thing, starting the gimbals early and all those things.

YOUNG — Actually, I felt pretty good about that. I don't mind the ground looking over my shoulder when I can't see what is going on back there, which is the way you are for LOI.

MATTINGLY — I thought that was a good plan. I thought the idea of the EMP was a good one, because fortunately, in preparation for the mission, we had all told the truth, and I told them what erasable programs I knew about. They told me what ones they had on the console that no one had ever heard about, and we wrote them all down and we had MIT look at them, got it all aboveboard so that these kind of things have been looked at and checked out. I felt real comfortable about it, knowing this very problem of a CDU and the way to work around it had been tested. I'd even flown this one in the simulator.

YOUNG — I'll tell you what did bother me about it was the fact that they'd been having CDU problems at Downey, and nobody had ever mentioned it to us before we were on the flight. Maybe it happened at Downey after lift-off, for all I know.

MATTINGLY — It may have happened last week. It may have happened on 012. I don't know where it was.

DUKE — No, it was spacecraft 117. It's the Skylab vehicle.

YOUNG — Okay, to make the thing go, you've got to have the feedback. When things go wrong, everybody who is going to be operating the gear should know about it.

DUKE — Except for that oversight, I think they did a super job of getting things ready procedurally.

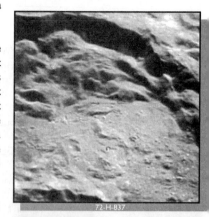

72-H-837

MATTINGLY — The execution and preparation for the burn were absolutely nominal. The other thing that strikes me as being a surprise (and everyone said that's the way it is) is that when we lighted the thing, it started on bank A and when I brought bank B on, I got a "chug." I'm not sure that's the right word. But there was a change in propulsive force and there was a decrease in chamber pressure when I brought the second bank on. And I would guess it was about a 3-psi decrease.

YOUNG — Sort of like the reverse of a nozzle closed burner light. It sort of sagged a little and it was a sharp one. It went "clunk."

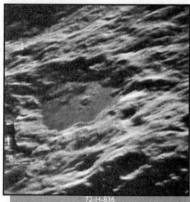

72-H-836

MATTINGLY — It didn't feel the recovery. It wasn't like a pop where it dropped down and picked back up. It dropped down and then subsequently, it picked back up. That was something that took me by surprise. I guess it gave us some concern during the burn. I remember considering that maybe I should turn this bank off because I felt something had gone wrong with it.

YOUNG — You said it was a 3- to 5-psi chamber pressure drop.

MATTINGLY — Yes.

DUKE — I think it was probably the helium or whatever it was in those lines between the ball valves and the injector.

72-H-630

MATTINGLY — The ground said that is normal. The first time you burn that thing, that's what'll happen. All I'm saying is that if that's normal, somehow in the last 6 years, I had missed all that. I don't remember anyone ever saying that they could feel any difference at all when you put the second bank on. I don't think you could have missed this. I'm just surprised. The simulator certainly doesn't show anything like that.

YOUNG — It is sure a noticeable thing.

MATTINGLY — Other than that, the LOI was just super. The countdown clock on the CMC was just lined up with the burn time prediction from start to finish. The rates were good and stable. Everything was just as nominal and perfect as it could be, from what you could monitor.

DUKE — As far as my side goes, skipping on through the sounds, I didn't have a sensation of sound, but of sight. The ball valves came open before you could feel the thrust definitely leave. Then you could feel the thrust build up. The PUGS, we left it just like it said, and didn't touch it, with the guarded switch down. It performed just perfectly, I thought. It regulated at exactly on the DSC. It was around 100 decrease where it had always been in the simulator, and then at crossover went right on up and regulated around zero. I think we ended up with something like 50 increase or 50 decrease at shutdown. But that's in the Flight Plan.

MATTINGLY — Did you see anything at crossover? I saw no chamber pressure changes.

DUKE — The PUGS jumps a little bit but that's all. The SPS pressures at ignition: The oxidizer side came down from 210 to 200 just as they predicted, and the fuel side went up from 165 to 170 just as they predicted. They were stable throughout the whole burn. It was just like they predicted.

YOUNG — Excellent prediction. The EMS, the time, and the guidance system in the CMC were in perfect agreement.

MATTINGLY — DOI, I'd say, worked exactly the same way. It was an automatic shutdown.

DUKE — Right on time.

YOUNG — And we had a plus 0.4 of a foot per second, which we didn't trim.

DUKE — A little underburn and you trim no underburns.

YOUNG — That put us in a little high orbit but that didn't bother us. Gravitational Effects on Spacecraft Attitude.

MATTINGLY — I didn't do any maneuvering with the stack manually, to detect it. I would hope that the flight control guys with their changes, in the biasing of the attitude, in looking at the different dead bands, can tell us. They can answer that question for us. I don't think we can.

YOUNG — I didn't notice any. You have to do manual pulse maneuvers and we didn't do any of that. We were all working P20s. SIM Experiment Prep.

MATTINGLY — There's absolutely nothing you can say about the SIM preparations; just followed the checklist down, and it all happened like you'd expect. It all went fine. The only thing that I was a little surprised at was that when I released the boom tiedowns and the solar monitor door, I heard those. Or felt them. It was a very, very low sensation, but when I threw the switch, there was something that took place. Maybe I heard the door

bang, but I was surprised. I didn't expect to hear anything. The gamma ray and the mass spec had written in one of the malfunctions procedures that if you don't get a boom retraction (and since this has been a previous problem), one of the first things you do is look and see if they're extended. So the first thing we did after the first extension was to go look and see where they ought to be, if they looked normal. You could see 2 feet of the boom plus the instrument. Both of them. I thought that was worthwhile, that we had learned where to look for them. They really were not where I had expected them to be. And subsequently, it paid off in knowing whether the boom had moved.

YOUNG — Communications. I didn't see any problem with communications at that time. The PGA donning problems have already been referred to. We talked about the trouble we had when we were donning, with the restraint zipper. Scrambling in and out of that thing is no piece of cake. With the LCG, with the WMS, the FCS, and the LCG on, and getting the suit up over a guy's rear end, what you have to do is start way back where the zipper starts and then go back and pull the zipper out, because the zipper invariably gets tucked under and you just can't get the suit on until you do that. It results in taking 5 minutes longer to do it in zero gravity or one-sixth gravity than it does in one gravity. I guess I'm at a loss to understand why it doesn't get hung up that badly. The problem is that you can't push on anything in zero gravity. In one gravity, you push on a guy's rear end and pull the zipper out; and in zero gravity, if you push on that, you just keep right on going. I don't think it's any problem but it just takes more time. On activation day, we started a half hour early on everything, primarily because I had trouble with Charlie's restraint zipper. We woke up a half hour early and just leaped right into everything. We didn't eat very much. We got that out of the way in a hurry and jumped right into donning our suits.

DUKE — I still believe that the reason that half hour paid off so much was that we got up and got our half hour's head start without talking to anybody, and as far as the rest of the world was concerned, we were still asleep. And I think we got more done in that half hour than we would have, had we let anybody know we were awake. I'm sure the EKG showed that we were up but nobody called us and we didn't call them, so it was a great half hour to have.

YOUNG — The reason I'd like to emphasize that is that it seems to me adding an extra half hour to the Flight Plan on getting ready to go into the LM day isn't the total answer, unless you're going to let the ground let you alone.

MATTINGLY — You've got to have time to do those things undisturbed. It seems that when the ground calls you on something, it takes far more time than the total amount of time it takes because they ask you to write something down. You stop what you're doing and you go get something, and you have to treat every word as being important. It just seems that when you stop and start things, it's just a waste of an awful lot of your time.

YOUNG — I guess the one thing that I was concerned about that we'd forget was some piece of equipment on the CM to LM transfer list, so we checked that and checked that. It's a fairly straight thing, but there's no single piece of equipment on there that you don't really need. So you better have them with you.

MATTINGLY — The tunnel worked like a champ. Everything was in there just right, and that little procedure that Jake worked out for how you can verify that you've got all the umbilicals hooked up I think was a good plan.

YOUNG — What procedure was that?

MATTINGLY — You open one of the latches on each system so that it will show a barber pole, and then you put power on the probe. It lets you know that you've done your thing right. The only thing on the tunnel business that I got a little concerned about was that we were slipping behind at the end. We were ahead of almost everything until you were late on the P52 or something. I was hung up waiting for you and you were waiting for me, and I was afraid we weren't going to get the tunnel bled down to zero in time for a doff. That thing takes 20 to 30 minutes. It seems like an eternity when you're sitting there waiting for that gauge to go down.

YOUNG — It takes a while.

MATTINGLY — Other than that, I thought we were ahead on everything we did.

YOUNG — IVT to the LM. Condition of the CSM Thermal Coating. LM Entry Status Checks.

DUKE — All of that from IVT to LM down to IVT to CM is all nominal and is adequately covered in the transcript and the activation checklist.

YOUNG — Except for the PGA donning, and we've already talked about that. I wouldn't change a thing either because it sure worked.

MATTINGLY — One of the things I want to comment on was the pressure equalization valve in the tunnel hatch and the way it was used during LOI. We altered the nominal procedures for that equalization valve being opened pre-LOI and remaining that way. And it seemed like by the time you do the burn, you would like to have the LM pressurized so you could go in there if you needed to for an abort. It seemed more prudent to have the two vehicles isolated during dynamic flight, so we ran the valve in the closed position after equalizing first and opened it up afterwards. I think this is a reasonable mode of operation. It differs from the Flight Plan. We ran the experiment booms out right after we entered lunar orbit according to the Flight Plan, and they checked out nominally. Subsequently, we ran the mass spec out after DOI, and on its first attempted retraction, we had no indication on board that the mass spec boom had retracted. The ground did receive a TM indication and showed an almost retracted contact, and we flew the circ burn with it in that position. The mapping camera went out nominally on the first deployment, and we operated the camera. On the first retraction, it took approximately 3-1/2 to 4 minutes to retract. This is an excessive number, and the ground played with that for the rest of the mission. When I checked the rendezvous transponder, I had obtained the CSM 113 calibration values for the transponder self-test. There was one value - test A on the transponder - which I had obtained as being 2.8 volts, and it turned out to be 2.1 in flight. All the rest matched up perfectly. The pan camera was turned on the rev after DOI. As soon as we turned the pan camera on, we got a main bus B - undervoltage, and the pan camera was turned off immediately. We let the ground take a look at it. Subsequently, they determined that the pan camera was merely coincident with a whole lot of heaters coming on. Following this, we put the pan camera to operate, and the main bus B undervolt light came on. I turned it off, and while we were talking about it, we had another main B undervolt light. It came on just momentarily and by the time we could look at it. all the voltages and occurrences looked nominal. We did nothing and I never saw another undervolt light for the rest of the flight. Apparently, it really was the heaters. The other thing that bears some comment is the windows. At the time we entered lunar orbit, we had a small amount of what looked like condensation on the inside of the outer windowpane on windows I, 3, and 5. Window 5 had a small residue on the plus-X into the window, and it made a strip which had a very sharp line across it about 2 inches back from the leading edge of the window. It looked like a salt spray residue. It was there to the end of the flight, and it seemed to me that it may have gotten a little bit enhanced as the mission went on. But the thing that struck me about it was that the trailing edge was a very sharp line like it had been masked at some time in the process. I couldn't tell if it was on the inside or the outside of the outer pane, but it appeared to me to be on the outside of the outer one. But I never was completely sure.

SPEAKER — During the boost phase, there was a lot of smoke flying under that protective cover. I guess that's been commented upon. I was watching and it looked like smoke trickling down the side.

YOUNG — I never noticed it.

MATTINGLY — I didn't see that. I didn't see that pattern on any window except number 5. The rest of them had condensation. They had a small amount of it inside which stayed with us the whole flight. As far as the UV photography out of window 5, we didn't make any comments about it because the area that the camera was looking through looked like it was clean. All the UV photography log sheets will say that window 5 was a clean window, which it was, as best we could tell through the area where the photos were taken. The systems people are going to have to tell us where the mass spec boom really was, since we ended up jettisoning it.

YOUNG — Yes, when we looked at it, it looked like the door was closed but it didn't look like it was all the way down. It didn't look like the door was all the way closed.

MATTINGLY — By the time you got next to it, both doors should have been partly ajar. We had both of them malfunctioning by rendezvous.

YOUNG — Do we have pictures?

DUKE — I'm not sure that'll show up.

MATTINGLY — I think it will, Charlie, if you know what to look for. You people familiar with the hardware can do that.

YOUNG — Activation Through Separation. Command Module Power Transfer. We got this half hour head start and we were all suited up, and we were running 20 minutes ahead of the time line and we asked them if we could go ahead and power up. They said, sure, go to it because we had plenty of power and everything. The first thing that happened is that Charlie leaped in there, and there was no grind from the yaw antenna and the high gain.

MATTINGLY — I think the only comment you can say about our part of it was that we got tight on bleeding the tunnel down. Other than that, I think we did everything fine on the undocking.

YOUNG — Tunnel Close-out. You didn't have any trouble with that?

MATTINGLY — No problems with that. Everything worked.

YOUNG — Maneuvering to Undocking. You didn't have any problem with it?

MATTINGLY — No.

YOUNG — Undocking and separation were nominal. I remember when you backed off, and you said that's as far as she's going. And I thought, here we go again. And you hit those other things, and we separated.

MATTINGLY — Well, the idea was to let it hang on the capture latches, and we did, and it gave a bigger "clunk" than I anticipated when it got to the end of the probe. There were no dynamics at all. It was just waiting for them to stabilize, and they were just steady as a rock. It just takes a little squirt of ACS and we were gone. I thought that all worked super. Somewhere in here, we did a VHF check and we had some trouble getting together on B duplex most of the time. It turned out to have better comm on A simplex, although the VHF generally sounded super. Maybe you understand what happened. I really don't know.

DUKE — It was either our B receiver or your B transmitter. They ought to be able to check the B transmitter in the spacecraft. We could isolate that. We had VHF problems before flight and they changed out both VHFs.

YOUNG — It was almost unreadable at first, and then it seemed to get better.

DUKE — We were clear to Ken but on our B receiver on voice ranging; and then on B simplex, it was a little garbled. It sounded like he was in the bottom of a cave. But it did get better. So it would be wise to have ASPO check that out.

YOUNG — Formation Flight. Looked like you were doing a great job there.

MATTINGLY — The time line after undocking. I didn't do any formation driving around after we undocked. We went ahead with the Flight Plan, and went to do our landmark tracking and the P52 and all of those things. Everything there worked super. We did skip the transponder check.

YOUNG — We skipped the transponder check because we just couldn't fit it in. It wasn't necessary anyway because we were going to go whether it worked or not. It may give you a warm feeling, but if your mission rules say you go without it, and it's something that you're tight on, to me that's the first thing that you leave out.

MATTINGLY — LM Inspection Photography. I took a couple of pictures of you as we separated, but we didn't do any maneuvers or anything. You could see the gear and I guess that was really all I was looking for. I didn't see anything hanging off anywhere. I'd like to comment on the fact that we had two comm loops, and I think that was really super. That made life so much easier because Charlie could get his pads when they were convenient for him, and I could get my pads when they were convenient for me. I don't know what it cost the MOCR guys to run two comm loops, but on our end, it was really a big help.

YOUNG — It really is bad when you're trying to cough up two sets of …

MATTINGLY — That really helped. I can't say enough nice things about that. Stu really pushed on that.

YOUNG — Is there anything you can say about the fuel cell purge?

MATTINGLY — I can say I didn't do one. Okay, we skipped the radar check. I also skipped the COAS alignment, again, because it looked to me like I wanted to have the extra time to get ready for the circ burn. The only thing we were doing different from normal was that we were going to start it at 8 minutes, instead of 6 minutes, so that you would have time to take 509 out of the computer. I think I probably started it at about 830. I was really intent on making sure that the circ burn went on schedule. When we got to the secondary yaw gimbal check, it was unstable and diverging as soon as I touched the thumbwheels for a trim check. There was no question about it being a real live thing, because the spacecraft was wagging its tail just like the indicator, and it was diverging. I shut it down and in succession I tried it in Auto SCS, then I tried it in Rate SCS. I failed to try it in ACCEL COMMAND. I tried it in TVC. I tried it in CMC and got exactly the same response. Whenever you turned the gimbal on; it was stable. The first thing that moved the gimbal excited a diversion oscillation, whether it was in SCS or CMC. We got a change to the mission rules document about a week before launch and I was almost too mad about it to read it. Now I don't know if I was glad I read it or not. One of the things was you had to have four servo loops to do circ.

YOUNG — Ken asked me about it and I said I didn't think you should burn.

MATTINGLY — The thing that had bothered me about it was the point that Charlie raised later, and the only thing we had done differently was to run those gimbal motors for 20 minutes before LOI and 20 more minutes before DOI. We had always been so careful. We researched this point and we talked about it. One of the SIM problems we talked about was what do you do if your gimbal motor gets stuck on. Do you kill the bus or do you let it run? They researched it and came to what I thought was a surprising conclusion. Just let the thing run. It's unlimited in lifetime. My concern was that this might not be a random failure of the yaw gimbal, but maybe running this extra time had heated something up. I might have temporarily been a hero if I had let you go land on that rev, but if I didn't bring you home you weren't going to like me very much. That's got to be the toughest decision I had to make.

YOUNG — You were not by yourself, because I didn't want you to do it

MATTINGLY — Let me say something about what we did after that. Again, it was one of those things where, fortunately, we'd been down this road before. We had talked in data priority about what you do if you don't do a circ burn. It was one of those things where you never could really get your heart into it because we could never figure out how you were not going to do circ. But we had talked about it and they had come up with a TPI rendezvous scheme, which didn't seem to make any sense to me because of the navigation at those ranges. So we tried it and we flew some of these things on a one-rev, no-circ, re-rendezvous. And with the kind of separation maneuver we have you can expect to be somewhere around 1200 feet separation. You just wait until you're that far apart and then brute force it. And it costs you a few percent, 5-foot-per-second DELTA-V. In retrospect, I think we probably made a mistake, because without any direction from somebody, I should have rejoined on the LM at the first opportunity.

YOUNG — Yes, Ken asked me what to do and I told him I didn't know.

MATTINGLY — I didn't do it, and all I can say, in retrospect, is I think I should have. At the time, I decided, well, there is still a chance we might get to land and I don't want to waste any gas that might cut into that RCS budget until we have to.

YOUNG — I guess the thought that was sticking in my mind was that when we would come to that time that we would be go for PDI, and if they could figure out what was the matter, we'd go ahead and land without having to re-rendezvous. And 1200 feet wouldn't make any difference in the phasing at that point. When we came around that side we were ready to go for PDI because we had done everything that we had missed before.

MATTINGLY — We knew we couldn't do PDI with me in the same orbit.

DUKE — You couldn't do that.

MATTINGLY — We knew that point. Let me just say that I think I should have joined up on you without any call from anybody and I didn't do that. I didn't know you couldn't do PDI with us in the same orbit. Yes, sir, because you don't have any abort targeting.

DUKE — We were at least at PDI-2 now.

YOUNG — I can get to that LM from any place in the sky, if it's got the DELTA-V to get there.

MATTINGLY — There is no mission rule that covers this specific case. I think there ought to be. Okay, then we came around and we didn't do anything. Then they said we've done it. Stu had flown one and I had flown one at the Cape and they said do it just like that when your closest point of approach is at some time, which I fail to remember. And it turned out that at that time, which was supposed to be our closest point of approach, we were still opening. So I lost confidence that they really had a good handle on our relative trajectory. And they said just brute force it. Well, I put in 3 feet per second towards you, at the time they gave us. And we started opening. We were opening at the time I put it in. I put in 3 feet per second and the opening rate increased. It turned out that the 3 feet per second was retrograde. You pointed that out and I got to thinking that that's retrograde and we're closer to apogee than to perigee, and is that really the thing to be doing? I tried to figure out a clever way to do this and let orbital mechanics work for us. I finally took it back out just to get ourselves safe. About that time we picked up the ground and I felt that the ground never really understood the question I was asking them. Did I have a good state vector? I showed a 6.3 perilune. I figured that if I had a good state vector when it started I was still safe and that's no sweat. But their numbers were showing me in a 9-mile perilune and I began to wonder if maybe I had a bad state vector to start with. Finally, they had us brute force at the bottom. I was reluctant to brute force from anywhere. I felt that we had opened up to over a mile then and, at that point, I was ready to go and do a regular rendezvous, if they wanted to put us together; and let us use orbital mechanics instead of fighting it and still save the gas. When they told us to brute force it, I guess in my own mind I felt that we'd lost the ball game. We were going to get back together for good. The flight guys are going to have to give us the data on how much gas it would cost. It was a pretty expensive thing. It turned out that using John's rendezvous radar, for line of sight, gave me a faster jump than I could get out the window. We just never let it get away from us. We did whatever we had to. I had wondered what it would be like getting into the dark and doing this and we quickly found out.

YOUNG — He was 600 feet out and he could see us because he was in the earthshine. He could see us perfectly.

MATTINGLY — You could see the outline of the LM. The tracker light is beautiful. You could see the outline and if I had known that I would get there before earthshine stopped, I would have told them to turn out the tracker light because it was so nice. I kept worrying about when earthshine was going to quit and what was out there. And throughout the entire thing, the only sensation of closure rate I had was John's reading them out. The EMS and the radar were uncannily close together. I couldn't believe how close they were. But it was somewhat easier for John to read me the range rate than for me to have to take my stopwatch out and measure the closure rate. And there was never a sensation, even when we were right together.

YOUNG — We're talking about closure rates like a foot a second. It's fabulous. That's really fantastic gear.

MATTINGLY — I turned my spotlight on so I could see him. And you guys saw the spotlight, and knew it was on. I turned it on at 500 feet and we were still in earthshine and I couldn't tell it was on. So I said, well, probably earthshine is too bright. Well, then the earthshine disappeared and we were somewhere between 300 and 500 feet, I don't remember where, but the spotlight didn't do anything. We had to be inside 300 feet for the spotlight to be effective. At 300 feet, I could tell that the LM was there. I could see a shape. And then, inside of that, the old spotlight was just fine. And they finally turned the tracking light on and I could see it.

YOUNG — That's the old Gemini light we used on Agena.

MATTINGLY — But I was surprised that you had to be that close to get anything from it. Station-keeping? If flying airplanes in formation could be as simple as flying two spacecraft, the whole world would be aviators because you could kill a rate. I don't know how long we sat there without my touching a thruster. We just came up there and sat there and that was it. Station-keeping in the dark, they asked us to dump the water and I really had no qualms about leaving the couch and going down and dumping the water. We started that in the dark and that was when you noticed that we had been dead in the water with each other. I put the dump on and that spacecraft translated away. Did you see that thing? Right out the side.

YOUNG — That's a real gap of water coming out. Yes, just moved right out.

MATTINGLY — Really did.

YOUNG — I couldn't believe it.

MATTINGLY — Not fast, but it was unmistakable that that's what it was doing. And didn't take much to counteract it, but that was sure there, so I guess I'm more of a believer in FDO data now, than I was.

YOUNG — Yes, we saw it, relatively speaking.

MATTINGLY — I guess this is an appropriate time to make a comment on if you ever have to do a night rendezvous. Docking at night is going to be an absolute piece of cake. It just couldn't be simpler. The rendezvous, I think, because of the absolute lack of any sensation of motion along the line of sight. Now, we didn't see big rates. But 5 feet a second, I'm sure I could not have seen. And that's kind of uncomfortable in the dark. You've got to have one of those rate systems going. Radar preferably. And without that you don't really want to commit to this. Because I think what will happen is you'll be worried that you aren't closing fast enough and then all of a sudden the angular size of that thing is just going to grow and with that big marshmallow you'd never stop it.

YOUNG — I agree with you. You'd get the little marshmallow out of the way, hopefully.

MATTINGLY — Yes, that would have to be your technique. You'd have to dump. Because that big marshmallow just glides right on by. We did our separation burn on the next rev, after we'd gotten into station-keep and all that. And again I think we got out of sync and maybe should have done. I was trying to save gas at all costs and when they gave me the separation burn, they gave me an attitude to go to. And I got the feeling that maybe they were unhappy that I elected to burn at three axes and burn in the attitude I was in.

YOUNG — I think that with the small differences you can afford not to do that. Somebody's got to keep somebody in sight.

MATTINGLY — When they told us the time was coming up, we had just a few minutes and the only way to get to the attitude they gave me was to go real fast and I just felt like my obligation was to save gas at that point. I went three axes and kept you in sight. I would do it again.

YOUNG — I think it's a good head, man. You don't want to run two of those things together out there.

MATTINGLY — The circ burn after that was just super nominal, didn't bring the secondary yaw gimbal on until after the gimbal check and then brought it on for redundancy. And we did roll 90 degrees so that I had the two good gimbals in the yaw in the middle gimbal plane. I made that one change to it, which was no problem at all. We had some residuals which were higher than I anticipated, in fact, I got quite a bit of attitude excursion on the burn initially. I was surprised that it didn't run as stable as any of the other burns. And you could see the gimbal swing and the pitch attitude. Pitch and yaw were both moving.

YOUNG — You may not have changed your weight, because of the RCS maybe.

MATTINGLY — Well, if that's the case, I'm surprised how sensitive it is. But we ended with some residuals as a result of this attitude excursion.

YOUNG — You had to roll to get them out.

MATTINGLY — And I got both residuals in the two axes I couldn't burn. And I had to take one out and then roll and take the other one out again, to keep from firing over the SIM bay. I was just surprised at the magnitude of the residuals in all of the three axes. I didn't anticipate that. After that, the rest of it was all nominal. Why don't you pick up your side?

YOUNG — I wanted to say I thought Ken handled that thing beautifully. I think I'm the guy that coerced him into taking that DELTA-V out that he put in, retrograde, because I was really nervous about that. I didn't really know. I think I remember you saying 3 feet a second but I wasn't really sure and we were too close to apogee to be doing that, and I just didn't want to see the command module disappear behind one of those hills. I don't think it would have ever happened but I didn't want to see it.

MATTINGLY — I'm sure if we had done it we would have been safe. My intent was that once I committed myself to that path, I didn't care how much gas it cost, I was going to finish it. But, I was worried about sagging below you, thrusting up toward you and going underneath. You talked me out of that.

YOUNG — It was a good rendezvous. I don't see how you could have played it any better. The closure rate was so slow, but the inertial line of sight rates were just perfect. All the way.

MATTINGLY — I guess it's worth saying that we have different impressions here, but it's my impression here that inside of half a mile my COAS gave me better line of sight rates than the radar. Outside of that, your radar was good.

YOUNG — If you correct for what the needles are telling you to do, you could be doing the wrong thing. It wouldn't amount to a hill of beans in the LM because you just take it right out again. On the command/service module, you would be squirting out gas to fix something that really didn't need to be fixed. You're going to fix it anyhow.

MATTINGLY — And then you're going to take it out later.

YOUNG — So I reverted to the COAS when we started diverging and I think it's the right way. There, we were 20 minutes ahead and I figured we had that time line just beat to shreds. I figured we were going to be ready to go down to PDI minus 1.

DUKE — Everything worked except the S-Band steerable. Eighty-two was "nominal except the steerable and it would not move in yaw. So that threw us behind on the up-links. And we had to manually do the P27s, which impacted the P52 dock. We missed everything else we did on order here. Everything else was nominal, except for the RCS; the RCS activation we had a double reg failure in system A.

YOUNG — And the first words were, "Open the ascent feeds." Charlie didn't hear that so I opened the ascent feeds before Charlie realized what I'd done and he said, "What'd you do that for?" Because we just had a SIM that blew the whole mission when we opened the ascent feeds.

DUKE — So I closed them.

YOUNG — So Charlie closed them and it turned out to be the right thing to do. If we'd left them open any longer we'd have probably overpressurized the APS, or gone too high on the APS pressure. People don't realize that. And then after we'd done that they called up and said don't let the APS pressure get above 180, which we hadn't even looked at.

DUKE — But those were the only two problems we had, the S-Band steerable and the RCS system A.

YOUNG — We tried working the P27 together. Charlie would call it out and I'd punch it in, but we were getting in each other's way; so Charlie finally took the P27 and I went and pressurized the RCS. We were working the time line in parallel, something I could do on my side I would do at the same time Charlie could do on his side. I think that's the only way we stayed up on it. The PGNCS Activation and Self Test was nominal. The Suit Loop Checkout was nominal. Everything was nominal except that yaw steerable. It was like somebody forgot to take out the "Remove before flight" pin or something. That's the way it looked to me.

DUKE — The Landing Radar Self Test, we did that after undocking and that didn't work the first time.

YOUNG — It didn't work the second time either. We were in a communications attitude and we were at perigee and I don't know what it was but it was reading erroneous numbers. The first time it read on the tapemeter, it read 8000 feet and it's supposed to read something like 494 on the range rate. It was reading totally erroneous on the range rate, and the next time, the VERB 63 radar self test where you set the gnomon flag and everything, was totally wrong, and then we tried it again and it read the right range rate, but the wrong altitude. And then, we got to an attitude where we weren't pointed at the ground and the thing worked like a champ.

DUKE — It was locked up, but we had such a super radar, that it locked up on the ground. Because on PDI we were locked up at 50 000. So we think that's what happened.

YOUNG — I don't know what happened, but it sure flunked.

DUKE — We didn't do the Lunar Landmark Recognition because we were behind. The MSFN Relay we never exercised. Their comm uplink was loud and clear the whole time, but I guess poor Jim had his hands full hearing us on the down-link on the omni, until we got the 210 up.

YOUNG — DAP Loads were nominal. The Rendezvous Radar Checkout we didn't do, we tried it one time.

DUKE — Now wait a minute, this was just before undocking, John. This was the rendezvous checkout. That worked fine.

YOUNG — Deployment of the Landing Gear was just a clunk and it was down and there's no doubt in anybody's mind. Undocking was nominal. Separation was nominal. Like you say, we didn't do the Landmark Recognition and the MSFN Relay.

MATTINGLY — We skipped the landmark tracking that we did after DOI. I guess that deserves a couple of words. When we get through, I'll come back to that. I guess it was significant that we were able to get both the training landmark and the landing site on the same rev and track them both. The auto optics and all that really worked like a champ. The thing that surprised me and, I don't know if anyone has mentioned it and I just didn't listen, but when you look through those optics at the Moon, you're really getting a dose of heat in your eyeball. I think you should never do more than two landmarks together, and you really should do only one, on any rev, because your eyes were very sensitive to looking at that bright concentrated light. And you could feel the heat when you put your eye to the telescope.

YOUNG — Let me tell you what Ken did the day before we started all this, into low Sun angle. He tracked landmark 16-3 the day before, on OJT training day and how many marks did you get it - four, five? Ten? He'd been set up to track the OJT landmark. He tracked that one and then without changing attitude he put in the new landmark for the 16-3, which is right at the landing site and tracked that son of a gun, the day before we were supposed to go down and land at the low Sun angle. I don't know how the ground will use that data, but if they had been able to collect that data instead of the OJT landmark data, man, we'd been in Fat City. I don't know if they did collect it.

MATTINGLY — The reason we didn't plan to do it that way, was that when we set up the Flight Plan originally, the landing site was going to be in the dark on DOI and then with the month slip the landing site slipped out in the daylight so that it was infeasible to do that. That was a pen-and-ink thing I wrote in because I wanted to look at the landing site early and we stuck it in there.

YOUNG — My suggestion would be if the landmark is going to be in daylight at all, get the landmark at the landing site, rather than fooling around with something. Even if he doesn't get the data that will give him a chance to do some landing site recognition, which he's going to need for the next day. After he gets the data, there's one big block you could forget about worrying about.

MATTINGLY — That P24 sure is a smooth running program. The auto optics did not put us on any of the targets, directly. We always had to correct them, but once you got going on it, that thing was really nice and smooth. Super easy thing to do; the only difficulty you have at low altitude is target identification.

YOUNG — I'd be willing to see what your landmarks tracking did for the state vector.

9.0 SEPARATION THROUGH LUNAR MODULE TOUCHDOWN

YOUNG — We're just going to talk about the lunar module. Preparation for PDI was nominal, except we didn't have any comm.

DUKE — We had to go to a different attitude than we expected, to maintain high gain. We never did get to see the ground.

YOUNG — DPS/PDI Burn and Performance.

DUKE — I think it was nominal. The radar locked on at 50,000 feet which is amazing.

YOUNG — Let me say something about that. As we were coming around the corner for PDI, we didn't have the high gain and our yawing was 20 degrees for the descent. Charlie says we were on the omnis and we couldn't get lockup for the state vector that we needed to do PDI. About 12 minutes prior to PDI, Charlie suggested that we roll right 20 degrees. We rolled right 20 degrees, and it improved the comm margins considerably. I think it's something that everybody who's going to work omnis should think about, putting the omni antenna toward the Earth as best they can to get that update, because it was 12 minutes before we got that update. I figured we were almost in another wave-off situation. That was a good suggestion and we did PDT from it. Off-nominal, we did it from zero yaw all the way down.

DUKE — The engine worked just as advertised; throttledown came right on prediction, and the profile we started high and to the south, about 16,000 feet high as a matter of fact.

YOUNG — 16,000 feet high?

DUKE — Yes, we started about 66,000 feet.

YOUNG — And we were about 16,000 feet to the south also.

DUKE — Right after throttledown, we were back on standard profile.

YOUNG — Yes, it was really beautiful. Adequacy of Procedures and PGNS Performance during PDT was nominal.

YOUNG — Spacecraft Trimming and Systems Status During PDT. Nominal.

DUKE — Yes, a couple of things here, John. We didn't do the COAS calibration or the LPD calibration due to being behind, thanks to the comm problems on that first rev. As it turned out, everything was perfect.

YOUNG — Yes, and as a matter of fact, for any reasonable rendezvous or any reasonable landing, the COAS calibration and the LPD calibration are just something to take up time in the time line. Those things are done on the ground, and they're done perfectly adequately, based on our experience. And it costs you gas too, because they're 2 degrees per second maneuvers to get to them and they perturb the orbit.

DUKE — I'd just like to add a comment on it. If you consider trimming RCS maneuvers in the LM, there's no question in the real world that the thing is maneuvering. You can tell what jets are firing. You can in fact hear the jets when you have your helmet off. At 2 degrees a second in the lunar module, you can feel the rotation rate. Like Ken said yesterday, I never felt the rotation rate in the command module, but in the LM at 2 degrees a second, you can sure feel it. And no question in your mind which way you're going and which jets came on.

YOUNG — The Radar Tracking Attitude was not as planned.

DUKE — Not as planned because we were late doing the rendezvous radar checkout. But once we got the attitudes on that first rendezvous with Ken, the radar worked.

YOUNG — That preparation for PDI was sort of off-nominal in that we had a wait and a GO for PDI. We had - was it two revs?

DUKE — It was after the no circ burn.

YOUNG — It was the third rev. Six hours. Okay, so it was three revs. During that time, we did a couple of P52s to keep the platform warm. In addition to the one that we did, we missed a rev of doing P52s in there somewhere. But there weren't any torquing angles. That's really a good platform.

DUKE — There was no gyro drift compensation required no PIPA bias. In fact, we never had any update, even after sitting on the surface for 73 hours in power backup, they didn't change anything.

YOUNG — The VHF Range was in good agreement with radar. The radar tracking attitude was right on, but we did it on the back side of the Moon just prior to the circ burn. The first one that we did, somehow, I didn't get a main load block. It was in the dark and I was pointed at the tracking light, but it wasn't like the usual tracking light. It probably had something to do with Ken's attitude. I just got a very faint light out there, and I didn't have a main load block and I couldn't get a lockup. So we did another one for the rendezvous. We knew after we'd run it for an hour and a half that it was going to work.

MSFN Acquisition via PCM High and Update Pads.

DUKE — We had a pretty tough time with the high bit rate, due to the loss of steerable.

YOUNG — Charlie, I don't know. They weren't getting lockup coming around the PDI; man, I was really sweating that when you said, roll, let's take the yaw out. And we did it and they got lockup. That was a cool move. That put the omni antenna right at the Earth. That may have been just the margin that we needed, 12 minutes prior to PDI. I thought we were in trouble then.

DUKE — Me, too. I thought it was really gone. The up-link was always clear, and we didn't have any garbled transmissions. So I had no trouble with the pads. The ground kept saying that we were really noisy, really horrible. So I used a louder voice. I know on Apollo 10, we had a couple of revs of that and it's really miserable trying to get things passed back and forth.

YOUNG — We sure did, didn't we, come to think of it.

DUKE — The only thing I can say on Attitudes and Position is that the preflight time line that we flew until circ, we did all those attitudes and maneuvers except for the one to the landmark tracking or viewing, and we had to get a high gain lockon so we were pitched up and didn't see the landing site.

YOUNG — Yes, if Ken had done the circ burn, we wouldn't have had any trouble doing PDI on that rev, so I think that the decision to cut a rev out of the time line was perfectly safe. It didn't work out and there may be other reasons why it wouldn't work out in the future, but if you think positively, there's no reason the time line itself doesn't constrain it to doing the PDI after the first rev.

DUKE — Even with all our communications problems, in doing the manual updates to P27, we got a little behind there. But by the time we got to circ, we were caught up. And I agree with John. I think that the time line was adequate and I recommend it for 17, if they plan to get out first.

YOUNG — One thing in favor of getting out first is you've already got your pressure suit on, and you don't have to go through all that Mickey Mouse again. Boy, I tell you, that taking that rascal on and off, you got what I consider the worst part of your prep already done. And I think that ought to be considered as a factor in this business. The LPD Attitudes and Accuracy (Calibration). I think the LPD was perfect. I don't have any gripes there whatsoever. When we pitched over, we were north and long and you could see that. I was just letting the LM float in there until I could see where it was going. I took out the north because according to our preflight maps, the north country was a little rougher. There were more contour lines up north than down south, so we took those out and when we got in close, we backed up a little and put in some rear updates. I don't remember how many there were. But at pitchover, you could see (just like preflight) Gator, Palmetto, and Spook, and the inverted deep shadow pattern through Stubby, Wreck, Trap, Eden Valley, and Cove right into Spook, although at 15-degrees Sun angle, it wasn't as apparent. Of course, we had already seen the landing site on two other occasions when we were flying over it because of the three-rev slip. There's just no doubt in your mind when you're at pitchover, and the first thing you see was South Ray. There was some question about whether we'd see the ray patterns at the low Sun angle, but there's no doubt that we were seeing the ray patterns from South Ray at pitchover, and there's no doubt, at least in my mind, as to where that machine was flying to. And it was a simple matter to redesignate to the south and back up a hair.

DUKE — We'd agreed that I was going to look out since I had two good craters on my side, and it looked just like the L and A.

YOUNG — In fact, it was working so well I was tempted to let it do the thing all by itself, but the trouble is we got down low, and I could see that we were going to land in that pothole down there. We took over, I guess at about 300 feet, and pitched forward a little, and we could see the surface all the way to the ground. Right close in there out of my window, I could see that crater down there, so I went forward a little bit and landed. I counted one-potato after we got the contact light and shut the engine down; even so, I think that we fell about 3 feet. I think we're very fortunate that the landing was so flat because I really couldn't judge the slopes. We just lucked into almost zero roll and a couple of degrees or 3 degrees pitchup, and of course we'd taken the yaw out. When we redesignated to the south, we must have had 30 degrees of yaw and took it back out. Like I say, at that Sun angle, we could see the rocks all the way to the ground and I think that was a great help. From 200 feet on down, I never looked in the cockpit. It was just like flying the LLTV; your reference is to the ground outside. You had another thing that nobody has ever remarked about before, and that was the shadow. I really didn't have any doubt in my mind how far above the ground we were with that shadow coming down. I had no scale of reference to the holes but with the shadow out there in front of you and coming down, it really takes all the guesswork out of it. For that kind of a Sun angle, if the radar had crumped, I don't think you'd have had a bit of trouble in just going right in and landing just like a helicopter. First, we could see the thing up all the way to the ground; second, the shadow was right there to help you with the rate of descent. When Charlie says you stopped and you're hovering, there wasn't any doubt in my mind that I was hovering. I could look out the window and see that we're hovering just like a helicopter. We were well into the dust, maybe 40 or 50 feet off the ground, when we're doing that.

Change in Appearance of Features, Distance Estimation of Landmarks. I think because we practice so much with the L and A, we had a pretty good hack on how far away we were from things. And the only change I'd make (and I don't know how much of a change it is) is that after we got out and went up on Stone Mountain and looked back, you could see Double Spot and you could see the lunar module, and it looked as if we were maybe 70 or 80 meters further east than I said we were originally. That's just a guess because we're sitting up north on a hill and looking back, and you could see Double Spot. And the lunar module appeared to be sitting in a hole over behind Double Spot, and it's almost in a direct line from where we were on Stone Mountain. So we must have come very close to landing exactly where we were scheduled to land in the first place. And I emphasize again, the only reason for landing there was to get a spot that wasn't so hilly. Preflight, that region around Double Spot was the only flat place on the map as far as the contour lines go. I think it was a mistake.

DUKE — It turned out, I think, the flattest spot that we saw on the whole traverse was up to southeast of North Ray Crater in that valley past Palmetto. It was a broad smooth valley and hardly any craters in it at all. No rocks.

YOUNG — No rocks at all, and on the contour map, that looks pretty bad.

DUKE — But it was about where Dog Leg was mapped, in that area, back off to the right there by the traverse. I had a good feel at pitchover of exactly where we were; once we got to the ground and I looked out John's window, I felt like I could reach out and touch South Ray and Stone Mountain. They just looked that close to me. I had a tough time estimating and I knew they were

5 kilometers away and I just had a tough time estimating distances of big features, once we got on the ground.

YOUNG — On a clear day, the mountains are 40 miles away and it looks like you'd be there in 4 minutes. It's the same thing on the Moon. We kept going over rises and I kept thinking, here's Stone Mountain; and then we'd go over another, and it wouldn't get you there.

DUKE — There were lots of ridges between us and that mountain. It wasn't apparent when we first looked out the window.

YOUNG — I know people have remarked on this before, but there's a lot more light in that vehicle than there is in the lunar module simulator. I don't know why they want to keep it so dark in there, but we didn't have any trouble reading any of the gauges. Sometimes we get in the LMS and everything is all turned out in there, and it's just like normal light. We never did have any trouble seeing any of the instruments. My subjective opinion of the light produced by those utility lights was much better than it is in the simulator.

DUKE — In the dark, we had the integral lighting up and the side panels on, and that was about all you needed.

YOUNG — I couldn't judge slope out the window worth a hoot, and that's the truth, even down low. The ground looks flat but I'm sure it would look flat if it had been a 6- to 8-degree slope too. I don't see any way around that. I've done a lot of helicopter flying, looking at slopes and you can't judge slopes in a helicopter from 100 feet on down very well either. And that's the way it is. But I don't think you'd be in any trouble if you touched down within 10 or 15 degrees of being straight up and down. It's going to bother you some in deploying gear off the lunar module, but it isn't going to bother your performance.

Hovering and Blowing Dust. We did a hover for a short period of time there at about 40 feet off the ground, and the rates were practically zero and there was blowing dust.

DUKE — It started at about 80 feet, John.

YOUNG — Yes, 80 feet. Certainly, it started there and it got a lot worse, but you could still see the rocks all the way to the ground. The surface features, even the craters and with something like that - which really surprised me. I was expecting two things: either the dust would be so bad we couldn't see anything, or there probably wouldn't be as much dust as there was. Possibly, it's the 15-degree Sun angle that did all that. Because there's certainly plenty of dust down there to blow, and there's nothing thin about that regolith around the LM.

DUKE — How about that Zero Phase? I never noticed it.

YOUNG — First, the thing starts out as a Sun coming and that turns to a shadow. That was zero phase getting better all the time.

DUKE — I was excited at that time.

YOUNG — Yes. When you said, here comes the shadow, that was before I had seen it. In fact, you were watching it out your window and I looked up there and I saw it, and I said, "Yeah, man, there's no doubt," and that baby got bigger and bigger and dropped right down in front of us, and man, that's a good gauge.

DUKE — Was it?

YOUNG — Yes, we saw it in a film.

DUKE — It looked just like that one in the LMS.

YOUNG — I don't know if you know this, Deke, they had a shadow for us, and I guess we've been using it. In fact, we'd been practicing without radar a couple of times just to see what you could do. The thrust-to-weight ratio in our lunar module is such that if you're at 100 feet and you have

20 feet a second down, if you go to full throttle, you'll have that all killed before you hit the ground. So we're looking at very high sink rate descents off-nominal with the shadow. Man, it really makes a difference when you're looking for it. A crater from 50,000 feet looks like a crater at 5 feet, that's the bad part of it. But with a shadow, as it gets bigger, you know you're getting closer.

Touchdown. Estimate of Vertical and Horizontal Velocity. It wasn't much.

DUKE — When the computer said level off, we leveled off at about 40 feet off the ground. I think you had a good feel that it was leveled off. There was no question in your mind that you'd stopped coming down. I remember seeing them at about minus 0.7.

YOUNG — And then we pitched forward again. I didn't want to go backwards at all.

DUKE — But the needles were great.

YOUNG — And from looking at the probes, we must have come pretty much straight down.

DUKE — Every one of them had folded straight up.

YOUNG — Significant Impact and Vertical Motion Sensations. When we got the contact light, I counted the one-potato and shut the engine down. The thing fell out of the sky the last 3 feet. I know it did. I don't know how much we were coming down, maybe a foot a second.

DUKE — I don't remember exactly, but about 1.8, I think I saw, right before touchdown.

YOUNG — Engine Shutoff and Probe Contacts Pad Cue.

DUKE — Dave had the same slow descent too, and they had the same sensation that when you shut the engine down, the LM really free falls. I think a softer landing would be a 3-foot-a second descent and allow the tailoff to cushion you as you continue down because the gravity really had us. I had a definite sensation of falling when you shut down.

YOUNG — I don't see what you're talking about. If you got 3 feet a second and you fall when you shut the engine off - -

DUKE — If you're leveled off, 5 feet above the ground and stopped the engine, I think you'll accelerate faster than if you came down at 3 feet a second when you got contact and just kept coming at 3, then that 3 feet a second is constant. And by the time you can shut the engine down, you've already hit the ground at a 3 feet a second versus the gravity acceleration type fall.

YOUNG — I'm not sure that's true.

MATTINGLY — Seems like the only way that could be softer is if you're saying the engine would be running when you hit the ground.

DUKE — Yes.

YOUNG — You could do the same thing by waiting a little longer to shut the engine down.

DUKE — That's right.

YOUNG — With all those bloody rocks around there, I don't see - we had a lot of choice. Just want to make sure you were on the ground.

DUKE — Yes, it's a definite warmth, but there's no doubt that the LM can hack it.

SLAYTON — Did you get any stroking of any of the struts?

YOUNG — I don't feel that there was because that first step was a biggy.

MATTINGLY — You were talking about if you back into that crater behind you, and that gear is only going to help you if you can stroke it, isn't that right? It seems like there's some advantage to hitting the ground hard.

YOUNG — No, Ken. I wouldn't stroke that gear, man. I'll tell you, that would really jar your teeth. I'd rather have them cut off a couple feet just to get me closer to the ground. I'm only talking about the slope.

DUKE — If you land on the upslope, the rear gear as you fall down could very well stroke.

YOUNG — If the rear strut had been over that hole, we'd have been just like the Apollo 15 guys. We would have been landing on the first 3 and the engine bed. It really dropped.

DUKE — We had about 4 percent on the fuel and 6 percent on the oxidizer, and we had just gotten a descent warning light. Things never matched up the whole way down. They were running 2 to 3 percent low on the oxidizer side.

YOUNG — We had 4 percent fuel and 6 percent oxidizer. That's about a minute of hover time or better.

DUKE — I remember their saying descent 1, and the quantity light was on when I looked up to check the systems after shutdown. I don't remember their saying a minute, giving us any call.

10.0 LUNAR SURFACE

YOUNG — The postlanding powerdown was nominal.

DUKE — We had a few changes because of landing 6 hours late. They had us power down the AC bus and also the LGC DSKY breaker to save power, and all that worked just great. I thought those procedures came up in good shape. We just floated right to the checklist for those changes and then we were right on.

YOUNG — The trouble with pulling the mission timer is you lose a clock, and that makes the emergency lift-off times absolutely meaningless. You don't have any idea what time it is. We kept getting a block and lift-off time and I didn't know what to do with them. I didn't know what time it was.

DUKE — We had no idea what GET was, but we had a Houston watch. They ought to read them to you in Houston time, then you can use your wristwatch.

YOUNG — What you would do in the real world if you had to do an emergency lift-off at one of those times, you had contact with the ground, you would roughly use that time to get you powered up on time and have the ground count you down to the second for the lift-off.

DUKE — But these are no-count times.

YOUNG — That's correct.

DUKE — Then, they were worthless.

YOUNG — It seems to me if you lose your timer, you aren't going to hit it on the second - you may hit it on the minute.

DUKE — You're right. I highly recommend the procedure that we used for parking that platform in gimbal lock. And we parked that beauty for 71 hours, and we didn't even have a PIPA bias or gyro drift update. And, it just worked like a champ.

YOUNG — Venting was nominal. Landing site orientation was about as flat as you can get. Like I said, "It's more luck than skill," (laughter). I hate to admit it, but it's true. Twenty-five meters either way and we would have been on a 10-degree slope.

DUKE — It's like landing on a carrier.

YOUNG — Add 15 meters to that, and we would have been in a hole.

DUKE — I had a good feeling that we knew just about where we were when we touched down. That big ridge was on our maps and had Smoky Mountains right out my window. John had called Double Spot going by, and I could look south and see all the landmarks down there. We had a good idea we were within a couple hundred meters anyway.

YOUNG — We could have almost triangulated off Stone Mountain with some kind of compass. You could have said, out there at 10:30 I see a ridge line and a crater. You could have done that and from one window you could have triangulated yourself in. There wasn't any reason to do that.

Here is a problem that was annoying to both of us. The night before, we filled the drink bags full of orange juice in the CSM; and the next morning prior to suit donning, we put them in the suit. Every time we bent our head, the microphone would get caught in the drink bag and put some orange juice into the air in zero gravity or would squirt the side of your face. Charlie really got covered up with it. It really was an annoying problem.

DUKE — My valve was really bad.

YOUNG — Mine didn't work all the time, and I was really being careful. I'm sure it got all over us because once we got on the surface and looked up at the lunar module, the travano cover had orange juice all over it. It was in dots, less than 5 percent, but there was a lot of orange juice on the travano cover. I'm sure orange juice is something you don't want to float around on wire bundles. I think we need something to stop up that hole in zero gravity and in one-sixth gravity until you are you could pull with your teeth that would work. I think it's essential when you're going out for a 7- or 8-hour EVA, you have to have something in that suit to drink.

DUKE — Yes, that really saved me out there.

YOUNG — I took my suit off and didn't put the drink bag in right for the first EVA. I didn't get anything to drink while I was out on the Moon and that was bad. I sure could have used a drink about half-way through. You do sweat a lot while you are out there. You sweat in your hands, you sweat at the back of your neck, and you sweat on your feet where you don't have water cooling. We should have one that doesn't spend its time wetting you down. And there was another problem associated with this. Before we went out the next day, Charlie had to clean the orange juice out of his microphone to get it to work. He wasn't transmitting at all.

DUKE — On VOX.

YOUNG — He had a comm carrier with one mike gone because of a busted wire and had to suck the orange juice out of the other mike to get it to work. Now that's a pretty marginal operation (laughter).

DUKE — Every time the left microphone hit that valve, the juice sort of migrated up that microphone in under my helmet, and this whole side of my head was just caked with juice.

YOUNG — Charlie looked like he had been shampooing with juice.

DUKE — It was really terrible.

YOUNG — The whole side of his face was just one big mass of orange juice. We got it on the helmet seal between the second and third EVA. We cleaned the orange juice off the helmet seal because we couldn't get the helmets unlocked and off. I thought we were going to spend the night in the pressure suit.

DUKE — It really wasn't on the O-ring; it was where the two surfaces mate together.

YOUNG — Yes.

DUKE — The stuff had seeped in under there.

YOUNG — The vacuum dried out that thing, and left the glue there. When it was time to take the helmets off, I couldn't get Charlie's off and he couldn't get mine off. I tell you, I thought we were going to stay in the pressure suit. (Laughter) I couldn't pull the button out, and I couldn't get it to slide.

DUKE — The button would come out, but I couldn't make it slide up or down.

MATTINGLY — If that's the case with both of you, then is that really a case against the orange juice, or is that something else?

DUKE — It's the orange juice.

YOUNG — Mine was leaking, too. At least, it wasn't leaking as bad as Charlie's.

DUKE — It was enough to solidify when he stepped out on the surface.

YOUNG — Where you get the problem with the orange juice is during the prep. It's not bad once you get on the Moon. It's not bad because you're not bending into it all the time. While you're doing a prep, there's a lot of looking down you have to do, and every time you bend your head forward and wrap your microphone around that thing and pull back, that works the plug and it squirts in your ear. It's already under pressure, because you have 32 ounces in there, and you're bending forward so your chest is pushing on it. It's just like a pump that pumps orange juice right in your mouth, your face, or ear.

DUKE — Maybe you could design a valve like the one for Skylab.

YOUNG — Design one that works. Well, I'll tell you, I really believe that by having a lot of something to drink in a pressure suit is a way to go. I think it sure helped me and Charlie out on the surface, but it certainly got to be a problem with orange juice floating around the cockpit as an electrical conductor. With it floating all over you and getting in your comm carrier, it's a problem; and then floating down in the neckring or worse yet would be getting it on the neckring seal where you couldn't lock that helmet. In training, we had orange juice get on our neckring and the only way they could get the thing locked was to go back and take the neckring apart and clean the residue out of those locking dogs. They took the whole helmet apart and cleaned it out. That's the only way we could get it to work. That would bite you in lunar orbit, because I don't know how to do that; I don't know how you take that neckring apart.

DUKE — Even with all those problems I'm glad we had something to drink.

YOUNG — Yes, I am too. Now whether it has to be orange juice, I don't know. Maybe plain water would do. In fact, for the first EVA, water was what I had in mine. I drank the bag the day before. Maybe they could fortify the water with the potassium, if they insist on that being there - or maybe there would be a
pill you could put in there. I don't have any idea whether it would make any difference whether you did it or not.

SPEAKER — They should be able to make that valve so that it doesn't leak.

DUKE — They overdid it.

YOUNG — It does exactly what it's supposed to. The trouble is every time you catch your microphone in it and pull back it pulls the valve forward and it works just like it's supposed to, and when you let up on it, it stops. But I mean it's sort of a rock and a hard place. If the microphones came around your nose you wouldn't have this interference problem with the thing but that would be a big redesign.

SLAYTON — I don't think it would be worth it.

YOUNG — I think they could make a little soft cap that you could pull off with your teeth because you sure don't want it leaking on you at zero gravity.

DUKE — That was really terrible.

YOUNG — It could drown you. Charlie was in there with a helmet full of orange juice when we were coming down to PDI.

SLAYTON — Was it your plan to leave the helmets on once you'd landed or go straight out for EVA?

YOUNG — No, no, we were going to take them off.

SLAYTON — So we could put a cap on there that you could take off after you take your helmet off.

YOUNG — Take it off, just prior to donning your helmet for the EVA. Yes, with your hands.

DUKE — It's a piece of cake getting out of those suits. It sure fills up the cockpit; they seem to be fatter than all those old training suits. We stowed them back on the engine cover, but we still had enough room to do everything.

YOUNG — When I unstowed the hammock and climbed on the top of the suits I wasn't sleeping on a hammock, I was sleeping on a bed because the suits were right underneath me. The suits were up into the hammock about 3 inches. So I wasn't suspended, I was laying on top of the pressure suits. It's kind of an unusual position because it gets right up under your back.

DUKE — As I reported, the first two nights I took a Seconal. It helped me. I slept really well all three nights on the lunar surface.

YOUNG — The first night I was really warm and I had taken all the gear off and hung the FCS and the WMS up to dry and slept in the sleeping bag with nothing on. I woke up in the middle of the night and my feet were freezing. So I turned around and put the ISA over my feet and went right back to sleep. Worked like a charm. But the next couple of nights I slept in the LCG because it was really cold at night.

DUKE — Chilly at night. You needed a sleeping bag. Even with the LCG on, you needed a sleeping bag. You didn't need it when you started to sleep but by the middle of the night, at least, I'd wake up in the middle of the night and I was cold and I wanted to get in that bag. The hammocks were great.

YOUNG — Yes, I thought so, too.

DUKE — We had too much food. Dave said that they ate everything. But we couldn't have possibly eaten everything.

YOUNG — We did our best; we did pretty good on them. I thought we were only getting two meals a day and I thought we did eat pretty much of everything, as the log will show. I think we left out the frankfurters or something else like that.

DUKE — I'd like to say "atta boy" right now for John Covington coming up with those new

procedures. Since we slept first at least I think it was, John. The Lunar Surface Checklist, his part of it was just outstanding. It was reorganized in real time and we never had a feeling that we were pillar to post in that
checklist. We had to flip pages but it was all flowing well and once we got in it, it led us in right back where we were back in sync by the time we got the prep card that we used, the cue card. He had done a lot. Those things were well organized and I thought those set of procedures for the preps and posts flowed smooth as glass.

YOUNG — Yes, I do, too. We might as well take the prep and post all in one bunch. Every time we took the suit off - it's real handy that the thing is standing up by itself in one-sixth gravity. It's really handy for you to close the zipper up to lube the pressure zipper and get those connectors before you put the thing away for the night. On the second and third EVA, because everything was really getting dirty, and I don't know whether it's a real problem or not or an imaginary problem, we were really getting concerned about whether we were going to be able to do things like fasten the connectors. So we were taking special care to lube everything and, therefore, we ran out of lube.

72-H-572

DUKE — We had one left for the zipper on EVA-3.

YOUNG — I think you should have some more lube in case you do get to a situation where as you're doing your last donning, something is not working right and you need to go back and lube it again

DUKE — No. Not only did we use the lube but we used water and the towels to wipe around the outside of the connectors. And the wrist rings - it wasn't in the O-ring part that was so stiff - it was the mating surface between the suit and the sliding ring.

72-H-573

YOUNG — Yes. That's been remarked on before. Somebody said they taped their wrist ring but that seems to be like a cluge. I think they should come up with something that keeps the dust out of the wrist ring. Maybe an overflap that you Velcro on the other side of it to keep the dust out of there because I just don't think you should have a problem donning and doffing. We really got a lot of dust and I don't see really any way out of it when you're picking up a bag on the Moon and you're holding a bag and Charlie's dumping the dirt in there, the dust goes all over the place and it's just as easy for it to go down your shirtsleeve as not. The fact is we had both dirt and rocks underneath the flap that you raise to get the glove open.

72-H-574

DUKE — Well, to pinpoint, EVA prep, PLSS donning and checkout went great. That little beauty is just what we expected. It worked just like the procedures, followed straight through. We felt as if we were on time during the whole donning.

YOUNG — Yes, we didn't think we were behind anywhere.

DUKE — We didn't have a clock to see.

YOUNG — What we had planned to do preflight was to have the mission timer running and we had our time line blocked out, so that for each sequence that took, say 20 minutes, we'd know where we were on the time line and we wouldn't have to keep bugging the ground. They wouldn't have to keep bugging us either to speed up or slow down. We didn't have to do anything over, but, on the first one, I forgot to put the drink bag in until after I got my suit on. And you cannot put the drink bag in with the suit on. Charlie put it in and I helped him stuff it down and it wasn't in good enough.

DUKE — Well, it went in, but you couldn't drink out of it.

YOUNG — I turned my head while we were outside, pulled the valve over, got it down in here somewhere and never could get to the valve. I had it down in my neck ring somewhere. I mean, I tried. I was down in the suit scrounging around for it, but I could never get there.

DUKE — Cabin depress - We used the overhead valve and it was horrible to reach in training, but in the one-sixth gravity it was nothing; just right up there and got it. I really felt familiar with those procedures. That square filling we did really paid off, I'll tell you.

YOUNG — Sometimes Charlie and I could reach the valve on the PLSS and sometimes we couldn't and there didn't seem to be a reason why. But, in the main, we tried to help each other in the vehicle to turn the oxygen on and off. And as everybody has remarked, once you get the gear on the only thing you can do is get out of the spacecraft because you've run out of room to do anything else. Let me say something about donning and doffing. We used a different procedure than that in the time line in regard to coolant. We used only lunar module for cooling during donning and doffing and we only used the LCG pump. We got the other hoses out of the way and sort of semistowed. We didn't say anything about them changing those procedures in real time, but it kept us significantly more hydrated than we would have been had we been on air coolant. What we would do is we plugged into the water and got it running through our suit, cooled the LCG down, and then climbed into the pressure suit, which took a big hunk of energy. And then immediately plugged the water in and pushed the pump in and gave it a squirt of water. At various intervals when we started to feel like we were running out of cooling we'd push the pump in for maybe 30 seconds worth, give it a squirt of water, and not use air coolant all the time that we were donning and doffing. I'm sure that's the best way to work that system because it keeps you from sweating. It keeps you from doing the kind of cooling that must dehydrate you.

SLAYTON — You didn't run the pump continuously, though.

YOUNG — No, I think that I'm going to recommend that they change the procedures to do it this way. I first noticed this when we were running long durations of suit runs and we only had air coolant. You really feel bad after you finish a long duration of suit run with only air coolant. If you run with water coolant you feel okay. Now I don't know why that is, but it sure is a fact. Because 5 hours of running in the suit with just air coolant you suffer the next day. Five hours of water coolant down at the Cape training you can go the next day and run another one. You don't feel like it very much but you can do it. I think it may be a difference that if you can keep your body from sweating you sure should do it, and running that water coolant for donning and doffing was ideal, plus it had the added advantage of getting those two big long cumbersome hoses out of the way, which were right in the traffic area. That's really a son-of-a-gun when it comes to doing things like changing comm out in the lunar module and we made that change in real time because it was actually like we practiced in the mockup down at the Cape. I sure think it paid off. I think it's a better way to do it. I just don't think long-duration air coolant in the pressure suit, which we were in a lot of the time, which makes your body sweat and you have to replenish the water by drinking it. You don't always have time to think about doing that; it may be a significant factor in getting the crew back hydrated, as opposed to being dehydrated. Okay; cabin depress, beautiful.

DUKE — In 2 minutes we could open the hatch just like they said.

MATTINGLY — Did you have a lot of stuff come out of the hatch equalization like we had?

DUKE — No, we saw a few dust particles fly out but that was all. To do the actual depress we used the overhead valve and just left it open. I never noticed much floating that way. The LM was extremely clean. You know how many screws and little washers and things we found floating in the command module. I guess maybe on the whole flight we found five in the LM the whole time.

YOUNG — It was extremely clean until after first EVA, and then from then on, it was really dirty.

MATTINGLY — Yes, I was thinking more about the subsequent depresses. Did you have a lot of rocks and crud flying through there?

YOUNG — No, actually it cleaned the floor off pretty good. When I opened the door, the dirt would go "zip" right out.

SLAYTON — Do you think one-sixth g is enough to keep that stuff from going out that top hatch.

YOUNG — You know all that Velcro on the floor, it just gets caked with dirt. You can't stand on the floor. I guess it didn't hurt anything, but I know when we donned the suit, we had our jettison bag down to stand on like everybody said, but our feet and hands and our arms were all full of dust when we put the suit on. So it was all going into the suit. And it didn't seem to bother anything. You don't know how much it's going to bother. You don't have a feel for whether it's going to give you a problem or not. There's just no way to avoid it. The second EVA, we had in places that much dirt and dust on the floor and that's after cleaning each other real good.

DUKE — The place where most of that dirt came from in the place you can't clean was the strap-on pockets we had.

YOUNG — We got smart after EVA-1, and before we got in, we closed the flap. But the first time, I got in with that flap open, and my pocket caught on a hatch sill and when I came in with that right leg, the dust just flopped out. You had a pocketful - you had a contingency sample right in your pocket. Once outside we were talking to each other. I guess Charlie and I really were the only two people who really had a good handle on how the prep was going because the ground didn't have high bit rate. We felt perfectly comfortable. When our coolant became adequate Charlie talked me out of the door. I guess I had a little more trouble with the line up than you did.

DUKE — It is because of the hatch. The hatch is only about three-quarters open. I can't back up any more. Once you get the hatch full open, you can get centered in the hatchway. I could do it but you couldn't.

YOUNG — I never had any concern about getting in or out. It was just a question of knowing what I was going to get caught on. Once I got outside I had our new LEC strap. The adjustment feature on that strap is at the top of it. We marked it preflight how we wanted it adjusted. I don't think it was adjusted that way because when I lowered the ETB to the ground, it landed on the ground, and we were trying to avoid this. We didn't want any dust on the ETB so we could keep the dust out of the cockpit. We had to adjust it on the later EVA. What I recommend is that they put the adjustment strap on the bottom so if you do land on a slope or if you land and you stroke a gear and you want to readjust the strap you can adjust it to your eye level on the ground, and not while you are hanging on the ladder with one hand.

MESA deploy was nominal and it appeared to me to shake the whole vehicle, but Charlie said he didn't notice anything. When I got down on the Moon the environment was just as good as I thought it was going to be. The second thing I did was reach down and pick up a rock just to see if I could do it and sure enough that was a piece of cake. So I really thought we were going to be in business with that suit mobility. I went around to the MESA and the first thing that I noticed on the MESA was that the height was too low, but I didn't do anything about it at the time. I loosened the TV blanket and opened it. Charlie came on outside about this time. We didn't have to move the TV to the tripod because we weren't getting TV. We deployed our antennas. It had been easy for us to deploy our antennas when we were on the surface, but once we were in one-sixth gravity the only way that I could reach up to deploy Charlie's antenna was to have him come over and grab hold of something like the ladder or the Rover and bend over so I could get my hand up to it. I just couldn't get a hand on it. That took a little more time because we had to move over to a new position. I don't think the communications would have been bad if we'd left the antennas stowed. I ended up breaking one because we forgot to restow it. I'm not so sure we shouldn't leave them stowed.

72-H-601

DUKE — My Egress went right by the checklist - with the breakers turned off the lights and open the hatch full. I came right on out. It was even better than I expected and easier to do. I felt right at home the minute I hit the ground. I just felt right at home. Then I went over to the MESA. My first job was to take out the drill and the core stems and I couldn't. The way the MESA was hanging you never would have gotten them out. It was almost like you were looking at it flat. Normally it sits up, but it was almost all the way down.

72-H-604

YOUNG — The MESA was supposed to be adjusted to the green line and it wasn't adjusted to that green line. It was about 18 inches lower than the green line so the MESA was lying right on the ground. Maybe this is a preflight problem. If it had been adjusted to the green line, which is where we adjusted it to, it would have been in perfect position.

DUKE — It was hanging down on about a 60-degree angle. It looked like the spec high case to me. That is where the vehicle is high and you have to drop the MESA down to reach things. It looks like you should pull the black strap to adjust it because it has a pulley arrangement, and you think that is the mechanical advantage. You pull, and pull, and pull and it is locked down. The strap you want to pull is the green one up above. This strap has no mechanical advantage at all except the gravity field. We finally figured that out after about 5 minutes. We wasted 5 minutes.

72-H-611

YOUNG — I recommend that they adjust it to the green line where it belongs and that late in the EVA training program we adjust the MESA. We had been checked out on how to raise and lower the MESA height, but I forgot how to do it.

DUKE — We did it a long, long time ago.

YOUNG — It was too far back from when we did it the first time.

DUKE — Maybe in training Stoner and his people could give us various cases, instead of the flat floor case, which we always get to keep you familiar with it or put a decal on it that says pull here for adjustment. That snowed us.

YOUNG — It not only snowed us, it snowed the ground, too. It took them a while to figure out how to do it and there's 5 minutes down the tubes. Inspection of the vehicle showed that none of the things had predeployed to which I may add a hearty "thank God." What had happened is that both walking hinges were open. The walking hinges were released just like on 15. We had to put those back. We didn't have any more trouble with it until we got the front wheels deployed. On Charlies's side the left rear wheel was knocked down and locked so Charlie gave that a pull, and pulled that down. When we got a little further down, the wheel came down and locked on the forward chassis so we locked both of those. When we got it on the ground two hinge pins were a little extended. I think they were partially inserted, but we had to insert one on Charlie's side and one on my side. I don't remember which two they were but we had to insert them.

DUKE — The wheels popped open just like they had done in training. There's a gold sleeve collar arrangement that has a couple of pins in it so that when the wheels are fully out those pins lock in to hold the wheels in place. That was what was not locked. All you have to do is push on the wheel to extend that mechanism and it locks right in place.

YOUNG — Other than that, the deployment was easy.

DUKE — Sorry we didn't have it on TV.

YOUNG — When we got it on the ground Charlie and I just picked that baby up and moved it over so we didn't have to back it up.

DUKE — One thing I'd like to comment on here, is that I think they overdesigned the Velcro holding the seat down on the LRV. A 2 inch or 1 and a half inch wide piece of Velcro is wrapped around on themselves to hold the seat in place. On my side I started pulling the Velcro to try to get it off of the
outboard handhold and the seat and all I succeeded in doing is picking the vehicle up off the ground. The Velcro wouldn't come loose. Then I tried a couple of snatch loads and by snatching it I eventually got the Velcro off. It was a hard job getting that Velcro off. They really over killed that one.

SLAYTON — I gather you thought all the Velcro was difficult to work or excessive.

DUKE — Yes, I did.

YOUNG — For what we were using it for it was. Although we had a couple of cases of these in-house devices that we will talk about later, where the Velcro was burned off. The glue that attaches the Velcro melted or something, like on the padded bags, and on the TV sunshade. And there was something else.

DUKE — It was just like you'd expect, Deke, when you wanted the Velcro to work, it wouldn't work and when you didn't want it to work you couldn't get the son-of-a-gun off of there. If you're going to do that PLSS harness thing, you need a bigger piece of Velcro and an easier way to attach those bags, because we lost two SCB because the Velcro came loose, which allowed the bottom to flop and then it fell off. One of them ended up on the Rover, wedged in between the rear wheel and the aft pallet and we recovered that one but the other one with the little sample SCSC fell off and we never did see that. I was surprised we didn't see that going back, John, on the return traverse, but I never did see that thing.

YOUNG — Oh, I know why you never see it, it didn't fall off. When we got back, the SCSC bag was lying on the footpad of the lunar module.

DUKE — Oh, it was?

YOUNG — Yes, it wasn't on the back of the Rover like we planned to put it. I don't think we'd have been able to use it. Okay, LMP inspect LM and pan.

DUKE — Well, I got out and looked at everything and it didn't look like the struts had stroked at

all. The engine bell was in great shape and had not impacted anything. Although there was about a 50-centimeter block just to the right of the engine barrel, and it extended above the engine barrel, the bell was still off the ground and had not hit anything. I did the pan as per checklist and also added some photos that Houston wanted us to do of the peeled paint, the shredded wheat, and on inspection we had that one panel on the minus Y side above the APS propellant that we've already commented about. There was another little spot above the below quad 4 that I took a picture of, and then we took the pictures of the steerable, but I never saw anything on that steerable that would give me any hint of what was wrong.
Apparently just the servo electronics was gone on it.

YOUNG — Your pictures came out good too, Charlie. It just looked like that thing was just welded down.

DUKE — I was really expecting to see a "remove before flight" pin up there but there wasn't. It was perfectly clean, but it just didn't work at all in yaw. I had to go back and take a picture of the cosmic ray and I forgot to look at the decals that time, but the next time we came around and looked at them they were all black. And then later on they decided to have John move the cosmic ray.

YOUNG — Okay, the far UV camera. I didn't have any problem with off loading it. I was expecting trouble from the bags. It came loose and I expected trouble picking the thing up and getting it off and I expected problems with keeping it out of the dust. I was able to get around the front of the vehicle and hold it over my head. It was easy to carry that weight around. The place that we had to mount it in the shadow was in a small subdued crater with about a 3 to 4 degree slope, and probably it was more than that. The only way I could get the camera level was to really step down on two of the legs, push them clean down out of sight in the dirt, with the other leg sitting right on the surface. And that was the first problem I had with it. The second problem, which was a continual one was the battery cable. Even though I didn't deploy the battery too far out in the sunshine the battery cable had a mind of its own and insisted on staying about 4 inches off the ground around the camera where I was walking, even though I pulled the whole thing back in there. So, every time I walked around the camera I had to pick my feet up to avoid the battery cable and 2 or 3 times I tripped over it, but fortunately it was the battery that moved and not the camera. To move the camera in azimuth completely destroyed any leveling. As the mission went on, to move the camera in azimuth got harder and harder and finally it got so hard that every time you moved it, it would pick the camera up off the ground and destroy the level. I had to relevel it after every setting and that took a lot of time. And in some cases, because we were really in a hurry I didn't get the level where I would like to have had it, as far as being perfectly level. Adjusting the level, using the wheels, just couldn't compensate for both the slope and picking; the thing up every time. From that standpoint, it was pretty bad. There was something wrong with the azimuth - it just got stickier and stickier. It didn't work at all like on a training model or either a qual model. I don't know what it could have been. But, I would devote a reasonable amount of time to leveling the camera after each setting. I think we have probably devoted three to four times as much time as we'd allotted for each setting. I got behind in doing that. Charlie recommended that we put the UV alignment in both checklists so that either guy could do it.

DUKE — I had to interfere with John's conversation with Houston to get that going. But it's a place to save time, because the LMP really doesn't have that much to do on the load up. Get the cameras configured, and the films stowed, is about all. So, it's good to have some kind of cross talk within the checklist so you don't have to interfere with the other guy. Apollo 17 probably won't have that problem with it since they don't have the UV, but sure enough something will be there, and so they really should have those TV and LRV power-up procedures in both checklists.

YOUNG — There's something else I don't really understand about the far UV camera setup and the alignment that we're doing in real time. Except for the geo-corona and the Earth, we changed every target in real time, I do not understand that. After doing it for months, we changed every target in real time. There were at least two targets that they called up that were pointing and we got the alignment with this tough azimuth change and they were pointed right at the lunar module and you know, when I turned the thing around I was taking a nice picture of Charlie looking out the window. In my opinion, that's an inexcusable waste of time on the Moon, doing that kind of thing, but we did it any way. I just can't believe we're doing that. There was one setting, when we turned that son-of-a-gun around and it wasn't even clearing the ladder good. I don't understand that. I'll never understand that, but that's what we did. Golly! Changed every setting after we practiced these things for months and we specifically reviewed them and, in many cases, when we started all this business back about a year ago. About one out of four would be pointing at the lunar module and we'd say go back and research this as we don't want to be taking pictures of the lunar modules, surely. So, they had them all changed, and I figured that before we launched, we had them all down in real time and by golly if we didn't change every target. I can't imagine that 6 hours could do that to you. It would on most targets but I find it hard to believe. Plus we had to move the camera after each EVA, and I don't understand that either. It was in the sunlight after the second sleep period. The bottom half of that square box the spectroscope looks out of was in the sunlight and after the second EVA, the upper 3 inches of the cassette handle was in the sunlight. Of course, that could clobber your film because if the heat goes down that cotton-picking barrel like it probably does, and heats up that film, that wipes you right out. I don't understand that either. That would be a tragedy to lose that for something like that. But we may have.

DUKE — Everything went well on the LRV load-up until I tried to get the power connector on from the LRV to the LCRU. The flight gear is really stiff and the cable had a set in it and it was just tough for me to get that Astromate connector on. It took a lot longer than it had in training, but it finally locked in and that was really the only problem that I had on the front end of the load-up. The antennas went on slick, the TV, the GCTA, everything checked out correctly.

YOUNG — The cables were what was stiff.

DUKE — You'd get the connector aligned and to see that you were aligned, you had to put it in place and get over to the side, and when I'd do that, the cable would spring out again.

YOUNG — On the back end, the old pallet just fell off the quad 3 and fell onto the LRV and went right in place and locked right in place and I don't understand it. Just a piece of cake. And the only problem I had was that on one of the penetrometer pins, the wire pulled loose so that instead of having a loop, I just had a long piece of wire. I was able to wrap the wire around my hand and pull the pin. The pin was pretty hard to pull and it was being pulled at an angle since the loop had broken. But, nevertheless, it came off. That's the only problem I had loading the LRV. Unlike the training equipment, both the shovel and the rake were easy to lock on because it tells you which way to turn the thing to lock it and how to put it on. We didn't transfer pallet number I at the beginning of EVA, we did it at the end.

DUKE — Yeah, at the end, we did the Bio-canister. It was easy; man, that beauty came right out of there at one-sixth gravity.

YOUNG — In one-sixth gravity, it comes right out and in one g, it's a son-of-a-gun. I can honestly say I had as much trouble putting the flag together in one-sixth gravity as I did in one gravity. My main concern was with the TV sitting there watching us, that we'd end up with the flag in the dirt and us standing on it. As soon as I picked up the lower leg of the flag, I dropped it and I was in the dirt. So, I was bent over holding the flag up with one hand and I picked the thing up and put it together. Okay, the ALSEP off load, Charlie!

DUKE — Yes. It was right below eye level, both pallets.

YOUNG — Is that where it was in one gravity?

DUKE — It looked exactly like it was the same level as in the training building. I flipped the switches for the descent ECA temp. monitoring, and the pallets just came right out with no trouble. About this time I think you were doing something with the far UV. Anyway, we got them all on the ground and put the pole together and by that time you had come back. We got the pole together and RTG tools out and dropped that down. I didn't quite get it locked on before I started changing it. I thought I'd lost the dome, but I started over again and made sure that the tool was locked in and then the dome came right off with no trouble. Offloading of the fuel was just like we'd done it in the trainer. It wasn't red like it's painted in the training model, it was black.

YOUNG — Was it black, Charlie? I was wondering what color it was.

DUKE — I couldn't feel any heat coming off of it. One problem I had on carrying the antenna mast, it apparently looked like it was locked on the RTG side and I could not pull it up any more and it looked locked to me though there was about a quarter or a sixteenth of an inch between the little half dome and the mast. But it still looked locked, but as I was bouncing out there I had gotten to about 25 meters and the RTG package fell off. It bounced into a crater. I thought I'd blown it then because of those very fragile fins on the RTG. But, I looked at it and it hadn't been damaged at all. In fact, it was hardly dusty. I put it back on and made sure I had it locked the next time. And the only thing I can say about that to make sure that thing is locked. You don't want any gap between the post and that little half dome. If you do, the thing is not locked in. You cannot feel it snap in like the training gear snaps in. Some of the hardest work I did was carrying that beauty out to where we finally deployed it. I highly recommend that you put that thing off to the left side of the LM if you can, if your experiments will allow you to. Because on lift-off, that MESA blanket we had went sailing right straight out front just like it did on 15 and impacted about a 100 meters out in front of the LM. It could have been another wipeout on the central station, like it almost was on 15, so that's probably a good idea to put it off to the left.

YOUNG — Yes, and I don't know why that MESA blanket did that. Maybe we should take the MESA blanket off and stow it inside. I don't know which MESA blanket it was.

DUKE — I don't either. So, I think you're wasting your time pulling those blankets off. If you just put the ALSEP off to the left you don't have that problem.

YOUNG — There's no way it can get there.

DUKE — When I got there I separated the packages and lined them up.

YOUNG — There sure were a lot of rocks out at that site, and it wasn't the world's greatest ALSEP site. After practicing for months on a flat terrain, with no craters, we ended up with a lot of craters and a lot of rocks and a lot of holes. You sort of had to thread your way as to where you were going to put each piece. I'm not too sure it was a good idea to have a rock between the central station and the PSE right out there at the end of the ISM. I'm not sure how that affects the data.

DUKE — They said it was all right, John.

YOUNG — They said they were having thermal problems with the PSE.

DUKE — Already! That's because there's dust on them.

YOUNG — Yes. And, because I kept falling over the rocks as I walked by it.

DUKE — That's the only flat spot I could find.

YOUNG — I know, Charlie. You did a good ,job. I don't see how you could have done any better.

DUKE — I wasn't about to carry the ALSEP all the way to Spook Crater.

YOUNG — I'm glad you didn't carry it any farther. If you had, when we did the ASE deployment, we'd have been down in a big hole. We just barely missed being down in a big hole as it was. Let me say something about the cosmic ray experiment. When I deployed it, I forgot to look at the temperature decals on them. I'm sure as a result of our three revs, plus our time enroute, all those temperature decals were black before we ever got there. I can't imagine that anybody would think they could put something on the side of the lunar module and expect not to see more than 140 degrees F. It saw 140 degrees F during translunar coast because the temperature was plus or minus 250 degrees. They could stand about 120 or 130 before the experiment is ruined. I never could understand why they were worried about thermal. When I pulled the ring, I pulled it down 3 inches and it pulled completely loose. And it was just the upper 3/32 of the hole that was visible. It was supposed to be completely gone. We got all but the 3/32s before the ring came off. On the next EVA, I moved the experiment down to the plus Y strut to keep it out of the sunshine. I still think that long before we ever got to the Moon, that thermal problems were a factor and, that experiment was gone. I don't know how you could have avoided that unless you put a lot of insulation on the detector or put it in the MESA. Let's say something about the LRV checkout. When we first turned on the LRV we didn't have any rear steering. I don't know why because all the switches were nominal. I didn't move any switches. We were on secondary power; 15 plus or minus volts was secondary. We tried it on both of them. After load-up, when we got back in the second time, it all worked. I don't know why it didn't work the first time. I wasn't particularly concerned about the rear steering not working because we sort of planned to not use rear steering.

YOUNG — In fact, if the rear and front steering had gone out we planned to go in a straight line as far as we could, get out, pick the thing up and point it in the right direction and keep right on going. On that particular site that would have been hairy. We'd have been doing a lot of getting out and picking up. But I still think if we had to do it for any reasonable amount of time it would have been a lot better than walking.

DUKE — The experiment came right off and the connector went off. This was something it hadn't been doing in the early part of training. It went right on. I unrolled it and took the probes out. We had one Boyd bolt stick. I had to use the tool twice on it, but, it finally came loose and broke the box apart. I thought I had a pretty good place for the heat flow. After I had deployed the probes, I went back and got the drill. I assembled the drill and the stems went on and the drill worked like a champ. I started out real fast and I think I ran into a little rock. It went right on through that rock, or whatever it was, and I had no trouble. I think I was way ahead of the time allotted to drill the hole. I had planned to use my foot to hold the stem while I took the drill power head off. That didn't work at all and I'm glad I had that new wrench. The wrench was just ideal. I could easily put it on and hold it with my leg while I twisted the drill head off. I got all three sections in real easy and I put the probe in with the rammer. Everything worked as advertised. About that time John said there was a cable loose over here. I looked around and the dadgum thing had broken off. So, I stopped on the heat flow at that point and went to the deep core.

YOUNG — My feeling is that kind of thing is almost unavoidable. If the cables are way up off the ground you never knew whether you were stepping on them or not. When you are standing in one-sixth gravity with a backpack on, you're looking about 3 to 4 inches in front of your toes, unless you make a positive effort to look over at them. Every one of those cables had a memory and were all at some distance off the surface. If you want to make that whole business compatible with the suit operation when you run into the cable, it will be strong enough so that it does something like pull the central station a little so that you know you're moving something. Maybe it should be such that it can stand a tangle and trip. That cable evidently was really flimsy. Some cables allowed you to do that. I was pulling the active seismic experiment around and that cable was on there so taut that I actually moved the central station and had to go back and adjust it. I didn't pull the PSE cable, but I had the feeling that if I had, it would have moved the central station. The RSM cable was very strong. But, that cable wasn't. I didn't know I'd done it. I had no idea. And I certainly didn't mean to.

SLAYTON — We should have helped you from the ground some on that. You can go back and look at the color TV and we could see it. The Cap Comm was looking at the black and white set and isn't very obvious on black and white. Besides the time delay, we had the wrong flight plate up.

YOUNG — It's an 8-second time delay for the whole thing to get through. You can probably do a lot of damage in 8 seconds.

SLAYTON — That's right. It probably would have been too late anyway. But, it might not have been.

YOUNG — It was sure a tragedy. If it had just moved the central station before it broke. I would have stopped right there and fixed it.

DUKE — Everyone of those cables had a memory. Every one of them were off the ground.

YOUNG — A guy really can't lift his feet too high around a central station. because when he does, he kicks dirt all over the PSE. It was a bad thing, but I still think it was incompatible with the kind of limitations that we are working with in the pressure suit. I blew it. I tripped over the whole thing, but I didn't, even know that I had done it. I was completely out as far as the active seismic experiment when I looked around and saw this cable following me.

DUKE — It wasn't your fault.

YOUNG — It was my fault. I didn't know I did it. The PST deployed normally. I had some misgivings about where we put it, but, we leveled it and deployed the thermal skirt. And, we padded it down so that it would be level. The last picture I saw of it shows that it is up off the ground a little bit and I don't know what to tell them about that because we sure made an effort to make sure it was flat. We did go back and check it three or four times during the mission, but it is up off the ground on one side. Maybe that will give them a thermal problem. With that rock between the PSE and the central station walking on one side or the other of that rock would tend to get a little dirt on the skirt. But, it was level and they should be getting good data. The LSM deployment was nominal in every respect, except for the problem that Charlie had deploying the curtain.

DUKE — The Sun shield flight curtain.

YOUNG — It was easy to align it. The ASE deployment was nominal. The way we did that was we drove the Rover out to the central station and then drove out on a heading of 290 degrees in a straight line for 100 meters. When I went back to reset the UV taking the Earth pictures of the geocorona I fouled up that procedure and had to do it all over again. Even if we had done it right, Charlie would have been out at the site and what we intended to do was take the Rover and run a little recon to pick a good place, if there was such a thing. I have the feeling that no matter what place we picked, it would have impacted one or the other experiments because of the blocks and because of all the craters out there. I don't think in a reasonable time that we could have picked a better site than we did. It's 20/10 hindsight for a man sitting on the ground with photographs to say, well, you should have put it over there. That's no good. We drove the Rover on a heading of 290 degrees, turned around, and came back. Then when we deployed the ASE. I'd say it was within 1 foot in 100 meters of being straight down that line. It was as the photographs show. The ASE went just like it was supposed to. The mortar package was no trouble to deploy or install. One of the legs on the mortar package hung up. I couldn't extend it. I can show whoever wants to know how that pin was hung up. If I had a pair of tweezers, I could have gotten in there and pulled the pin. But, with a pressure-suit glove, I couldn't get my fingers in there to do it. So, we deployed it in the ground with three legs extended. It was leveled with the ground on a heading of 333 degrees. Both it and the mortar package were pretty much level. I think it will work satisfactorily. Arming of the mortar package at the end of EVA-1 was no problem. That was a one-time item. I have some misgivings about being able to deploy the mortar package with any amount of slope. Fortunately, the place where the mortar package was sitting was almost level. And, it was the only place around there that was. If it had of been tough regolith like it was in a couple places, we would have never gotten it in the ground.

SLAYTON — Core sampling, Charlie.

DUKE — The first core tube I put on the drill went right on and locked in place like it was supposed to. I was able to push it in maybe half of a stem length. I started drilling and it seemed to auger in on me. It went in much too fast. I held back on the next two. The only problem I really had was when I tried to get the drill head off to add another stem. It seemed to be real tight and didn't want to unscrew. The drill didn't want to unscrew from the stems and that happened on all three sections. I really don't know why. I checked and it didn't look like I was gauling any of the threads. The stem threads on the stem side all looked clean and weren't gauled. When I put them stem to stem, they went together real easy. But, when I tried to get the drill head off, it was hard to get off. But, once I broke it loose, it unwound easily. I had to really make a conscious effort to make sure that the drill stem did not unscrew in the ground. It really wanted to back off. When I tried to put the wrench on it and unscrew the drill, the whole thing would turn and I really had to make an effort to stop that. It went into the ground great. Once I had it in, I did 15 seconds of clearing the flutes. While I was doing that, I tried to pull up and the thing just came right on out of the ground. I pulled up 4 or 5 inches. It was coming out easy. I said, man, it is going to be a piece of cake to get this out of the ground. I took the drill off and capped the top. I stopped the flute-clearing activity and then I tried to pull it out of the ground. Boy, I couldn't budge it. I took the drill head off, capped it, and used the jack on it. The jack worked as advertised, and it brought that beauty right out of the ground. I was amazed. I had to jack it out almost 7 feet before I could pull it out of the surface with my hands.

YOUNG — You said the bottom layer was white.

DUKE — I don't remember the details right now. Since we lost the heat flow, they said they wanted to see if the holes had collapsed. I took the rammer and dropped it in the hole. It fell out of sight. Only the top 3 inches of the rammer were visible. It was perfectly open all the way down. Breaking it apart over in the Rover was nothing. It worked just as advertised. We propped it up and retrieved it after we had ended the EVA. I'm glad we had that extractor because the extractor works great. I had the same problem with it that I had in training. Every time you picked it up to try to set the C-clamp back down again, the bottom plate would shift on you. It wanted to walk clockwise with you. What I did was put my right foot on the plate and Jack with my left hand. That worked great. It held steady then and it speeded up the process.

YOUNG — I tell you we weren't disappointed about this EVA. We'd been practicing with real deployments on the training gear. Every time we deployed it, we'd have some kind of problem that nobody had ever seen before. Well, we had the same thing in flight. I think all the problems we had with the training gear oriented us for our real-time problems.

DUKE — We had ALSEP deployments and that was really good training. I felt right at home with every piece of gear we had. We loaded up for the geology and I changed out the magazine on my camera. We got the bags ready to go and off we started.

YOUNG — Somewhere in there, the ring came off the 20-bag dispenser.

DUKE — That was right at the beginning. Several small screws that hold the aluminum plate and the ring that holds the bags to the camera backed out and the whole ring fell off. This allowed the bags to still be held to the camera but to dangle such that you just couldn't reach up and pull one off. We discarded that set of bags. Luckily we didn't throw them away. We put them on the Rover and used them later on, because we almost ran out of bags. I recommend we have several sets of those 20-bag dispensers, because if you have any failure like that, it really can slow you up. The best thing to do is just discard that set of bags and get a new set. But, the way we were using them, we couldn't afford to do that.

YOUNG — I dropped the bags off the camera out of that place. During the first EVA, we had them taped on, remember? On the second EVA we forgot to bring the bags back in and taped them on the same for the third EVA.

DUKE — I had them taped on both cameras, but the tape came up. The gray tape doesn't hold too good on that metal surface.

YOUNG — You need some mechanical latch to hold them on. Probably something that springs up and down. You need some way to keep them from falling off. Mine fell off on the second and third EVAs at least 10 times, and that really slows you down. Twice they fell off when I was driving the Rover. They just vibrated out. Fortunately, it fell on the seat or we would have lost a couple of 20-bag dispensers right there. I didn't realize that I dropped them until after they were gone. It seems to me like that's a simple fix. The trouble is, they fell off in training, Deke. I kept saying, "Is this flight?" Everyone said no, the flight one is really stiff and it won't do that.

DUKE — I think it happened more in flight than it did in training.

YOUNG — The system is supposed to handle that and it didn't.

DUKE — We had no trouble. After station 1, we saw Spook. We went by Spook and it was impressive. We went by Buster and Buster was a heck of a lot bigger than I had imagined; 50 meters. It was really a blocky crater. We kept going and we felt as if we'd landed a little bit west of where we were supposed to end that the distances would be shorter.

YOUNG — They didn't turn out to be.

DUKE — We got to a crater that...

YOUNG — It was really a big crater.

DUKE — ...turned out to be Halfway Crater. We weren't to Flag; then, we got back on the Rover again and started to Flag. The reason we realized that we weren't there was because the distance was only about three quarters.

YOUNG — Halfway Crater looked as if it was 100 meters across.

DUKE — The craters all looked bigger.

YOUNG — Flag Crater is 300 meters across. What we're saying is we couldn't tell the difference between a 100- and a 300-meter crater. And, that's the truth.

DUKE — I think it's the sharpness of the land and the degree of the subduing of the craters.

YOUNG — You know, I would have been willing to buy that for being Spook Crater. But it wasn't.

DUKE — When we got to Flag, there was no question. Man, this is a big crater. It was a lot bigger than the one where we stopped first. We found the right place and found Plum. Plum was smaller than Buster. Plum and Buster were supposed to be the same size. Buster was just gigantic compared to Plum.

YOUNG — I expected Plum to have a bright rim around it like it has in the folders. It didn't have a bright rim around it, but when we dug down 2 inches, there was all this white material.

DUKE — John took the scoop and pushed it down there and dug away on the rim of this little crater. There was about 3 centimeters of gray regolith and right under that it was just whitish white ash. It was white when we put it in the bag. I don't know what it looks like right now but, it was white when we put it in the bag; just ash white.

MATTINGLY — Some of the other things that looked white when we got them inside were pretty black.

YOUNG — If that doesn't look white, I'll eat every bit.

SPEAKER — It was sure marshmellowy when we got it.

DUKE — That was one of the unique things in the navigation system on the Rover. It's superb, absolutely superb. The map holder is worthless. You can get in there with a 16-millimeter camera, your Hasselblad, and your knee. By the time you get in, you're pushing the map holder out of the way. It's sitting there and you can't even see the maps. What I ended up doing was taking one with the headings and the topography on it - the 1/25 thousand - and I wedged it in between the 16-millimeter camera and the staff. That's a great place for the map because you can look up at it and see it. You can reach up and pull it out if you have to and just push it right back and wedge it in place. Unfortunately, our maps and photographs didn't look anything like the topography.

YOUNG — ... no resemblance to where we were.

DUKE — The Rover nav was working so good, and we could see our landmarks. So, we had no trouble navigating. We really didn't need the maps; but, where they were stowed, they were useless. If you need a map, then you better pick one that you think you will use and stick it up there on that camera and just wedge it in there. That worked great.

YOUNG — We couldn't handle the maps.

SPEAKER — Say something about Rover driving out there in zero phase.

YOUNG — Man, I'll tell you that is really grim. I was scared to go more than 4 or 5 kilometers an hour. Going out there looking dead ahead, I couldn't see Craters. I could see the blocks all right and avoid them. But, I couldn't see craters. I couldn't see benches. I was scared to go more than 4 or 5 clicks. Maybe some times I got up to 6 or 7, but I ran through a couple of craters because I just flat missed them until I was on top of them. And, I don't recommend driving in zero phase. They keep saying they want it included in the traverse and I specifically cautioned them not to include it on the traverse. But, there is no way for us to get to Flag Crater without driving in zero phase. It sure is grim. The other direction was about twice as good. I saw my tracks on the way back. We were doing 7, 8, 9, and 10 clicks. It wasn't any good during the traverses where we were going down-Sun. I was tacking a lot of times. But, when you got to a ridge, you couldn't tell if it was a dropoff, or whether it was a smooth, shallow ridge. In a couple of cases, you couldn't see there was a ridge. I didn't care for that much. It's kind of like landing an airplane aboard ship where you're looking right into the Sun and you can't see what you're doing. You just go ahead and land it anyway. It is not normal, but on occasion, you have to do it; but, you'd just as soon not.

DUKE — We found mostly breccias out there. Sometimes they looked like tough breccias to me. General rock type and the blocks, as John said, were numerous. We had two types of regolith, a white underneath and gray. We also found some crystalline rocks out there. We did a rake and it's a great big plus for that rake. That really is a good sampling tool.

YOUNG — There's no doubt that those rocks probably came down first. Then later the dust fell and covered the whole thing, because they were sure covered up.

DUKE — Okay, travel to station 2. Going back, we just followed our tracks. The slopes were small in that traverse. The local slopes approached 5 to 10 degrees, I guess. We were going up and around a couple of the craters.

YOUNG — You know, we never encountered any of these features on the geology map. They were mapped as scarps or steep features that they said we're going to have to drive around or over or maybe we could pick up outcrop. We just never ran into those. And I think that's because those guys were reaching for and pulling out features that weren't there. I mean, I looked for these things and sure enough if you really imagined it, you could see something there. But, I think with that scarfing we had, they were reaching for it. Because they sure weren't there in the real world. If they were, they looked like every other slope that was around there.

DUKE — I think the whole way out there, we were in this series of rays from South Ray.

YOUNG — You couldn't tell where one took up and the other left off.

DUKE — The rock types you could classify as just angular to sub angular and with very little filleting around them. There were some rocks from North Ray that had fillets developed and I think that might be a way to tell one from another.

YOUNG — The rock from North Ray was a little more rounded and not as sharp. We had rocks from North Ray and South Ray in the landing site. ... only by the angularity and the old rocks have more zapped craters and are rounded off.

MATTINGLY — You are saying that those are North Ray rocks which have fillets? That's a hypothesis.

YOUNG — Yes. That's a guess. But, I would say that is the best way to tell them apart.

DUKE — I bet you five bucks that's right.

MATTINGLY — I thought maybe you had traced them on the way up there.

YOUNG — Of course, some of them could have been from Buster, too.

DUKE — When we got to Spook, John had that LPM to do. I took the pan and went scooting on up to Buster and took a partial pan of the interior. That was really an impressive crater. We had some 3- to 5-meter blocks that covered 70 or 80 percent of the bottom of that crater. And they trended up to northeast slope and out the southwest slope of the crater. They were large meter-size blocks on the northeast rim. Around on our side, the southeast side, we didn't have anything greater than half a meter. I sampled some rocks there and did a radial sample of about half a crater in diameter and then about a crater in diameter back toward where the Rover was. We got an number of samples and hopefully they're the right ones. I'm not sure Buster was a secondary crater. If it was a secondary crater, it was really a big breccia in there, because it was a big crater. It was a primary crater and it had excavated into bedrock down there.

YOUNG — I think Charlie got some rocks from Buster Crater that came from the bottom.

DUKE — The rocks were very similar to the ones that we saw at North Ray. We sampled rocks out of South Ray Crater up on Stone Mountain, but the rocks down there looked very similar to the rocks up at North Ray and the rocks at Buster. So, there could be some underlying formation that goes across the whole Cayley Plain. That might prove to be totally wrong, but at least from a color inference, those black and whitish rocks were everywhere; they were everywhere, I think.

QUERY — Say something about cooling when you were driving on the Rover.

YOUNG — It is best to operate on minimum coolant, which is practically no flow at all when you're sitting on there. The rest of the time I was running between minimum and intermediate coolant. That was sure adequate for any of the work that I did the whole time, except there in the last EVA. I was on the Rover and I forgot to reset it, and minimum and intermediate seemed to be pretty good there, because I think we're getting a high Sun angle there toward the end.

DUKE — One thing about the Rover, which will come up more on the next EVA, but on these local slopes that got steep, you really had no sensation of climbing a steep slope. But, you really knew it when you were going downslope, a steep one. The thing could have gradually increased to a 25-degree slope and I don't think we would have realized it.

YOUNG — The pitch meter broke off almost before we started up Stone Mountain. The pitch meter face fell off. The only way you'd know it is that you'd only be making 8 kilometers an hour, you'd have the thing fire-walled. On a level, you could do 11 max; and, downhill, if you took the power off, you could do as much as you wanted to if you let the thing go.

DUKE — Going upslope, the pitch meter needle was working and it was pegged at the top.

MATTINGLY — And what was the scale?

DUKE — Twenty degrees. At one point, I looked over there and the thing was pegged at the top.

YOUNG — Charlie said it was pegged and I said, "Oh, Charlie." I didn't believe him. I didn't feel as if I had a sensation that it was being pegged. I tell you one thing, we wouldn't have gotten out and worked on a 20-degree slope. You just can't handle it. Although, I think maybe we did when we were standing in that crater. We might have been on a 20-degree slope.

DUKE — I think so, too. Rock type change - Buster - as I said, we sampled whatever the major rock type there which was in the crater and on the rim, plus soils and all.

YOUNG — Let me say something about the LPM. The thing that surprised me was that there were no problems reading it with the Sun. We also anticipated that it would be hard to wind up; but, it was easy to wind up. The problem was to unstow it. After each deployment, it got harder and harder to pull loose and I thought on the last one that the cable might bust before we got it completely unwound. Just shows, you never know what your problem is going to be up there. It just got harder and harder to pull free. But, it was easy to wind up and it's easy to set up and easy to operate and we'd probably have made a few more readings if anybody had been interested.

DUKE — That was all real time, too. We did some readings that weren't in the plan. And we dropped out some stops, which was no big deal. Let me talk about the 500-millimeter targets, John, just a little bit. I took those of Stone and the only thing that was significant about Stone was these lineations that I described and they trended southwest. It looked as if they started at the east and gradually climbed upslope and over the ridges and you could just follow them all around. They did not follow any contour lines or any of the bench lines. They seemed to transect all of that stuff. They were very closely spaced during the first EVA when I took the 500. But, later on, we looked back over there on each EVA and as the Sun changed, the spacing between these lineations changed.

YOUNG — Not only that, when we got up on Stone Mountain I didn't see any lineations.

DUKE — You couldn't see it.

YOUNG — I don't know what that means, but they sure weren't there.

DUKE — At Flag Crater in the undisturbed regolith, you could see lineations that were mostly northwest, southeast. It looked as if the regolith was loosely compacted and the particles were standing up and the lineations were formed by the Sun shining on these particles; casting little slight shadows. I think that's really what those little lineations are. I took some pictures of that. I hope it'll show up. That was the feeling I got. If you just kicked dust over it, it'd mask it and be gone. I think maybe those things up on the mountains were the same kind of feature but on a more gross scale.

YOUNG — On Hadley Rille, it might have been.

DUKE — But the 500 worked great. It was a little bit more difficult to stabilize it than I thought. I used the range sight and it fired off the pictures. We could see into Stubby and we could see into the southeast wall of Stubby. You could not see the apparent flow that is on the photograph. It was not apparent to me looking at it, I thought. We got some pictures of it. That's all I can say. I jumped off at station 3 with the camera and John started off. I got into position and I squeezed the trigger on the handle, which was easier for me to do than punch the button. I could tell the camera was running by the vibration and just watched John do his thing on the Grand Prix and I think we did 2-1/2 minutes worth of film and then called it quits.

YOUNG — I didn't get up to any great speed, maybe 10 clicks at the most, but the terrain around there was too rough and too rocky for that kind of foolishness.

DUKE — It was to do that.

YOUNG — I was driving around craters and a couple of times I did a brake out there on the turn to show them how it looked. Driving the Rover, when it brakes out, is no problem. All you have to do is cut back like you do in snow when the backend brakes out. The trouble is when you cut back, you overshoot and you may end up going the other way. But, at least you're not going, you stop. You're relatively slow when you brake back the other way. We never did - on a couple of brake outs, when we were in a hurry and we may have had only three during the whole time, but the thing changed direction as much as 90 degrees on what was still brake out like that, and then cut back into it. We ended up going back this way 100 degrees. But the thing would be stopped. I never did have the feeling that we're going to turn over. Although, one time we had a couple of wheels off the ground and went sideways. I wasn't too impressed with that.

DUKE — One time I thought I was going to be under the thing. That was on EVA-3 and we'll talk about that later. But picking up core tubes, I'd left them standing on the little tripod. I picked them up and ran back to the LM. I didn't bother getting back on the LEV, just put them in the bag. At first they were a little bit too long for the bag and I couldn't get the Velcro down on the bag. But, once we got them inside and tamped it in, the snaps snapped and the Velcro Velcroed. We brought those beauties back as advertised. The solar wind deployed just like in training, perfectly nominal;

shoved it into the ground. The cameras worked nominally, even though we got them real dusty and it was hard to see the setting after the EVAs. We wiped them down with a wet cloth inside and changed the film outside. When we changed the film they were extremely dusty and yet the camera never quit.

YOUNG — Not only that, I guess according to the photo guys, we got some dust inside on the reseau. The camera still worked although it left a couple or three streaks across the film. It ruins the PR value of the things, but it sure doesn't hurt the data. But the thing worked. I thought we might have had a hangup.

DUKE — I did, too.

YOUNG — But, it didn't.

DUKE — Once or twice I watched the film and when I'd squeeze the trigger it'd sit there and it'd hum and then it'd go.

YOUNG — There was only one time I had to change your camera, to change the red into the white. The rest of the time those things were working and it was great.

DUKE — We had good luck with the cameras, just great.

YOUNG — It sure worked a lot better than that training gear. If it worked like the training gear, we wouldn't have taken very many pictures.

DUKE — I went on AUX water during EVA-1 some time during the Grand Prix. I didn't feel the cooling change at all. During the ingress, I took the pallet up and I just jumped up on the ladder. I didn't feel stable enough to jump up on the ladder with the pallet in my hand. I jumped up to the first rung and John handed it up to me and then I felt stable enough to go on up with that big pallet. You could stick it inside and put it over against John's left-hand stowage area and I still had plenty of room to get in and take all the gear off of it and hand the pallet back out. John looked real stable to me coming up the ladder with the bags in his hand, the rock bags, and the SCBs, and the SRCs. We had no trouble with ingress.

YOUNG — The technique I used, I'd stand at the bottom of the ladder and bend down and spring and I could get up to the second rung of the ladder with either the SEC or an SCB in my hand. That is really the way to fly. You feel like Superman jumping up off the ground like that. The way I would do it is, I'd put a bag on the LEC down at the end of the ladder, take the bag in my hand and leap up the ladder, hand one of the bags to Charlie, and then pull up the LEC and hand the other bag to him. That saved me a trip back down the ladder. It was real easy to pull up the LEC. You just lean against the ladder, do like this, and the thing would come up and then you'd grab it down a little further and you could pull it up in 3- or 4-foot grabs, even though the actual weight of something like the ETB was probably a good 30 pounds with all the cameras and everything in it. It was real easy and I'm glad we went to that LEC because once we got adjusted, we never got any dirt in the cockpit from it. Although we got so much other dirt, I don't know if it made a lot of difference. Let's make sure we got all these items. Yes. I had some difficulty attaching the power cable from the central station, but, after finally fiddling with it and pushing and tugging, it went on. Okay, I guess the recommendation on the LEC is to put the adjustment strap down at the bottom of it instead of at the top. Charlie said we need more sample bags, that's true.

DUKE — Our Rover seat belts were great. The adjustments that we had made inflight in the zero-g airplane turned out to be just exactly right. So, I recommend that the Apollo 17 guys have one-half hour of parabolas and get that seat belt adjustment and have them mark it down. That's what we did and we had a little mark on them and it just worked perfect.

YOUNG — The TV operations in both checklists - The penetrometer cone fell off. Antenna alignment on the first EVA was no problem on that 180 headings. It was easy to position the Earth in there, but I'm sure glad we had that training that we had with the IESD over there in the stack looking at that thing to get a feel for the problem, because it really was a problem at high Sun to get the Earth in a picture because it's so dim. We'll talk about that later, too. My yo-yo broke.

DUKE — Was that on the second EVA or the first EVA?

YOUNG — Second one, because I was getting tired and thought I'd use the tongs and put them on the yo-yo. I pulled it out and picked up a rock with it and put it back and it came back about that far. We brought it back, so it should be around here somewhere.

DUKE — It's at the house. I forgot to bring all that stuff.

YOUNG — We should give it to them and see what's the matter with it. But I don't think with the soft regolith that you need a yo-yo. You just carry the tongs out there in one hand and stick them in the ground. It's more convenient to do that than it is to pull the yo-yo out. So I don't think we're in trouble there.

SLAYTON — What broke?

YOUNG — It just came back to about here and then quit.

SLAYTON — Oh, I see. The recoil mechanism.

YOUNG — When we were opening the battery covers, of course, we had to dust the LCRU (they got dusty all the time), the LCRU did badly when they opened the battery covers. We had to park the LRV such that it was rolled into the Sun a little. That may have given them a thermal problem they didn't know about. We parked it at the right heading, but when we opened the battery covers the dirt just flew up in the air and came right back down on the batteries. We had to dust the batteries. I could see why they got dusty. There's hardly any way to avoid it. That was on the first EVA before we lost any fenders. We were relatively free of dust in the front of the vehicle other than what we accumulated on the front of it as we drove.

DUKE — When I got into the Rover, I mounted the Rover, reached over and grabbed hold of the handle below the camera and jumped in. When I did that my arm would go out and knock the seat belt out of it's little loop and most of the time it ended up over on the console, which was no problem. But, one time it ended up down under my leg and I couldn't reach it. So, John had to get it. I don't know how we could tighten that loop up or something, but it was a recurring problem with me, four or five times, anyway. John never had that problem with his on his side. I think his loop was a little bit smaller than mine. And a little bit higher too, and that was probably the reason.

YOUNG — On three or four different occasions on the first EVA, we pulled the purge valve off my seat; either fastening it or unfastening it from the seat belt pulled the purge valve pin out. I didn't know how it was happening. Charlie never saw how it was happening. We never experienced it before. What we did on the second EVA was to turn the purge valve around so that the plug came out this side. I found during the suit integrity check, when we pressurized the suit prior to egress, I could pull it that way. So, I said "Let's do that and get it out of the way" and that's probably the best way to take care of that problem in real time. The ground was going to suggest that we tie the purge valve with a piece of string and that would have just been something else to pull loose and get in the way. We really had too much stuff to do. So, I think that was an adequate fix, a real time fix, for that problem. The other thing they said about the purge valve which I really didn't understand, they wanted to bring mine back for analysis. We don't know whose purge valve is whose, so I don't think that was a rational request to bring one back opposed to another one. But they got them both back, I think.

DUKE — Yes.

YOUNG — It was just catching on something. I don't know what.

DUKE — I think it was your seat belt, John. It happened when you dismounted.

YOUNG — Yes. It must have happened when we dismounted, because we found it lying beside the Rover.

MATTINGLY — That's surprising, those things are hard to pull out, unless you just get them lined up right.

DUKE — He did it three times.

YOUNG — I did it three times on EVA-1 and had everybody nervous. I didn't know what was going on. I thought we followed the time line on EVA-1 pretty good. We didn't allow for the extra time it took to do the UV camera setup and alignment. We just didn't have that in there. That put me behind Charlie when he was going out to the central station and also for things like ingress. It was just before ingress when we reset the camera that the son of a gun really was hard to turn in azimuth and I didn't think I was going to be able to turn it. Cabin repress was nominal. PLSS refurbishment was nominal. We were able to stow my PLSS in the stowage station with the harness on it by moving the harness up and stowing it. I'd say this about our EVA-1 predonning: The first thing that happened to us that really slowed us down on EVA-1 was that Charlie's tool harness fell off, and we both had PLSSs on. It was really tough to get the tool harness up off the floor or off the side and underneath Charlie and put back on him with both PLSSs on. We should be able to belt that tool harness on so it wouldn't fall off until such time as somebody makes a positive motion to get it off. That's really a drag, that tool harness. Not only could we never keep it from falling off, but we could never keep it on once we got it put on. We could never keep it in the vehicle; we always had to tighten it down. We tightened it down repeatedly inside the vehicle. Once we got outside the vehicle, we'd have to tighten it down for the last time. PLSS refurbishment was just as advertised. We managed to keep the PLSSs level because we were fortunate to have landed fairly level. But one of the toughest things we did was stow the PLSS with the tool harness on it in the Commander's station and get it to lock. There's no way to guide that thing in there. There should be some way to guide it into the stow position. You do not have any idea. It's a blind connector into that pin. It's just by hit and miss that you get it in there and stow it. What would be a minute job, if you could see what you're doing, takes 5 to 10 minutes. It's necessary because you want to get that thing up and out of the way and make sure it's secured. The Apollo 15 crewmen were continually turning around and knocking it off the bulkhead, and that's no good. The EVA-1 debriefing - I thought that the questions were pretty good, with the exception of the kind of questions where they asked you what rock did you pick up at what station. Nobody can ever remember the answer to that. After you've gone through eight stations, all the rocks look the same. You can't remember which rock you picked up at which station.

YOUNG — PGA doffing was the usual problem of getting the PGAs off. As we said before, we lubricated the zipper when we got the PGA off, and then fastened the zipper, and pulled the seals down tight. It was after EVA-1 that we noticed the wrist rings were getting clogged with dust. There should be some way to cover those wrist rings (the things that snap in and out), the sliders that keep them from getting full of dust because it makes them practically impossible to work. After EVA-1, we experienced a little stickiness with the helmet. Not a great deal, so we didn't pay any attention to it. When we took the suits off, they were all dust covered, up to our knees, even though we kicked our boots off as we came up the ladder. We took the suits off and put them into a jettison bag, pulled the jettison bag up over the legs, and laid them on the couch like everybody else has done. We put a bag down on the floor to stand on, but that did not keep us from getting dust all over the place. One of the problems was that we had dust on the bottom of the PLSSs even though we wiped it off, and dust on the side of the OPSs for some reason. They were lying on the floor. As a result, when we got in our LCGs, we were sort of standing around, like I had one foot on OPS and one foot on the midstep and was sort of leaning back against the shelf on my side. Charlie was sort of standing with his foot on the ETB and one foot on the midstep, and we were up out of the dirt. Our hands were black when we started taking each other's wrist rings off. We got our hands dirty and I didn't get the dirt off my hands until after we'd landed. Washed them up good. I don't think Charlie did either. We managed to get dirt on the bottom of the LCGs, on our sleeves, and on our hands that got into the suits. It was just a little dust. I don't know what problem it entailed, but it sure looked like it might become a problem. The only thing I can say is we stayed out of the dirt as best we could. It was all over the floor. Just hardly any way to get off of it. We even had some on the midstep where we'd laid the ETB up there. It was dust covered too from dropping in the dirt because the LEC was too long to keep it off the ground.

DUKE — Another 8 ounces on the stowage list in the form of a jett bag would be an outstanding addition to the LM stowage.

YOUNG — For the second jettison, we loaded up the buddy SLSS bag. In training, the buddy SLSS bag had been adequate. Unfortunately, nobody had counted on us landing, eating a night's food, getting up the next morning, eating a breakfast, and then trying to stow all those food wrappings in the buddy SLSS bag. We just barely jammed all that stuff in there. It was really overly full. To tie the bag up, we finally ended up wrapping tape around it. It was really a marginal operation. Plus the other jett bag would give you something to stand on. I don't think the buddy SLSS bag makes a very good jettison bag. It just wasn't big enough for the volume we had to put in it on the second jett.

DUKE — On the second jett we had already jettisoned the LiOH canister. We were supposed to have an LiOH canister in that bag.

YOUNG — I know, we'd never made that.

DUKE — Never made that.

YOUNG — The eat and rest period, I think was nominal. Charlie took a seconal for the rest period the second time.

DUKE — I had a tough time getting my mind out of gear. I knew I was tired. I had a tough time getting my mind unwound. That pill helped me do that. I don't really think it helped me sleep any better.

SLAYTON — Could you eat comfortably without feeling pressed?

DUKE — Yes.

MATTINGLY — One-sixth g makes a lot of difference. You got a gravity feel. You can stabilize stuff. We could put our bags down upon the console and let them for a while to get the air bubbles out of them. There's air bubbles in the LM. I'd say 20 percent of the volume in the food bag was air.

YOUNG — You could squeeze the air out and drink it bubble free. That personally helped my digestion a lot because I'm not much on eating bubbles.

DUKE — I thought there was adequate time to eat, 45 minutes or an hour.

YOUNG — That additional time we put into the Flight Plan really helped. There wasn't any time we were loafing during EVA prep and post, that's for sure.

DUKE — We sort of combined things. We'd start the PLSS, the O2 refill, the 10-minute charge. We'd maybe get that going and I'd cut a bag and John would be piddling with that, things like that. We sort of combined things. We ended up with one spoon. I put my spoon where I could be sure to find it and it took me 2 days to find my spoon. We had to eat in series. We only had one spoon for 2 days.

SLAYTON — It seemed like we over ran most of those EVA post-activities by about 1 hour.

DUKE — Did we?

YOUNG — Take a look at the checklist, Charlie.

DUKE — I think the debriefing took longer than we had planned.

SLAYTON — The first one did. That's why we pushed the other one back.

DUKE — We really weren't hustling.

YOUNG — Body elimination really slows you down. I mean to tell you it does. It's necessary.

DUKE — We had some real-time readups which took some time. The ETB load-up, I did the night before. That took 15 minutes.

YOUNG — That's right. Charlie did the ETB loadups the night before. Something else we were doing that we got out of the way the night before, we serviced the drink bag. That took 5 minutes, and we sat it aside to get the bubbles out.

DUKE — That was about a 15-minute job. That was pretty close to the ETB load-up time, 15 minutes, which was about right.

YOUNG — We also lubed the zippers during doffing, instead of on donning as we had in our checklist, because it was just more convenient to do. The suit was standing up there right in front of you and all of the connectors were visible. It would sure help to get the zipper lubed and, also, it helped for Charlie to load that bag up.

DUKE — It's just a relaxed time to do it. You can double check and make sure you got the right film magazine and everything you need in that bag.

YOUNG — It is pretty self-evident what magazines you're going to need. Like, if you don't use up the one from the EVA before, you know for sure you're going to take it right back out with you. So, you just leave it in there.

DUKE — They were updating that real time.

YOUNG — Yes. Donning was hard. I'll tell you, pulling that restraint zipper was really rough. After we got the dust in the zipper, closing the zipper and locking it was pretty, pretty bad.

DUKE — Give me a new restraint zipper.

YOUNG — Restraint zipper and, also, closing the gloves and locking once we got the dust in there was really bad. It didn't hurt wrist mobility, but it sure was hard to get them closed.

10.2 SECOND EVA   DUKE — EVA 2 Prep Activities. PLSS donning and checkout, I think, was nominal.

YOUNG — Yes.

DUKE — We had the updates to the Flight Plan to get us back into the proper checklist sequence for EVA 2 prep and that flowed real smoothly I thought. We didn't do the computer stuff. The eat period went okay. We knew we were nominal for EVA 2. Except for the donning problem already talked about the suit seemed to really get tighter. I shouldn't say got tighter, but it held the same. I had about a .15 the first time we did it. The third EVA I had the same thing if I recall.

YOUNG — .15 to .2.

DUKE — Cabin depress went okay. Water boiler startup and everything like that went great.

YOUNG — Certainly had a problem with resetting the UV camera on EVA 2. We had to move it out of the Sun and realign it and reset it.

DUKE — The equipment transfer to ETB went down okay, with the cameras. Jettisoned the buddy SLSS bag with the trash.

YOUNG — We had to take special care not to get the buddy SLSS bag near the UV camera, so we dropped it over near the MESA.

DUKE — LMP Pan. I forgot which strut I took it off of. No trouble with the camera. About this time I think we looked at the CRE, and it read off scale, HIGH HOT. They made John move it about that point. So we got a little out of sequence with our checklist.

YOUNG — We had the realignment of the camera. That's probably what deleted station 7 on our traverse. That's probably what did it. Were we behind when we started EVA?

DUKE — No. I don't think so. I was reading the transcript back in the room there. The reason they took 7 out is because they wanted to add more time at station 10, around the LM. I think we really threw them off on that. They were all zeroed in on that vesicular basalt that turned out to be glass. I guess we really never regrouped from that.

YOUNG — They really were looking for that vesicular basalt to support their theory.

SPEAKER — There isn't any vesicular basalt up there.

DUKE — Not one piece. I called - this looks like a piece of basalt under near the engine bell. They said - get it, even if you have to get down on your hands and knees to get under there to get that rock. From then on they were boresighted on that basalt. Wasn't nothing but glass. Equipment prep - We loaded up - no problem there, was there, John?

YOUNG — No. No problem with the load up.

DUKE — Except that I really recommend those SCBs have a positive lock on the top. In fact, I would recommend a positive lock on the top much like is on the back of the hand tool carrier, so you wouldn't have to worry with that Velcro strip at the bottom. That is really a pain, trying to have the other guy bend over, you've got to pull down on the bag and out, and then with one finger try to thread that Velcro strap in through that loop back there on the bag.

YOUNG — I recommend they put some Velcro on the pack where the tool carrier goes too. Remember the time when it fell off inside the command module, that really slowed us down - on me getting down and getting back up to get it hooked up on you, that took about 5 minutes. Remember when your buddy SLSS bag fell off?

DUKE — Yes.

YOUNG — I mean your tool carrier fell off?

DUKE — Yes.

YOUNG — We both had PLSSs on and, man, it was just very - it's very difficult to bend over and pick that up.

DUKE — That was the hardest thing in the whole prep - getting the SCB onto the guy's tool harness. I don't know what the Apollo 17 has to do, but you could cut the weight down on that PLSS harness if you wanted to. No need to carry the tools there any more. The hammer fits right in the pocket.

YOUNG — And it's more accessible to the guy that wants to use it.

DUKE — All the coring operations were done off the back of the Rover.

YOUNG — You can carry the cores in the bag.

DUKE — LRV Nav Initialization. We came right back with it. The thing just worked great. The whole traverse. I take that back, half the traverse. Once we left station 8, for some reason the thing didn't update in range. It updated in bearing but it didn't update in range, distance.

YOUNG — We left station 8 and our bearing back to the LM was 007. That was the last bearing that we got back to the LM. The LRV traverse down south across Survey Ridge and over and down again. It was downhill all the way to Stone Mountain. In fact, from the lunar module you could see all the way down to Stone Mountain. You could see the whole traverse route except for behind the ridges. The ridge lines really had a lot of blocks on them. On Survey Ridge we saw this blocky region down there, and I was really glad to get out of there. It not only had a lot of blocks but a lot of craters. We were hard pressed to make any really fast time down that way. In fact we were hard pressed going down to make any good time at all. Driving cross-Sun was no problem. You could see everything. At least you could avoid it.

DUKE — We were pretty well located all the way. You could see Survey Ridge, and we had intermittent hummocky ridges between us and Survey. But as you topped each one you could look down and see Survey.

YOUNG — All the way to Stone.

DUKE — You could all the time see the Cinco crater and Crown crater. So we knew where we were going and the trafficability looked like a piece of cake.

YOUNG — We started up Stone Mountain, and somewhere in there our pitch meter face fell off - on the Rover. And we weren't able to tell exactly what we were doing. Although the needle was still working, I didn't notice that till after we got up on Stone Mountain. Except for the fact that you slowed down to 8 or 9 kilometers an hour while you were going up the mountain we just didn't have a feel for the slope. It was only after we got out to station 4, turned around, and looked back at that hill we just came up, that we got the idea we might have bitten off at least as much as we could chew. That was a steep ridge in places. There were breaks in the ridges. There would be a ridge crest and it would just drop out of sight. On the way back down, even though I was following my tracks, I proceeded very nervously. We just didn't have a feel at all that we were going to be going down a slope like that.

DUKE — At station 4, we parked in a crater because we had the feeling that the Rover was just going to slide off down the hill like it did on Apollo 15.

YOUNG — So we backed around and parked in a crater.

DUKE — There we had mostly the rocks we sampled that were identifiable as South Ray ejecta.

YOUNG — Yes.

DUKE — And secondary craters. In fact, the Blocky crater was a secondary crater.

YOUNG — South Ray is classic, because the blocks were all distributed away from South Ray.

DUKE — We had some crystalline and some breccias. Soil samples - we picked up shovel fulls. Every time we sampled a rock we got some soil.

YOUNG — It was at station 4 that the penetrometer cone fell off the first time - when you went to get it out, the two-tenths cone.

DUKE — I stuck the penetrometer into the cone and pushed as hard as I could. To me, it felt like it seated. I locked it. As I pulled it out though, it was locked just enough to allow the thing to come out of its holder. Once it got out it fell on the ground. It was right next to the Rover, so I could bend over and pick it up.

YOUNG — I don't remember seeing any crystalline rocks at 4, Charlie. I thought they were all breccias.

DUKE — Maybe it was 5 when we took the first one. Yes, it was 5.

YOUNG — A pure one type rock, I didn't see any of those at 4. Even the crystalline rocks could be one rock breccias. Remember like that anorthosite we found at the San Gabriel mountains - where it melted itself against itself.

DUKE — They're figuring that out right now. Once we got the penetrometer working, John helped me get the thing on and locked, it worked as advertised. The reading's going to be spiked. You couldn't apply a steady force on it. You'd start leaning on it and you'd lose your balance. You'd come up off of it, or, it would give. When it would give, it'd give fast enough to allow the little spring, that you push on, to back off. Then I'd push on it again and it'd bottom out. It ought to be apparent. Every time I did it I tried to call it out. I think they'll see some spiked readings on the drum. There was just no way to avoid that. I tried two or three different little techniques, and every time it worked the same way. The double core we pounded right in. That was an easy operation on the back of the Rover, to assemble and disassemble. Fredo gets a great big case of beer for thinking of that way to stow that rake on the back. We didn't do the padded bag sample.

YOUNG — Not at 4. We did at station 11.

DUKE — Yes, 11.

YOUNG — Travel to station 5 was downhill all of the way with the brake on. Driving downhill, I did it with idle power and sometimes put on the brakes. That's all we had to use. I wasn't going too fast, maybe 4 or 5 kilometers. On a straight stretch might get up to 10. But coming up to a ridge, I'd slow back down again 'cause I had no idea what kind of slope was on the other side of that ridge. You could sure tell that if you let that rascal loose that she'd go down that hill in a big hurry. When you got the Rover up to about 10 clicks going down a hill, it's just like riding a sled on ice. No matter which way you turn the wheel the thing's just going straight. I mean it'd be sideways, but still be going in a straight line downhill. Lot of mass there.

YOUNG — Station 5 - station 5 was the first place we picked up what I thought was a crystalline rock. It was an angular rock. Was station 5 where we worked in the crater?

DUKE — We were in the crater, but you also had a grab sample.

YOUNG — It became apparent after we were at station 4 that we weren't going to get what everybody thought was true Descartes, because we kept picking up what looked like South Ray ejecta. The only way we could think of to do it was to sample towards the South Ray side of the secondaries, or find a primary and sample there. It just wasn't clear that any of those things were primaries because of the way those blocks were distributed in them. They're all on the side away from South Ray. Maybe there were a few primaries up there, but we sure didn't see any, did we?

DUKE — No.

YOUNG — It makes you think that there's a lot of craters on the Moon that must be secondaries. This is the first time that we've had a clear-cut example of it. A lot of those craters on the Moon must be secondaries as opposed to impact craters, because they don't look any different except for the block distribution. As many secondaries as the South Ray must have made in that region it is sure clear there must be a lot of craters that are not primary impact craters. I mean a lot.

DUKE — We sampled around the rim of this 15-meter crater.

YOUNG — We dug into it trying to run across some rocks, but it was all soft regolith: What we picked up looked like - with the exception of a couple of rocks which were probably from the secondary thing that got thrown back in there - were probably all dirt from Descartes.

DUKE — I think so. 500 Millimeter. I took that, not here but up at station 4. I took pans of Baby Ray, South Ray, and the west half of Stubby. Stubby was a very old subdued, tired looking crater and it really wasn't much for regolith. I got pictures of Stubby but there was not lineation, not evidence of outcrop, no evidence of anything except just old tired looking things.

YOUNG — I got the feeling from looking at Stubby that it was there after Descartes was because of the way the slope was off the mountain around Stubby. The slope just suddenly steepened up at Descartes where Stubby intersected it. That could be erroneous, but that's the only reasonable conclusion that I could come to. The rim of Stubby bisected or cut off some of Stone Mountain.

DUKE — Right. We did everything we were supposed to do and moved on to 6, still downhill.

MATTINGLY — You never said anything about when you went up there where you could tell when you went from the Cayley up on to Stone Mountain.

DUKE — No, couldn't recognize that.

YOUNG — Didn't even notice the slope.

MATTINGLY — No texture difference?

YOUNG — There was a slope difference, but we didn't pick it up. Going uphill you couldn't really tell, there wasn't much of a slope change.

DUKE — Going down, you knew when you were off the mountain, because it was a definite break in the slope, but you couldn't see any textural difference at all.

YOUNG — Same gray regolith, it wasn't black regolith. It's a gray regolith like it must have been at Fra Mauro.

DUKE — Station 6 - we had a fairly decent size crater there, 5 to 10 meters across, with some blocks. If I recall the raking came out pretty good there.

YOUNG — The surface was harder there too, remember?

DUKE — Yes.

YOUNG — We skipped station 7 for 8. That was on a block - South Ray. Ray, with some fairly big blocks. We tried to turn over one boulder and couldn't budge it. We did get some samples off one side of it. The rake soil was straightforward, the regular sampling was straightforward. It was here, about halfway along, we lost rear drive and rear steering. We didn't even stop. We went on to station 8 and then stopped and let the ground think about it for awhile. I got back in and started looking at switches. They called procedures they wanted to do, none of which worked. I got to really checking the switches, and sure enough it was in PMW-I, that was all. The message there is to see if the switch configuration is normal.

DUKE — I must have got it with my seat belt or something.

YOUNG — Your finger or something because that's a guarded switch. It'd sure be easy to get in there, clip with the seat belt and never even know it. Or for that matter, doing anything over there, reaching for the camera, etc.

DUKE — The only thing I can think of is that little "T" handle that you lock the belt with. When I reached to pull it off that loop, it could have hit that switch and not known it. The coring - the double core was a piece of cake. At one place it got a little hard, I've forgotten exactly where that was.

YOUNG — Station 10 - the one they let me do. I was really beating on that thing. They were hard to go in the ground but easy to come out - all the cores. Station 9 - we did quite a bit of sampling there. There was something that slowed us down at station 8. We did something that took quite a while. We traded the bags out.

DUKE — Yes.

YOUNG — That took a lot of time. We had an hour at station 8. We traded the bags out and changed the film. That cost us a lot of time. I don't know how you get around it. You have to do that. Only thing I could suggest would be Charlie's idea for the shopping bag wherein you don't have to change the bag out. You just pick up the bag and go sample with it or something. That really took a lot of time to change those bags out, I was really surprised that it took so much time. I don't understand why.

DUKE — Well, there's no way to get that Velcro on. That's the problem I had.

YOUNG — Yes.

DUKE — I couldn't get that little Velcro strap through the loop on the bag.

YOUNG — That's a manual dexterity test. Run the Velcro through the loop in a pressure suit glove. It never works the same way twice.

DUKE — I'd get the strap through there, pull on it, and it'd be twisted. The part I wanted to Velcro would be outboard and it wouldn't turn over.

YOUNG — That's one of those three-handed jobs in 1/6th g. You need a hand to stick it through, one to pull it down, and a hand to stick it through and keep it there when you reach around to pull it down and pull it tight.

DUKE — It's hard to do. We did the vacuum container core sample. That went okay.

YOUNG — Yes.

DUKE — The surface sample did.

YOUNG — I'm not sure that there's any material on the first one. That's the kind of thing where you sneak up behind a rock, stick it over there, never look at it, and let it rest on the surface. The first assumption you make is that it's resting on the surface. Nobody was ever able to verify that. I could see the imprint where the legs on the sampler had been but I was never sure that the plate inside had actually gotten down on the surface. You don't push it. You just let it sit there. You're taking somebody's word for it. The second sample, because of the unevenness of the ground, only 20 percent had something on it.

DUKE — We closed them up and stuck them in the bag.

YOUNG — At station 9, we were able to overturn a rock. We were able to overturn it, chip a rock off the top and a rock off the bottom, and sampled underneath. Underneath that surface, the soil was compacted very much like rocks on earth. You could see the outline of the material compacted down. It looked like it'd been sitting there for quite awhile. The etchings on the rock were all on the surface. However long South Ray has been there is how long that rock has been sitting there. After we left station 9, the last bearing we had was 007. I forget the clicks. So, I said, shoot, I'll just take up 030 and meander that way generally to make sure that we intercept our path so we don't end up west of the LM. We never did intercept our tracks but there was the lunar module just as we came up over the hill, about 300 meters out. Went right to it. We had pretty good pad built in. We had the traverse tracks of the 1.2 kilometers out to Flag. We really didn't have any trouble. We had a set of tracks we could follow back from either one.

DUKE — There was never any feeling that we were lost.

YOUNG — No.

DUKE — We knew exactly where to go. You could see Smoky Mountain.

YOUNG — Until we came up over that ridge we never saw the LM.

DUKE — Not once we left station 6.

YOUNG — Yes.

DUKE — After we left station 9, there was a crater we passed that was very old, tired, subdued. It was rimless. It was ridged at one end and it was very deep. It looked like a big sink hole.

YOUNG — It had a central crater in it that was elongated that sort of looked like a bend. We did a 360 and Charlie shot a pan while we were going around in a circle. Station 10. They eliminated the soil mechanics on station 10 because we semi-dug a trench up at station 4 - an exploratory trench up there. So, they eliminated that one.

DUKE — I did the double core at station 10.

YOUNG — No, Charlie, I did the double core.

DUKE — You did the double core while I was doing the penetrometer. I did do the penetrometer work there with the two-tenths cone, the five-tenths cone all the way out to the deep core site. Came back and then did two plates. Nothing out of the ordinary, the thing worked great.

YOUNG — When we got back to the LM, I didn't see any trouble with any of the close-outs, did you? Why don't you tell all about how you…

DUKE — On EVA-1, I got one of the sample bags - the end of it where it is thick plastic or Teflon or whatever it is - stuck in the seal. I couldn't get it locked. But when I opened it back up and saw what was happening, I just moved that bag back in. It didn't look like the seal was damaged and it locked right up. SRC-2 - we put that vacuum sample in and most of the core tubes, John's whole bag. Really packed it full and closed it up. It closed without any problem.

YOUNG — By taking the bags back and dumping them in the SRCs, we could get a lot more bags in. The bag that appeared to be full would generally fill up the SRC. You did that a couple of times. I think that's a good procedure because you'd do a couple of things - you save a bag that you don't need, and you get the SRC full. Furthermore, you get all that bag material out of there that we never quite licked because seems like when you put it in there with a full bag and you close that hatch, close the lid on it, invariably some piece of the SCB catches inside the seal, and messes up the whole business. We used the same procedure to haul up the ETB and the bags and it seemed to work straightforward and ingress was no problem. I think it was station 8 where we lost the rear fender and that was because I fell over it. I was coming out to help you and I tripped over the thing and it fell off. Avoid those fenders if you can. Every time that wheel came off the ground and went back in and dug in, it was just like we were watching rain. Dirt came over it, covered up the battery cover, and the instrument panel so bad that you couldn't read the POWER DOWN or POWER UP decals. When we got back to the lunar module, I brushed off not only the camera, but the batteries and the instrument panel as well. And that made the problem of dusting me and Charlie off even worse too. We had a lot of dust on top of our OPSs, had dust all over the place, dust on the helmet, dust around the neckring, what a mess.

DUKE — Raining dust.

YOUNG — Yeah. The message is don't trip over the fender. It didn't bother us any apparently, but it sure was dusty.

DUKE — No trouble with any of the transfer. We loaded all the bags up, and jettisoned the pallet. The ETB came up easy, John passed them into me and I'd pass them on up and put them on the ISS and then John got in and closed the hatch and turned off the waters as planned and repressed. EVA 2 Post Activities. Cabin Repress was nominal, PLSS Refurbish was nominal, Rock Samples Stowage and Weighing was just as advertised in the checklist. We did dry out the PGA every night.

YOUNG — I'm not sure it dried very much.

DUKE — No, it never did get real dry. The next morning we'd get up and it'd still be damp, but it wasn't uncomfortable.

YOUNG — On those evening meals we were eating everything that we could get our hands on. Sure got hungry after a day out there in the toolies.

10.3 THIRD EVA   YOUNG — EVA 3 Prep. There again it was straightforward. I think it was due to training. Man, I'm glad we trained as much as we did on that. Although that was the worst training we did the whole - I mean as far as keeping behind the power curve from start to finish - that's just a miserable way to spend 3 or 4 hours. But it sure paid off. PGA Verification, straightforward; Cabin Depress straightforward; Egress, straightforward. By that time I think we were maybe 20 minutes ahead when we got out. That's probably why they let us do the EVA longer. It was here when we got the UV camera pointings that we had to move the UV camera and the first time that we moved it to a place that they told us to move it to. Then there was no way to get the - get between the UV camera and the strut to work it so I had to move it back a little. We took some more time to change the azimuth and here again it was still on a slope that went down and it was impossible to get the thing aligned without really kicking it down in the earth on two legs and leaving one leg almost out of the ground. And every time we changed the azimuth - it changed the whole alignment of it. And they wanted to move the battery into the shade at this point, which we did. Okay, and I guess I had so much trouble with that every time that you had to do all the ETB stuff every time.

DUKE — Yes. It was in my checklist, right?

YOUNG — And that's a good reason for having those operations in both checklists. Because if one guy is operating the equipment, then the other guy's operating straightforward, like the ETB which everybody's done. Then if anything goes wrong with the equipment, then the guy who normally does it can sluff off and you still haven't lost anything. Load-up was the same thing. Between station 8 and 9, we lost the first SCB that fell off my back and lodged between the rear fender and the frame. It was full of samples but we didn't lose any. We didn't lose it.

DUKE — Yes, that was really lucky.

YOUNG — We'd have lost half a bag of rocks if that thing hadn't have hung in there. Pure luck.

DUKE — Really got to improve those SCB tie downs. Not only because you could lose the thing and lose the samples, but the way it is right now, it's just hard to do. And it's not secure what you're doing.

YOUNG — I like Charlie's idea of a shopping bag where you just have something with a strap on the side. Matter of fact we used the regular SCBs, if you just had a strap on the side where you could stick your hand in and if the bag was just a little more rigid so it would stand up instead of lay down all the time. You could carry that on the surface, set it down because of bending the suit and do things like independent sampling.

DUKE — Okay, I think the travel to 11 was just spectacular scenery the whole way and you always felt like you were right down on top of North Ray. In fact, I thought we were there when we crossed that first big rock that ended up to be station 13. Looked like it was - you could see it way up on the slope.

YOUNG — I thought we were up on the rim of it when we crossed that first ridge short of station 13. I said, boy, and there's a lot of boulders up there. As we got up to this ridge, there was a whole lot of boulders as we climbed up this ridge and I says, man, we're getting near the ridge, and they said you're 500 meters away. Sure enough we were. We climbed up over the ridge and it almost looked like a second ring around North Ray.

DUKE — No, Ken and I talked about this the other day. There was a ridge line north of Palmetto that came out south of North Ray and came back around again and where we crossed back over here, we started climbing again.

YOUNG — Well, it was a ray though, because it was just like Schooner Crater out at the Nevada test site. We were riding along in this regolith and North Ray blocks would be sticking out, I mean big blocks 2 or 3 meters across, the top of them would be sticking out of the regolith as we got over that ridge. And it's just very much like Schooner Crater where we had the big blocks that were all mantled by that powdery regolith.

DUKE — I think those might have been blocks from Palmetto, John, those might have been because those were really mantled. Really had fillet that looked like they were almost subdued.

YOUNG — But they could have been blocks from North Ray.

DUKE — Except most North Ray blocks were not that well filleted.

YOUNG — And they weren't as rounded either, if I think about it.

DUKE — Another unique part about that drive up there was that once we got up past Palmetto, there was a total absence of sharp craters. There were 1-meter to 2-meter sized craters.

YOUNG — There was a real absence of craters and the blocks were gone.

DUKE — Gone!

YOUNG — Yes, just practically gone much less than 5 percent.

DUKE — It was really like driving out in west Texas across some sand dunes.

YOUNG — There were some pretty big down slopes going out there.

DUKE — Yes, out to north of Palmetto.

YOUNG — Yes, and as we climbed up the rim to North Ray it was really a steep slope going right up to the edge of the rim. Of course, the old Rover didn't notice it - just went right up it. But when we turned around and came back down it we really noticed it. That's where we achieved Vmax and we achieved another Vmax when we were on the other side of Palmetto. I felt a lot better about achieving Vmax down our tracks than I did trying to set any going the other way. We did 17 kilometers coming down that hill for a short period of time. It doesn't turn. It just goes in a straight line no matter what you do to it. And I don't recommend that at all. All I do is slow down after that. You're just on a piece of mass that's going along.

DUKE — We combined station 11 and 12 and that gave us about an hour up there. We did the 500 millimeter of the interior, of what we could see of the interior of North Ray which was about, I would imagine, halfway down the wall, maybe two-thirds. You could not see the bottom. I wasn't going to get close enough to see in because there was no way you could have gotten out of there if you had fallen in. Well, anyway, I took the 500s there and I was looking back at Stone, I took those. I took some of Smokey, of the ray going up the flank. The Near Polarimetric Photo they threw out because we had a gnomon failure. And that happened on EVA 2. We didn't mention that.

YOUNG — Charlie pulled it out of the bag and the leg stayed in the bag. I never thought the gnomon would fail. I really was surprised. I figured what would happen was that it would bounce out of the bag and we'd never see it again.

DUKE — But we did do the Far Polarimetric Photo and it worked. I haven't seen the pictures but at least the filter worked like it was supposed to and I took all the right settings. And I let them know whether I was going from right to left or left to right. And I took three partial pans up near where we stopped the Rover. I moved North around the rim about 50 to 60 meters and took another series.

YOUNG — I guess we could have probably gotten down to the rim edge. It was about a 10 or 15 degree slope down there and I really wasn't too anxious to go down there and fall in that crater. So we stayed about 50 meters from the edge of that crater.

DUKE — I bet you there were some big blocks in the bottom of that crater, though. Sure would have liked to have seen that.

YOUNG — Yeah, I would have liked to have seen it.

DUKE — Did you ever go directly over it so you could see in the bottom, Ken?

MATTINGLY — Yes, there were some big blocks down there.

YOUNG — And that is a steep wall crater too, because I'm almost sure that that thing that looks like Talus over on one side has a linear orientation in it. What I'm saying is the...

DUKE — The blocks are disjointed but they're laid in there.

YOUNG — Yeah, they are laid in there like outcrops.

MATTINGLY — That's typical throughout. That crater inside didn't look any different than 10 million other craters around the Moon. They all have that same characteristic. And I don't know if that is a sublayer that's being exposed or not, but my impression was that it's not but that's the way the thing sort of slumps.

YOUNG — I don't see how it could slump uniformly like that.

MATTINGLY — Well, I tried to trace some of those I could see well with binoculars and you couldn't trace a continuous layer. The things bounced around, but it does at first glance - it looks like they've fallen in strata.

YOUNG — These particular ones sure did. Maybe they are a layer that got laid up there that slumped back down together.

MATTINGLY — It's still strange that they would end up in some kind of horizontal line.

YOUNG — Yes, it's weird.

MATTINGLY — But, looking down in North Ray I had absolutely no sensation of any strata.

YOUNG — Me either, me either - except for that line of blocks.

MATTINGLY — South Ray had material in it that looked to me like it was in place, but North Ray just didn't.

DUKE — We couldn't see inside South Ray except the upper part of the southwest wall.

MATTINGLY — But, I really didn't spend a lot of time looking in those craters. Because I thought pan camera pictures from right straight overhead have got to do better than I could do.

DUKE — Yeah, that's great. I can't wait to see those.

DUKE — Well, we saw the house-sized block. There was one there.

YOUNG — Couldn't keep Charlie away from it.

DUKE — And it was a biggy, but it was just a two-rock breccia. And some of the clasts were meter size. It was predominately black.

YOUNG — I bet there was more than two.

DUKE — Well, all I could see was two, let me put it that way.

YOUNG — I never really would have looked to see how many.

DUKE — John saw a shadow cone on one section of it that apparently had broken off.

YOUNG — We got a picture of it.

DUKE — And we got a north-east-west split and got a sample which I don't know whether it is going to show anything.

YOUNG — It's a good east-west split, Charlie.

DUKE — I took a flight line stereo of the thing.

YOUNG — The trouble is that Charlie couldn't back away enough to get the whole thing in there. Did you pan up and down by any chance.

DUKE — I don't think so. I don't really remember.

YOUNG — I don't remember you doing that either, but if you backed away too far he would have been over the edge of the down slope.

DUKE — But we got some sample off it in place and got some of the contact in the breccia, both clasts.

YOUNG — That was a big rock. There were some big white rocks up north of there we got.

DUKE — There was a huge white rock in the swale between Smokey and North Ray that we didn't get to.

YOUNG — We didn't go to, right?

DUKE — This was the densest boulder field of the whole EVA.

YOUNG — Right between Smokey and North Ray.

DUKE — North Ray, but we never got over there. We started going up there and it was just a glistening white rock out there that was very angular - looked like maybe it was 5 meters across. And it looked like one great big piece of crystalline rock to me from a distance. But it could have been a breccia. But it sure was angular and it was a biggie. But, we never did get over there to sample it, which was tragic. Okay, we did all the regular sampling. We had some problem up there. John's bags fell off and my bags fell off. We ended up wearing the bags on our little finger. We were carrying the bags like that and pulling them off. We probably wasted a few minutes independent sampling. Then we got organized and we started sampling together and that was a good move. And the regolith was very, very thin up there. You couldn't even rake the rake through it. It bent the tines on the rake. It was only about a centimeter or so thick and under that it was hard.

MATTINGLY — This is at North Ray?

DUKE — Yes, on the rim. Yes, what we ended up doing was I'd hold the rake and he'd kick some stuff in to it and we ended up kicking enough so we could shake out a few samples.

YOUNG — We practiced that on the ground. I didn't imagine we'd ever have to do that.

DUKE — I didn't either but we sure did - because it bent the tines just like it does in the training exercises down at the Cape.

YOUNG — It was at station 13 we achieved V max there coming over that rim - 17 kilometers and I sure don't recommend that. It was near the bottom of the slope and we had a straight path with no blocks. I just took my hand off the brake; I didn't add anything. But it must have been an awfully steep slope.

DUKE — I think on the whole way I was underestimating slopes and overestimating size of rocks and percentages for some reason.

YOUNG — Yes, I think I was too.

DUKE — I think that when they look at all the traverse pictures that we got on the way they'll see where I said was covered by 60 percent of the surface, it'll probably end up being something like 40 percent. It was covered with cobble size. 13 was really a big boulder. We stopped at the big boulder - when I say big, I mean 4 or 5 meters across. It was a breccia. On the east side of it, it had some very strange holes in it that looked like drill holes to me. They were perfectly circular. They weren't vesicles and it just looked like they were bored out of the thing and in only one side of the rock. In fact it looked like it was drilled out for blasting.

YOUNG — Yes, that's what it looked like to me too.

DUKE — We got some pictures of that.

MATTINGLY — Did you run anything inside it and see how deep it was?

DUKE — No. I started to do that with the pencil but the Rover was sitting over there and I just didn't. But we did get a permanently shadowed sample out from under there. And that was a true, permanently shadowed sample.

YOUNG — Yes, Charlie got way back up under there and no doubt in my mind that it's a permanently shadowed sample. In the meantime, I took the LPM reading.

DUKE — Yeah, John was tangled up in the LPM cables. Scratch, 14, 15, 16, 17. Ten prime was out northwest of the ALSEP, 20 meters, or so. And we got some nice crystalline samples there, plus it looked like it was ejected from South Ray material but there were some nice angular rocks we got.

YOUNG — Yes. We had the big rock bag on and I scooped up a big rock at North Ray. And I got another big one, maybe 2 big ones at 10 prime. I think they're fully documented.

DUKE — The crystal rocks that we found throughout, the whitish ones had a sugary texture to them. And the crystals were big enough to see crystals in them, meaning a couple of millimeters high, some of them. There was just a light - various shades of gray, I guess. There were no salt and pepper texture to the rocks.

YOUNG — You know on the bottom of that rock we turned over, I swear that looked like quartz - that white crystal on the bottom of that rock. I hate to say it but it sure looked like it.

DUKE — I think so.

YOUNG — The same old vitreous luster mashed quartz crystal - either that or white glass, one of the two. You could see right through it.

MATTINGLY — Did you get a piece of it?

YOUNG — Oh, yeah.

DUKE — The rocks really looked a lot like earth rocks.

YOUNG — They really did, I agree with you Charlie.

DUKE — They didn't look like the typical lunar volcanic rocks.

YOUNG — In most every case where you wanted to see a fresh surface you had to bang it to get it.

DUKE — In 10 prime we did all of what we did there, and we got a core as I recall.

YOUNG — Yeah.

DUKE — It went pretty smoothly and we were looking for some basaltic rocks. We never found them, but we did find some crystalline rocks. We scratched the Grand Prix, the LRV offload was nominal, and John drove it on out to the parking site. I was surprised. I hit one rock that was really a hard crystalline rock and it broke.

YOUNG — You hit some pretty hard there. Charlie, you hit them a couple of times there. When you didn't hit on a fracture surface, it was pretty tough.

DUKE — It was tough. Yeah.

SLAYTON — Were there lot of fragments flying around.

DUKE — No, they just pulverized, really.

YOUNG — Really surprised at the size of the chips we got every time. I thought we were going to get little bitty chips and we ended up with pretty good sized rocks.

DUKE — Some nice fresh surfaces, too. Sugary looking. I found a crystalline rock for John, for the LPM. It was about baseball size and just fit right into the bag and I was really lucky to do it. It was definitely crystalline rock, it was not a breccia. Had that sugary texture to it. The LPM stuff we sent back, there sure were some funny readings on those things. I don't know what Palmer's going to do with that.

YOUNG — I don't know either, that's strange stuff. Almost every place that we had put the LPM was where there was either a crater or an old subdued crater and maybe if those old craters were in primaries, maybe just the fact that you've got a real primary whomping in there would change the magnetic field. I don't know. There was hardly any place that we didn't put the LPM, that wasn't a crater of some kind. Like at the big rock where I put it up, I put it right into the edge of the crater. There's just no way out. An old subdued crater. It'll hardly show up.

DUKE — The LRV SETUP was no problem. Went as advertised. The SWC, just like in training. I took that dadgum thing off and tried to let it slowly roll up and it slowly rolled up and was about 2 feet long. So, I pulled it back down again and when I pulled it out again, it partially ripped on the left hand side. And then I tried to let it roll up. I had a little bit better luck with it this time it came up about oh, a foot long. It was about 3 inches or so in diameter, so I just crushed it with my hands and stuck it in the bag and it went right in the bag. We threw it in the ETB for transfer up. Then I didn't have anything else to do then until John got the cosmic rays done and we'll let him talk about that.

YOUNG — Like Charlie said, we didn't have any trouble with the LRV setup. We opened it all up, pulled all the circuit breakers except for the AUX circuit breaker and AUX switch which we pushed in, but, we still had to dust it real good, and set up the LCRU with a blanket folded over it, and all that went real well. There was no problem associated with doing that. The Far UV camera cassette was the next to last thing we did. When we went to get the Cosmic Ray Experiment back, I picked it up and brought it around to the MESA and pulled the last ring. The strap that was hanging down from the bottom of it, the nylon strap, was very much like the one in training. I grabbed hold of it and gave a little pull and it didn't move. I gave a harder pull and it didn't move, gave a real hard pull and tore out the bottom of the nylon strap. Then Charlie came over and helped me. We got the pliers out and put the pliers on the bottom of it and gave a real hard pull with that. And then it managed to start to come out. As soon as we broke it loose it was free and it came on out.

DUKE — I was holding the top part, the frame, and I was pulling and John was pulling with the pliers on the bottom of it.

YOUNG — If we hadn't had those pliers we probably would still be there. That doesn't mean we wouldn't have brought back the whole thing. We could have brought back the whole thing but it would have been an additional thing we'd had to think about stowing.

DUKE — But that cover was hot. That was the only thing on the whole EVAs that I touched and held onto that I could feel through those gloves. By the time John got through pulling that thing out of there I could feel the heat on my fingers.

YOUNG — Let's just finish this up. Cosmic Ray- Experiment stowage was no problem. Retrieving the Far UV camera cassette was a piece of cake. And ingress, did you see any problems with that?

DUKE — No.

YOUNG — We had a lot of bags. The last bag we got up was the big rock bag and let me just say one thing about that. Once, we got it inside, we had too many rocks in the big rock-bag and we had to reshuffle some of them by hand. We had to make the big rock bag weigh less than 45 pounds.

DUKE — It weighed about 50 something and we took two rocks out of it and got it down to 40.

YOUNG — The main rock we had in there was that big Muehlberger rock and one other one.

DUKE — Watermelon size.

YOUNG — EVA 3 Post Activities. The repress was normal. Again we tracked in a lot of dirt with us. Weighing and stowage was normal. I don't remember any problems we had with stowing any of the boxes.

DUKE — We left our helmets on and broke out the scales. We started tying down, just like we're supposed to before the jettison time. We weighed them and then we had to wait about 10 minutes until they decided whether we could keep all the rocks.

YOUNG — One thing we did that was necessary, we had to get the weight of the ISA down to 45 pounds. I had to reach in and pull out the Muley rock to get a rock out from underneath it. I think we took two rocks out of that bag, and put 'em in a half-full SCB to make the weight more balanced. I ended up touching a rock with my bare hands. I really didn't plan to do so. There must be some stray hydrocarbon on those rocks that I touched. They were big rocks. We did a suit loop check, depressurized the cabin, and dumped out the bag. I opened a hatch as soon as I could pry it loose, and a lot of the dirt went over the side with it - but certainly, by no means all of the dirt. The power up was nominal, wasn't it Charlie?

DUKE — We didn't do the P22, but otherwise, we just breezed right on through the launch prep.

YOUNG — One thing that I noticed, the Earth was in the window, and so was the navigation star. Which star was it?

MATTINGLY — Altair.

DUKE — Achernar.

YOUNG — Achernar?

DUKE — Yes.

YOUNG — That's really a B. And it's right in there. Where was Altair?

DUKE — It was the Achernar.

YOUNG — Achernar.

DUKE — Détente one.

YOUNG — It was in détente one, and the torquing angles were small after the gravity align. It was really a good alignment, both of them. I was really surprised to be able to see the star. We didn't even turn the lights out in the cockpit. We put the window shades up but we left the lights on in the cockpit and had no trouble seeing the star. And the platform torquing angles were small.

DUKE — The night before when I went to bed, I looked to see how the alignment was going to be and to see if I could see any stars. I and 3 were gray, 2 was a pretty good détente, except the radar antenna was there at the time but I could see something over the top. There was something white in both 4 and 6 I don't know if that was reflected Sun or not. I think it was reflected Sun off the steerable antenna in détente 4. I don't know what was in the other one but I thought those two détentes were useless.

YOUNG — Two and four?

DUKE — Four and six.

YOUNG — We had a couple data stars in those détentes.

DUKE — That's right.

YOUNG — So the next man that picks stars better think about those kind of things.

DUKE — Launch Preparation. We didn't have any problems. No, we didn't. Everything went well.

YOUNG — Even though we were completely powered down, as soon as we received the state vector, had the clock, and mission timer started, we were right back in business. We checked the TEPHEM did the computer check, and we were already to go.

DUKE — We were way ahead. At liftoff minus 35, we held for 20 minutes. The next thing was at liftoff minus 15, we had about 20 minutes with nothing to do.

YOUNG — That's right, we sat there with helmets and gloves on.

DUKE — We started a little bit early on the helmets and gloves because of a wrist ring problem.

YOUNG — I'm glad we did, because we had a problem. Every time we put the wrist rings on we didn't know if we were going to get them on, and then once we got them on we didn't know if we could get them off again. We knew we were going to get 'em off, but it sure wasn't the normal click click, push pull. I think we need some protection against dirt getting into those wrist locks.

MATTINGLY — If you'd wrap a piece of tape around...

YOUNG — I thought about wrapping tape around it but would the tape stick?

MATTINGLY — That gray tape would do it.

YOUNG — I'm not sure it would stick in a vacuum very long. I think you need something more than tokenism on that cover. The systems were powered up, and the launch preparation was nominal. I'll tell you what happened to me on the launch prep. About 3 minutes prior to launch as I was looking out the window and the Sun was bright overhead my left eye started crying. I had the window down all the way. I normally kept the window up. For launch you want the window down. I couldn't see out of my left eye. It became worse and worse, and it was as bad as it got just prior to launch. It was really bad. I closed my left eye and it was tearing badly. I think it was just due to the brightness.

DUKE — So it made you like snow-blind.

YOUNG — Yes. I just couldn't believe it. It was almost like I had something in my eye. I'm sittin' there and I can't see out of my left eye and we're going to launch in 2 minutes.

DUKE — I couldn't believe it either. I looked over at him and he looked like both eyes were closed.

YOUNG — Well, there was nothing I could do about it. I sure didn't plan to abort the launch for that. I had my right eye, and I was going to fly it using one eye if I had to. I was going to fly on instruments anyway. You couldn't fly on it out the window on the ground track without really being in good shape. We had at least five guidance systems and four different control modes going for us before I had to use my left eye, so I felt pretty confident. I think it's something you need to think about. Looking out that bright window all the time with a helmet on leaves your eyeballs with no protection.

MATTINGLY — Could you get any relief by holding your hand up?

YOUNG — Yes, I put my hand over my eye and that relieved it. Charlie couldn't figure out what I was doing.

DUKE — I thought you had something in your eye.

QUERY — Sunglasses?

YOUNG — You don't want that under the pressure helmet, because you have to look inside the cockpit and you can't get 'em off. That surface was some bright, and we were looking upslope. We were looking up the rim of this crater. I don't know why it didn't bother Charlie over on his side, but it sure was – sure got to me.

11.0 CSM CIRCUMLUNAR OPERATIONS

MATTINGLY — In general because of our altered time line, due to the delay in circularization, the entire time line seemed to get rejuggled. The primary activities and comments, I think we had to address out of all this, are going to be the way you handle that kind of an operation in real time, and what is a reasonable way to do that. The Operation of the Spacecraft - I think the command module is a real fine one-man spacecraft in orbit. It's a pain in the neck to fly a simulator in one g. You're climbing over and under things, but when you get in orbit it's much more efficient for one man to operate it in space than it is for three men. And one of the mistakes I think we've made in our planning is that we've assumed that with three men in there you could really pursue parallel efforts and it didn't turn out to be that way at all. I had the feeling that throughout the entire solo operation that I was getting more Flight Plan operation done with less fuss and bother than we did in any time when we had the extra people in there. Navigation, we scrubbed several of the P52s to save time, because the platform drifts were so small that it just didn't justify taking the time with it. And it was during this period that we looked at the scanning telescope and I kept looking for stars on the back side. I had in mind the Apollo 15 problem of not being able to recognize patterns. And this is where we picked up the fact that when the telescope is out of focus you just don't get enough light to paint a star image. Once you get it back into focus, the star patterns become quite obvious and it looks just like it did in Earth orbit. It's an excellent device. And I think we've already covered the fact that these things change their focus due to vibration or whatever it is that allows the jam nuts to back off. The systems operation actually is no problem. The ground did it all. The thing that we did differently, and I think bears comment is that for all the Flight Plan objects, that were done on the front side of the Moon are within communications with the ground. We gave the time line responsibilities to the CAPCOM, and that gave me a chance to sit back and do whatever it was I wanted to do. If I wanted to get a drink of water, I could do that. If I wanted to look out the window, I could do that. I didn't have to just sit there and stare at the clock. I could tell there was a big difference; I could relax when I got AOS and I could stop staring at the clock, which is your primary activity any time you're running the responsibility for the Flight Plan and the time line. And those guys on the ground just did a super job of keeping us on step. They would call and say it's time to turn the pan camera on and I would just float over there and flip it on. That way I could take the Flight Plan and take about 10 minutes out of each rev and look ahead at what was going to happen on the next rev and get an outline in my head of the things that were coming and the sequence of things. I had the proper phase of the experiment checklist out and I had the right film magazines adjacent to the cameras. It was really no problem to stay with the time line. The places it got to be a little bit sticky were when I didn't have that luxury because we had altered the Flight Plan. There were a couple of revs where we finally resorted to taking a blank piece of paper and Henry read up to me the sequence of events for the next 4 hours. I jotted them down and we worked from there and that was actually a lot easier than the times we tried to go through and alter the Flight Plan and delete the item in 99-10 and add some other item. And all the deletions turned out to be taking up an awful lot of your time.

We had the same problem in operating solo that we had before. It really didn't allow adequate time for taking time to go to the bathroom and a few things like that. Inadvertently during the development of the Flight Plan we allowed things to creep into the eat period. They look innocuous when you write eat period down and it shows gamma ray gain step 1 or something, and as long as there is some guy on the front side to give you a call you can afford that. But when you're on the back side and trying to eat, trying to mix your food, trying to keep your eye on the clock and run this, it becomes a messy thing. And we discussed this preflight and agreed that since it had slipped in there we recognized that we weren't going to make a big issue out of it. We'd go ahead and do it this way.

I would strongly recommend that eat periods be exactly that - they have got to be sacred. This includes no updates from the ground. We were continually getting Flight Plan updates during the eat periods, which requires you to read them back to the ground and you just can't talk back to the ground and choke off the food bag and do all these other things at once. I found that the VOX circuit, which I never thought much of in the command module before, really worked great. That freed two hands so you can hold a pencil and a Flight Plan and write and read them back at the same time. I don't know how well it came through on the ground, or how much clipping there might have been but it sounded to me as though the VOX circuit was performing superbly. I just turned it to maximum sensitivity and left it there. I think I ran in VOX almost the entire period. It was not clear to me then why we altered the time for the plane change. The flight planners did an outstanding job real time of giving me an outline of the items that were coming up so that I could have some plan for it. We protected against the yaw gimbal motor by going to the plane change attitude then rolling 90 degrees for redundancy. This attitude is a good redundancy mode for the middle gimbal direction, so if you did lose it you wouldn't go into the gimbal lock region.

SPEAKER — You did the plane change then a 90-degree roll?

MATTINGLY — Yes, that seemed like a very reasonable thing to do. No problem, a matter of retrimming the gimbals to make sure that after you calculated one angle that you retrimmed it again before you let it off. Both the circ burn and the plane change had a larger attitude excursion than I would have anticipated. They may be nominal. I would like to get the system guys to explain.

SPEAKER — Is there a short burn logic?

MATTINGLY — Yes, what it tells me is that the gimbal angles we put in for trims really weren't as smooth as they could have been. Perhaps that's the normal response. The deployment of all the experiments seemed to go nominally. The mass specs, as we'd already stated started not indicating on board that it was retracting fully, but the ground TM showed that it was retracted to a safe distance for the SPS maneuver. The mapping camera on its first retraction had shown an excessive retraction time. We deleted several of the extension retraction cycles trying to minimize the problem. There was one time when we got off and operating in this open loop fashion where the Flight Plan was being read up somewhere between 5 and 30 minutes ahead of the time that the maneuver operation was to be executed. We ended up going to full calibration and the ground read me in great detail all of the SIM bay switches, and did not remind me of configuring the jets for the SIM bay operation. I had the feeling that there are two ways you can handle real-time Flight Plan changes. One of them is to tell the flight crew we'd like for you to operate a particular experiment at a certain time. And let the flight crew be responsible for configuring it and doing it all. The other way is to have the ground figure out each switch and each activity that has to be done just like you would the normal Flight Plan and read those up to you one at a time. And then, let them bear the responsibility. The ground never had sufficient time to work ahead so they were reading me detailed steps they hadn't had a chance to verify that something wasn't left out. They left you alone in the sense that you had all the data that you needed to do the job when, in fact you really had to go back and scrutinize the procedures that were read up because no one else had had a chance to verify them either. They were sort of waiting for me to cross check it and as you get a little tireder you start thinking, well, they've done it. And this one time I know we caught ourselves off guard.

YOUNG — Well, if you are going to call the data up to a single crewman a short time before he has to execute it; he doesn't have time to think it through and make sure you've done the right thing.

MATTINGLY — Now, my whole theme on real-time Flight Plans are that we certainly did a lot of it. I guess I'm against revamping and reordering your priorities, burns, and trying to optimize things. It seemed to me that we spend months and months building a Flight Plan to go fly. We try to screen it and make sure that maneuvers don't go into gimbal lock, and that you have sufficient time to get to the new attitude and really thought of all the things that go wrong. Then you go and do all this thinking on the spur of the moment. It seems to me the chance of doing it wrong is so much greater that instead of getting slightly optimum data you probably are going to mess something up. I would feel much more comfortable when we discussed this preflight, that if you ever get behind the time line, we should drop what it was that we missed and press on with the rest of the time line. So that everybody knows where we're going and in which direction and why we do the things you have practiced and know how to do them. And we did not do that and I hope all the data works out. We'll just have to wait and see when it gets back.

YOUNG — I agree with you, Ken. I really feel strong that once a Flight Plan is written down it shouldn't be changed. I just don't think the science people should be allowed to redo the whole thing. I don't know what they were doing it for, but I just don't think we ought to do that. That's asking for trouble.

MATTINGLY — To me that's the time when you're likely to extend the mapping camera, while the door is still open and things like that. You get in a hurry and you need to get the camera on and you hit the switch and you forgot all about that door that's covered or something. I felt uncomfortable from that aspect before I didn't often get far enough ahead that I felt like we crammed the bistatic radar around and I think many of our problems were caused by regrouping. The ground knew that we were going to cut the mission a day short and tried to get all their major objectives in 1 day less.

SLAYTON — It was all supposed to stay the same and they were knocking a day off the end period.

MATTINGLY — But we didn't. We rewrote the whole thing. Apparently some science priority team got together and redid the whole business. Because that's the way it was coming up to us. That's the only thing I could think of that was going on down there and that is hazardous. As a matter of fact, on landing day when we were 6 hours behind they didn't do that. They just picked up 6 hours later and took the Flight Plan from that point on. I knew where we were and what we were doing. I went back and I started at the nominal landing time. I went back to see what things we had missed and by the time I went to bed I had picked up and done everything that was to be done on the three revs that we didn't do, except track the LM with the sextant, and that meant maneuvering. But all the rest of the photography and experiments, I went back and picked up. No, there was one exception and that was the photographs of the Moon and earthshine. And I was in the process of that when ground called and said knock it off and go to bed. We closed up shop on that one. But we had picked up everything else. And I think I was able to do that because I understood where I was and I knew what was going on and I had the feeling that the ground was in command of the situation. Okay, I think that's really enough of that. But that problem persisted throughout the rest of the orbital mission.

YOUNG — The bistatic radar was a source of continual confusion to us because we never knew what was coming next. It's kind of difficult to be right in the middle of doing something else and all of a sudden have to change your mind to something else. The crew has a Flight Plan which you can go by, and the ground changes it just for 2 percent more data of one kind or another, is asking for trouble. And I think we proved that on the LM jett day. I don't know where the mistake was in the LM jett day but there are three places it could have been. It could have been before it ever got off the ground if nobody ran through the procedures. I assume they did. It could have been in running through the procedures and not having the configuration right when we shut the thing down. It could have been that running through the procedures, the communications loop could have gotten messed up, in that somebody has to transcribe the procedures, somebody has to check them over and pass them to the CAPCOM and then he has to read them up to us. Then, Charlie is there standing, in the LM, in his skivvies, copying those things down and I'm copying them down and we may have misread some of the things that they gave us. And then, of course, in our execution of them we could have made a mistake. I really had serious misgivings about doing something that we hadn't practiced. We practice all the things that constitute abort-type situations like Apollo 13. We practice a contingency checklist. But we never practiced the phase of jump around, and grab here and there for LM jett. I was real nervous about being able to do that right.

DUKE — It ended up we didn't get to bed any earlier than if we had jettisoned it right at that day. It caused about two or three times more work the next day. Of course that ought to work, but it's only going to work if somebody on the ground has time enough to go over and run through the complete thing end to end in the simulator and make sure his base is right when he starts and his base is right when he ends. Then, he gets the procedures written right and there's a lot of changes that you have to make to do that sort of thing right. We practiced LM jett a lot of times. And it always worked.

SLAYTON — In that respect, we should have let you get tired that day and kept going instead of shutting down.

DUKE — I think so.

YOUNG — I would have sure felt better.

MATTINGLY — It turned out that we got to bed later than if we'd gone ahead and jettisoned it. We had more work to do the next day and I never felt like I was on top of what was happening.

DUKE — Nobody talked about that much after we pulled all the breakers and while we were waiting on that we were doing the same thing we would have been doing anyway.

MATTINGLY — Now you had a plan to store all that stuff in this order. But you can't do it because you have to review each item you have to stow and see how many of these things you need to use tomorrow. And can we get them in places that are easy to get to. As far as the SIM bay operation was concerned, the ground did it all and I just responded to their commands and what little I know about the experiment operation is what the ground told me, so I'm not going to comment on any of that. Dim Light Photography, I'm going to comment on that right now only to the extent that says how we performed it. The dim light photography that we did do was accomplished just exactly like the PI wanted it. Much to my dismay, we did two solar corona sunrise sequences. I used the tape recorder countdown for both of those and for the zodiacal light. They really worked out super. The countdown tape would come up the sunrise and say turn the camera off at sunrise. There would be the old Sun, and I think we got all that done. There's one section in the zodiacal light sequence where it was changing every 10 seconds. We were taking two exposures, changing the filter, and changing the exposure setting. I dropped my hand off the filter or something and lost track of where it was. I ended up having to skip two exposures. That's noted in my experiment checklist. I think I cover that later. The conclusion of a star chart that shows the pointing targets for each of the dim light sequences I think is a good thing. One time we caught ourselves pointing in a different direction. It turned out, that what had happened was that they had changed their mind on what they had wanted to take a picture of. But as a rule I always felt very comfortable and I knew I was pointing in the right direction because I could verify against the star patterns. I didn't trust the light shield around the 35 millimeter camera and consequently I darkened the spacecraft completely for all dim light and photo sequences in lunar orbit. Now, we did some on the way home with the dump sequences where we left a couple of lights on in the LEB flashlights that Charlie could use to read the checklist. All the ones I did in lunar orbit I did entirely in a darkened spacecraft. The operation of those things in the dark is really no problem at all. The magazines used were recorded and read down in real time as far as, which magazine, which frames went with each experiment. I have those things noted here in the experiments checklist. I think that really covers all the comments that are necessary to be said about this zodiacal light, solar corona, galactic light, Gum Nebula.

Attitude excursions during the CMC free periods were recorded. There was one set of pictures that required a free control mode and I missed that and left it in auto, but I knew that as soon as it started. I finished the sequence and I did not see any engine flashes, so I believe the data is correct. That experiment is noted in this checklist also. Earthshine - I don't think we did that one justice. Much to my surprise, when we first got to the Moon, looking at the Moon with a nearly full Earth, earthshine, you could really see an awful lot of detail. I had always wondered if earthshine photography was worth the effort. Had we done it on day 1, I'd have no question that we'd have learned an awful lot about features that are on the western limb of the Moon, which we'll never get a chance to photograph in sunlight during Apollo. Each day the features became less and less sharp. That was the reason that after PDI day I wanted to run that strip off, because I could already detect the difference in the earthshine sensitivity from the previous night. It was scrubbed and I tried to get it in the first thing after I woke up the next day. It kept getting pushed back in priorities. When we did get around to it I'm not sure that we really picked up very much at all. That's a shame because there really were some interesting things to be seen back there. You could see the outer rings of Orientale. You could see the Mare Rille between two of the rings. You could see the radial fractures, you could see some of the radial ejecta patterns. It was really a beautiful thing. I would suggest that if Apollo 17 has a chance to look at the moon with really full Earth that they really should expend a magazine and take a full strip back there.

UV Photography was accomplished primarily translunar and transearth. We did some in lunar orbit. We took some pictures of the Descartes formation in UV. I speeded up the times and centered around the Descartes light material. When I say Descartes, in this case, I'm talking about the crater Descartes and the light material that is just to the north of it. Skylab Contamination Photography was scrubbed in lunar orbit. Orbital Science Photography, I took as many of the strips as I could. Just having reviewed them, it appears that the prints I was looking at are accurate. Our exposure settings were rather poor. We did not expose all of our film. We only had one roll of film that was available for crew option photography. That was magazine Victor. I kept trying to pace myself on that, and save it. I figured that we had 160 exposures to last us 6 days at the Moon. And I kept trying to ration myself instead of taking a whole bunch of pictures the first time I saw something. As it turned out, we ended up at the end shooting it up just to finish it off. Not only did we spend as much time in lunar orbit, but a lot of the time when we could have been taking photographs out the window we were copying and executing the Flight Plan updates, which is really dead time. That's wasted time in lunar orbit.

I did a lot of work with the binoculars. I compared the binoculars and the monocular by using one against the other - by trying to observe, binoculars by just putting my hand over one of the barrels. Ten-power binoculars are, in my opinion, about the maximum magnification that a man can hand hold. I found that to really see resolution I had to lay one end of it up against the window in order to stabilize the optics, to keep them from bouncing around. I really felt like the binocular size was proper. It would be nice to have a wider view on them, because target acquisition was a little difficult at times until you learn how to do it. You'd look out and see a feature in your unaided field of view, then you'd want to look at it with the binoculars. The field of view with the binoculars was so small that it was hard to find the sequence of patterns that would lead you into the small area you wanted to look at. Or worse yet, you'd want to see something in there, then you'd look outside and try to put it into the perspective, and again it was kind of difficult because of the small field of view. The binocular small size meant that you could stick your head up in the corner of the window and look at it. I really don't think you could handle much larger. I found that the Hasselblad with a 250 lens on it was really too big to get between my eyeball and the window and see most of the things I was looking at. The SIM bay attitude that you fly, window 5 generally, includes the nadir, but not in the middle of the window, it's over in the corner. And for all the things you'd like to do you really feel like you're cramped looking out of window 5. You get a crick in you neck. Window 3 is a good one for looking away from the nadir and up towards the horizon. Got a good look at the northern part of the horizon because when I was awake we were generally in plus-X forward in SIM bay attitude. I could see up to the north very nicely. I had very few chances to look to the south of the ground track, because that was always when we were minus-X, and that generally meant that we were in a sleep period. So we didn't see nearly as much to the south of the ground track as we had anticipated. The overlays that are in the flight data file, used for preparations of the visual targets and overlays that you put on the maps to see what your field of view is from each window, turned out to be surprisingly close. We questioned that they were perhaps too small, that they had assumed an eye positioned too far back from the window. But in reality I actually had the feeling you had to decrease the overlay for window 5. Window 3 was probably about correct. Window 4 is much larger than is useful.

We took the gegenschein. Took a sequence of photos at the antisolar point, the midway point and the mobile point. Did that twice. If the star charts are right about where to point, we were, in fact, pointing in the right direction. The long exposures should show anything that happens to be out there. We did a water dump on the transearth coast. John pointed out if you look down-Sun at it you had a lighter spot in the sky. I looked out and saw the same thing. I think if we had taken densitometry photography of that area we'd have found that there was a little bright spot looking directly down the water dump. That may, or may not have any bearing on the gegenschein, but I think it's similar phenomenon and ought to be considered. One more thing about the binoculars, several times I kept trying to compare binocular and monocular vision and I could never find a discrete test wherein I could say the binocular saw things that you couldn't see with monocular vision. But I always had the feeling that I saw a fuller picture, and that there was more information, and that I saw more things when I looked through binocular vision. Several times through the mission I found that I got my fingers on the lens, or something, and I'd pick it up later and one of them would be in focus and one of the lenses would be out. I felt like the picture was missing. I'd start squinting with one eye, or the other, and find that I had a bad lens. I'd clean it off and all of a sudden everything was better again. Once I finally got the lenses cleaned off, and realized you had to work on it, and once I had adjusted the binocular I really didn't have to mess with the focus again.

Plane Change I was essentially a nominal burn. There was no problem with it. Comm was good. In fact the communications throughout the whole period were excellent. Supporting lift-off maneuvers were a piece of cake. We tracked the landing site landmark. We never did try to track the LM itself. I saw a glint of sunlight off of something bright. Sort of like the kind of a reflection you'd see from a wave out over the ocean. One time when I was looking with the binoculars, at the landing area, I believe I saw the glint off of the LM or maybe the ALSEP. And, another time I saw a glint over on the flanks of Stone Mountain. Right after that, Hank said that was in fact where the Rover was. It was nothing I could identify or pinpoint, but it was a flash of sunlight reflected off of something that looked entirely unlike any other features that you see around the Moon. As far as looking for the LM or the Rover or anything like that in 10-power optics I think it's really a waste of time and should never be pursued.

Rest and Eat Periods - rest periods are really okay, but the eat periods were continually being violated with Flight Plan updates of one form or another. It turned out that my favorite experiment in orbital science was the bistatic radar. That meant the ground couldn't talk to me for an hour and a half. I had a chance then to go to the bathroom, eat dinner, and get an exercise period or look at the Flight Plan. I think you really need those kind of periods every now and then throughout the day.

TPI Backup we talked about in the rendezvous portion. Midcourse backups were completely nominal and right down sequence. There are a couple of items I noticed in the Flight Plan leading up to the rendezvous point. First of all, we chose several photo targets and visual targets that were too far off the ground track to be appropriate. The photo targets may well turn out to have value taken near the horizon. They're hard to reach from the SIM bay attitude. The visual target, to be useful, really ought to be confined to those things that can be done within probably a 30° cone of the spacecraft nadir. The targets again will be biased toward the side that SIM bay will support. Because of the SIM bay attitude, you generally don't have a chance to look straight down and you can't use the full 2 minutes from plus or minus 45 degrees elevation, even when you fly directly over the target. Generally, you end up getting about a minute or a minute and a quarter, because of the SIM bay acquisition problem. You either see it early and you get to look at it a long time before it gets close enough to see, or, you fly over it and then you lose some time due to target acquisition, and the target is at best viewing position when you first see it and you may not recognize it unless you're really on top of the patterns that come in with it. I found that one of the things that saved me a great deal of time in doing the visual strips, and in just general operation with the photo targets and so forth around the lunar surface, was the rather large amount of time we spent preflight, learning the ground track and learning the significant features every 10° across it. This really paid off, in that you could just walk over to the window look outside, and you knew where you were. That would give you a good handle on where to pick up the cameras, how to set them, and what you ought to do with it. I think you can waste a lot of time trying to cross check between a map if you don't have that kind of familiarity with it.

A couple of times during the mission we recalled that we had a high O2 flow. Invariably it turned out that we had the suit circuit return screen clogged up again. I cleaned that thing off every night, and I was forever pulling off just piles of junk. It would take me 4 or 5 passes with the tape to pick off the screen on the main ones. When we went out and looked at the screens on each of the individual hoses there was always a lot of material in it. This was true even before the LM came back and brought all of its dust and dirt with it. This was true of just plain old spacecraft particles, little nuts and just little pieces of debris.

The use of visual targets at the subsolar point is probably a mistake in use of time, in that you just can't see very much. One of the things that bothered me a great deal the first two days in lunar orbit was the intense brilliance of the sunlight. I generally ended up with rather sore eyes or just tired eyes at the end of both the first and second days. I tried to wear the sunglasses but the problem you run into for looking inside you need to take them off. You want to pick up a set of binoculars and look at a detail, now you have to take the sunglasses off. The nuisance of handling one more thing between your eye and what you're trying to see convinced me to do it with my naked eyeball and grin and bear it. I never had anything happen to my visual acuity as a result of this, but it was a real annoyance having your eyes get so tired. After a couple days it seemed to me that my eyes had acclimated to the situation and it didn't bother them anymore. Someone really needs to think about the fact that this is going to happen to you. If there was some way of covering the windows so that you can look outside and not be fighting the sunlight, and at the same time use your naked eye inside, I think that would be a big help. And I would delete all visual subjects except of a very specialized nature in the vicinity of 30 degrees either side of the subsolar point; Because the features are so badly washed out and it's just a toll on your eyeballs to look at it.

When I changed the lithium hydroxide canister at 144:50, it was the first time I noticed that we had a swollen canister. It was just slightly sticky. Later in the flight we had one that was extremely difficult to get out and we'll discuss that when we get to that point. But the first one was 144:50 where it was slightly swollen, but that was one I jettisoned in the LM. The VHF communications with the LM, I think we commented on earlier as being pretty outstanding. I could hear the LM comm right up through landing and through T1. Part of their comments were going on with the checklist and so forth following T1. The tape recorder we had talked about using it for these time sequences. One of the problems that had been raised on previous flights was the idea that the tape recorder seemed to run at non-uniform speeds and things that were recorded on the ground didn't sound right when they got in the air. So, I made the decision to do our tape recording for the time sequences in flight instead of on the ground, then, whatever biases we had due to being inflight would be the same that the recording and the playback would be the same. It turned out that there weren't any differences they could find. The tape recorder seemed to be very stable as long as you Velcroed it down and left it sitting somewhere. But if you picked it up in your hand and shook it, or did anything with it, why you could hear the speed change. This was particularly noticeable when playing music. If you were touching it, just trying to hold it in your hand, it would cause the speed to oscillate. I ran several checks against the clock. In one 15-minute check, it was off just about 1 second, which I thought was pretty repeatable, so I continued to use it. Some time during the mission we found that we could not depress the little red button on the Sony recorder to make a recording. I was going to record some RCS sounds while I was in lunar orbit. I went to do this and was unable to get the button to depress. It had worked during the translunar coast because that's when I made my solar corona and Zodiacal light tapes. During transearth coast I noticed that once again it was working. There must have been or must be some foreign particles floating around inside the tape. The batteries on the tape recorder lasted much longer than I would have anticipated. We have one that seemed to last for only two tapes. And it really sounded terrible, like it needed to be changed. I subsequently played six tapes on one battery. I changed the battery not because it wasn't working, but because it was coming up on one of the low light level passes and I wanted to have a fresh battery in there to make sure I hadn't jeopardized the timing of it someway. I found that the number of batteries we carried was far more than sufficient. I wouldn't cut any off, but it was certainly enough. I had suspected preflight that there wouldn't be enough. I played all of the tapes in flight at least twice and probably half of them another time around. We used probably half the batteries. The next item was on the landmark, F2, I think, that I tracked on rendezvous day. I was given a crater in Mare Smythii. The crater was so large that it seemed to me it was an inappropriate target to be tracking for landmark correlation, so, I selected a small feature on the northern rim, which shows up marginally in the sextant photography. I was marking on one small feature on the northern rim of the crater rather than on the center of the crater itself.

12.0 LIFT-OFF, RENDEZVOUS, AND DOCKING

YOUNG — Lift-off was nominal. The pitchover was on time and we flew the most beautiful profile I ever saw. Shutdown was on time. There was one thing that we did differently.

DUKE — We closed the main shutoff in system A before ascent.

YOUNG — Before we turned on the ascent feed, the main shutoff in system A was closed. When we got into orbit, we used system B and crossfeed, which effectively gave us two systems. It worked beautifully. I never had any doubt that this rascal was going to work properly. The control system is just a champ. The insertion was nominal. I forget what the residuals were, but there was nothing to trim.

DUKE — Less than 1.

YOUNG — Two feet a second and asked for 10 up. We pitched up and locked on to the spacecraft a lot sooner than I thought we would. We were about 150 miles out. They redid that table in the time line. Charlie saw the command and service module in reflected sunlight first. This was right after we pitched up, which must have been 2 or 3 minutes after insertion.

DUKE — We inserted 170 miles behind and closing at 492.

YOUNG — I could not see it through the COAS. If I moved to the side of the COAS, after we had radar locked on, he was right in the middle. I was glad to get that data point, because everybody had been asking, can you see him in the reflected sunlight? And I answered, "Yes, I can see him at 150 miles." We kept an eye on him until he went into darkness. That's a good way to be able to check the state vector and make sure you don't have a bent radar.

DUKE — Once we got locked on the PGNCS needles, the 50/18 needles, the AGS needles, the radar needles were centered. I forget how many marks we ended up with but it was about 25.

YOUNG — It was a lot more marks than we needed. In fact, they said we could proceed if we wanted to.

DUKE — So we did, 2 minutes early.

YOUNG — Two minutes early. The TPI solution for the PGNCS is the one we burned. I forget what the numbers were, but it was right down the track.

DUKE — We lost radar lock when we did the burn. It was right at the edge of the limit, and when we lit the engine, we lost radar lock. Until that time, we hadn't lost lock. And of course, all we did was call up P35 and reacquire. When it changed the W matrix in P35, I either hit the enter button too fast or hit another button, and the W matrix didn't get changed. When I called it up again to look at it, it was the same as it was before TPI. I redid it and that resulted in our getting a few less marks for the first midcourse. I think it was 5 as opposed to 7. Five is adequate. All the TPI solutions agreed perfectly. Our onboard TPIs and the ground agreed within a couple of feet per second, so we decided to burn the PGNCS. At the first midcourse in all three axes, the biggest one was 0.9. We burned the first midcourse on the PGNCS, and there was no disagreement between us and the command and service module. On the second midcourse, the numbers were so small that we shouldn't have burned, but we went ahead and burned them anyway. I would think that it wouldn't be that smart, but maybe it was. I did not make a line of sight correction until we

were at 7000 feet. Beautiful targeting. I probably could have waited until I was closer in than that, but I didn't really feel too good about it. The braking velocity was 29 feet a second and that's exactly what it was. Line of sight control was just beautiful all the way in, and what the needles were saying agreed very well with what was happening on the lunar module until we were close in. Then as you noticed, we went out to the north about 70 or 80 feet and brought it back in. When we were about 600 feet out, I could see down the side of the service module, so I knew we were out just a little to the side. As opposed to the usual Kamikaze brake that I usually make, we kept it very conservative. We talked about that before the mission. We decided that we would always keep the braking within something that the command and service module could do. This means that, contrary to the braking gates that we use in the LM, you sort of have to lead them. In other words, at the range that you want to be at, you almost have to be at the braking velocity to give the command and service module a fighting chance in case it has to do it. I never had any doubt that we would do it all ourselves because that machine was working so beautifully. We just closed in and it was so good I wanted to do it again. It was really slick. After you finished your maneuver, they wanted us to do the 360-degree yaw. We talked them out of that because I was already in position, and all I had to do was the 360 yaw. We did that first and then you did your pitchover to the attitude where we took the pictures of the bubbles on the command and service module. You did the roll over and then went to the docking attitude. I think that the 16-millimeter camera doesn't really reflect on those bubbles too well, unless somebody studies it very carefully. There were sure a lot of bubbles out there on the thermal coat and on the surface. Then we went to the docking attitude and came in and docked.

72-H-534

72-H-533

YOUNG — PGNCS and AGS were in perfect agreement. The nav was beautiful. We'll get the numbers for this at a later time.

MATTINGLY — The things that I did differently on the command module side, as it turned out, didn't really have any effect on the rendezvous. Because of the

72-H-642

concern for the glitch in the IMU, we did all the attitude maneuvers for all the attitude controls in the SCS, and the navigation was being done by the CMC IMU. They worked those procedures out, we talked about them, and I wrote down some notes that Stu gave to me. It was a very straightforward thing. It was very similar to the way we would handle a no-IMU rendezvous. We had practiced that kind of thing, and the only thing that was different was when you needed to know a precise angle, we only had to go to the IMU and fly to that angle. So you had the precision but you used the procedures that we had gone through on some of these failed IMU malfunctions. So it turned out that we could draw on background experience, and it worked out real nicely. We got a lot of marks. I had 20 and 22 marks prior to TPI. I really thought I was going to pick you up at insertion. I think without a better state vector I wasn't going to find them. You'd never find it in the telescope. With the sextant, I think I would have seen you at insertion with the Sun angle, had you been in the field of view. I'm not at all sure I could have tracked it because the auto optics chatter was just enough to make it difficult to track. And I kept searching but I never did find it. There was a little dot near the middle of the sextant, which apparently was part of the sextant optics, and I tracked that guy for a while. It was sure being super because it was always in the center, no matter what I did. This is before you did your tweak. And then I caught it.

YOUNG — While we were sitting up before we got into orbit, we heard the VHF radar lockup. Were you locking on us?

MATTINGLY — I talked to you on VHF.

YOUNG — But we heard the tone; I forget where it was.

MATTINGLY — Before we got into orbit?

YOUNG — Yes.

MATTINGLY — Yes, I locked up on you just prior to the insertion. I gave it a try, not anticipating any success. You were chattering and I figured I'd just do it for drill, and it locked up. I blocked the inputs until I received a comparison with your radar. But it had been correct. And you didn't stop talking. You were talking in the middle of my lockup tone.

YOUNG — I know it.

MATTINGLY — I didn't think it would work, and it worked just super.

YOUNG — I was really surprised.

MATTINGLY — And the correlation between the VHF and the radar I thought was amazing.

YOUNG — Yes, it was really beautiful.

72-H-644

MATTINGLY — Everything on my side was just better than I've ever seen it in a simulator. I received a state vector from the ground, and as soon as you moved into the darkness, I picked you up in the sextant. It was a flashing light. I noticed I didn't pick you up in the telescope until I had you in the sextant at 100 miles. I picked your flashing light up in the telescope at 70 miles. And that was the first time I'd been able to recognize it in the telescope as being your beam.

YOUNG — That old 400-mile beacon is pretty good.

MATTINGLY — With the sextant, it was just beautiful. It's like a star. And at 50 miles, I had the Earth in there along with you and it still showed up nicely, even though the Earth was right in the field of view, which some people had worried about. I had 10 and 13 marks for the midcourse 1.

YOUNG — Yours was better than ours.

MATTINGLY — Everything was working so well that the marks were just pouring in. And I made a correlation of the boresight between the telescope and the sextant and where you boresighted. In the telescope, you were located at the bottom and about one line at the bottom of the crosshair pattern of the sextant and about one line width to the right. And I thought that was super agreement between those two instruments.

SPEAKER — Something less than 0.2 of a degree.

YOUNG — Yes, I think down at the bottom is 0.1.

MATTINGLY — It really surprised me how well it worked. And everything else we did on the rendezvous itself was nominal. Docking - We got out of sync with ourselves on this business of taking the photo sequence. I think we really did the right sequence, taking you first. I got the impression that both you and the ground were getting very impatient about getting started on docking, and it had been my plan before flight, my stated intention, that 15 minutes before darkness I was going to start my approach. I would do the photos and all that up until that point. And then it seemed like everybody thought I was being superconservative, but I had 20 minutes before darkness and everybody was getting ginchy that I wasn't making the approach.

YOUNG — I thought we made our approach right away.

MATTINGLY — I guess I wasn't concerned about the approach in the dark after a previous experience. There was no reason to waste gas hurrying. So I didn't. And we got there about 5 minutes before dark. And the only thing I would change by doing it again, I would approach faster. It was very obvious that you could see the LM attitude dead band. I was trying to play it cool and not chase the line of sights. But I'd lose my courage. Every time you get a little bit off, I'd go right over and several times I caught myself moving back and forth, chasing your attitude dead band.

YOUNG — When you approached, I was looking right at you, and you were really lined up well through the COAS. I could see you through the COAS. And as a matter of fact, if they painted a line underneath the window with a mark down it, you wouldn't need that docking adapter. You could dock with the line and make it every time.

MATTINGLY — I had that impression. I had the impression that if the target had fallen off the LM, it would have made very little difference in docking. The one thing I'd do differently is I'd approach faster. It'll make your line of sight rates less expensive to control. They aren't going to be big; they're going to be dead band kind of things. The other one was that I got contact and I was anticipating hearing it like everyone had said you would. Apparently, I approached slowly enough that I didn't hear anything. I saw the LM jiggle a little bit, and I don't know if you heard me contact.

YOUNG — I didn't hear it either. We didn't feel it either.

DUKE — I felt it; it rattled a little bit.

MATTINGLY — Nothing happened, so I thought, guess we're going to do something else. I decided I'd thrust at it but nothing happened. I didn't feel any motion and I was just ready to thrust again when the barber pole came up. So I must have just had it laying there, and it just needed a little push to get the capture latches in. And again, I think a faster closure rate would have made more positive docking.

YOUNG — You could have pushed the LM backwards.

MATTINGLY — But there was no rebound on the part of the LM. And they always talked about that, don't hit hard enough to capture it, you'll knock it away. But it didn't. It just lay on the nose there, and all I had to do was push it a little bit.

YOUNG — As soon as you said barber pole, I shut the PGNCS off and we were there. No thruster, no extra thrust applied, none of this fighting each other.

MATTINGLY — I went to free as soon as we contacted. The retract was just like the original one.

YOUNG — Very slow until it got there, a hangup on the edge, and all of a sudden the docking lights just appeared.

MATTINGLY — Same pattern. For some reason, we didn't take photographs of this. I don't know why. The ground called and said don't take photos.

DUKE — We did one thing procedurally at insertion. We had a lot of dust and pebbles floating around in the cockpit with us. We did turn on the cabin fan and left helmets and gloves on until docking, because we had so much dust in there.

YOUNG — That didn't clear any dust out because you have to open the inflow valve to get any of that stuff in the suit loop to clean it out.

SPEAKER — It just circulates it around. It has a filter behind it.

SPEAKER — Does it have a filter behind it? Well, it didn't clean much of the dust out.

13.0 LUNAR MODULE JETTISON THROUGH TEI

MATTINGLY — After docking, we went through our transfer items even though we knew we were going to be retaining the LM and going to bed. From my side, the time line entering the LM was a little bit slower than I had anticipated even though I pressurized the cabin prior to rendezvous. Taking things out and finding a place for them just seems to take a little bit longer. Perhaps that was because I'm methodical about it. The first thing we did was pass in the vacuum cleaner. I had checked the vacuum cleaner operation only to the extent that I turned it on and it worked and I turned it off. I didn't try to vacuum clean anything. I didn't try to verify that it really was sucking anything up. There's some question in my mind whether the vacuum cleaner really ever worked properly.

DUKE — It did. The screen was covered with dust. It probably was so covered that it stalled out, and that's what failed it.

MATTINGLY — In any event, some time later, I went into the tunnel to get something, and the vacuum cleaner was laying there making some funny little noise. I noticed the switch was on, but it didn't sound like it was running so I turned it off. It didn't interest me enough to see if it was still working. I think it had probably failed then. We tried it later and I heard a hum. I suspect it failed at the time I found it the first time. That was within an hour of the time we started with it. The tunnel operations were just like you'd expect them to be; all the latches were made, including latch number 10. This time, latch number 10 fired properly. We had no anomalies at all with the docking latches, nor with the probes. One other comment on latch 10 that I don't think we included before was that - when we went to undock, latch 10 cocked itself on the first stroke. It felt like it took about half the effort to cock that first stroke that all the rest of them took to cock on either the first or second. So apparently, it had not fired at all.

LM equipment transfer - when we started bringing that stuff over, our original plan had been to stow everything in this EVA stowage or entry stowage case, whichever was more convenient at the time we brought it over. Take our time and do that and then come over and do the LM ,jettison. We altered that because of the plan that retained the LM and having to put your suits on for LM jettison the following day. We just dropped the whole plan of trying to stow things item at a time. In general, we just sort of stuffed the stuff on board and decided to worry about it later. I don't think we really had any other choice considering the fact that we couldn't stow the suits and things. The command module filled up with LM dust and rocks and things almost immediately. Within an hour, it was very noticeable that there was a coating of dust on all the instrument panels and all the surfaces. You'd see little rocks float by in front of your nose. I was surprised how rapidly that stuff all had diffused in. It came over as soon as we brought the first bag or the first suit, or whatever it was. That stuff was just coming off of everything and it never stopped. The command module cabin fans were on at the time of docking. I turned them on right after docking and before removing any tunnel hatch equipment. They were working properly at that point with the cabin fan filter on. They failed some time EVA morning. The material we brought in we just stashed away.

DUKE — We tried to vacuum the suits and some of the bags that were dirty like the big rock bags and found it was almost totally worthless. You could do a little bit, but the best method was to take a damp towel to wipe things down. We were able to get some of the dust off this way. Fortunately, most things that were dusty went over in DCON bags. That was a lifesaver. Once we opened one of the DCON bags just a little bit to see what bag was in there. The dust floated out and we closed that in a hurry. That was a real mistake. I think Apollo 14 did the same thing. The transfer of the equipment was expeditiously done just according to the Time Line Book from my side. Ken was doing a great job taking care of everything as I passed it over. Had we been on a nominal two-rev-to-jett time line, we'd have been adequately prepared to jettison the LM at the time. I think it's a very loose time line. I think Ken does have time to stow things, at least temporarily, and we don't just throw them in there. The samples did not need to be vacuumed except one bag before I put it in the DCON bag. That was the big rock bag. Everything else was in good shape. Since we were going to retain the LM, we had to copy up about 10 pages of changes to the checklist. We started with the dock deactivation unstaged in the Contingency Book, drying out the water boiler. We reverted back to the Time Line Book the next day when we got back into the LM. The configuration on the LM side for the rest period was a complete powerdown. I felt like it took us 90 minutes at least for the water boiler to dry out. I really don't think personally that we saved any time by making that decision to retain the LM. I feel like we could have had an 18-hour day, or whatever it turned out to be and could have jettisoned the LM by keeping our suits on. But that was not the decision. The decision was to keep it. So, we had to take the suits off. John and I both doffed the suits in the LM. I thought it was quite a hazard over there floating through the LM with all that dust and debris. A number of times I got my eyes full of dust and particles. I felt like my right eye was scratched slightly once. I think the mode you want to operate in is get that gear transferred and get out of there as quickly as possible, get that beauty closed up and jettisoned. The ground could tell us exactly when we got it closed out. I would be willing to bet you that it was almost the same time, if not later, than it would have been if we had gone on and jettisoned the LM. The suit doffing was the same problems that you had. We ended up working up a new technique at least with the A-7LB. Getting it down and pulling your feet out first before you pulled your head out. That worked just great. The next day we got up, IVTed to the LM, and started the power-up before we got suited.

MATTINGLY — We started the power-up first.

DUKE — I never had any confidence in my set of procedures that we really had everything in the right configuration. We skipped from page to page and book to book. We seemed to have gotten everything going. The computer came up; the up-link came up okay. We didn't bother with the AGS. It looked like we had a good P30 in there. The state vector was in; the no-DAP light was off. All the gear seemed to be running, the COMM seemed to be up, we didn't fire the jets. We didn't do an RCS check, but looking back over the procedures it looked to me that we had all the necessary breakers in to fire the PGNCS RCS ,jets. One switch I did miss was the PGNCS mode control; it was in ATT. HOLD rather than in AUTO. The no-DAP light was out, so it should have held its attitude, and the ground should have had no problem getting the RCS burn off to the deorbit burn. However, when we jettisoned the LM, it just gently floated away with a slow roll and pitch, without any jet firings at all. Looking back over the procedures I don't know where the error was, or if there was one, or if we had a mechanical failure. We'll just never know as far as the mechanical failure. I think that we got the switches and the circuit breakers all pushed in. But it was a set of circumstances any way in which we didn't have a chance to review with any great degree of accuracy on board. In that kind of operation, I think you're just setting yourself up for something, which happened. Suit donning that morning was okay. We had the same amount of problems we normally have.

On close-out, the LM was still as dusty and debris covered as ever. We had the same problems with the wrist rings as we had on the lunar surface. We managed to get buttoned up all right and on time for the jett maneuver.

MATTINGLY — On the LM jettison, we really got behind at the last minute. I thought we were doing pretty good. When it came time for the suit integrity check, we got behind. We didn't have sufficient time to let the whole thing play out and stabilize.

YOUNG — Let the suit flow come down to 0.2.

MATTINGLY — We had to rush the integrity check in order to get it done and get off at something reasonable. At that, I think we got off about 30 seconds late. Even with the suits on, the tunnel pyros make a very obvious noise.

YOUNG — You said that the LM was maneuvering in pitch and roll?

DUKE — As it backed away, it started a slow yaw for the LM. It was LM yaw and a slight pitch up if I recall. But it never fired a jet, not one jet.

MATTINGLY — Very slow rates; we must have imparted an extremely small impulse. I don't think there's any question on whether it fired any jets because when you folks rendezvoused, I could see the jets firing. They fire in the daylight. The ones that were pointing at me looked like little flashlights going on and off. The others, I could still see some effluent coming out of the exhaust nozzle. I don't think there's any question that it never fired an engine at all in any axis.

DUKE — I agree.

MATTINGLY — Separation maneuver was a nominal thing. They changed the maneuver from the preflight value, but it was executed nominally. It was small; 2 foot per second. We deleted the shaping maneuver on plane change 2.

DUKE — Right after that, Ken, we had to jettison the mass spec boom.

MATTINGLY — Yes. The lithium canister that I changed at about 181 hours was stuck. It was number 12 that came out of the V chamber. We had to jiggle it an awful lot. This was on the night of rendezvous. We had to jiggle that thing an awful lot to get that thing out. My concern at the time was tearing the little cloth strip that pulls it out - jamming it and then finding that I was unable to get it in or out and thus would be unable to change the canister in the opposite compartment. Ground told us later that there was no way we could tear that little strap off. The canister did not look much different, but it certainly was swollen good and tight. Somewhere along the line here, we were unable to get the mass spec to come in. It was before LM jettison. You saw it in the LM window.

DUKE — Yes.

MATTINGLY — Out the window. They told us to retract it for LM jettison. I went to retract and they said it stalled out. I had a barber pole. You looked out and could still see it. So, we went to extend once and it went just a little way and stalled again. We cycled it in and out a couple of times, and Charlie was able to tell that the mass spec was neither at full extension nor was it coming in. It was just moving back and forth a little bit. So they gave us a pad to jettison it and that all went quite nominally. The thing that was most significant about the boom jettison was that, when it left the spacecraft, it had practically no rates of any sort except in the translation. Just as stable as it could be. It was moving out rather rapidly. It just looked like a big arrow going out. We should have some pictures of that. But that stuff came off without any trouble. I don't remember even hearing the thump when that thing went. I don't remember a sound or anything that was associated with the mass spec test. The satellite jettison, we did on time. It was perfectly nominal as far as we could tell. It was done in the dark so we didn't see anything. I don't remember any sounds associated with that either. It did its thing on time and barber pole changed state just as advertised. During all of that time, we were getting Flight Plan updates almost by the minute. The ground had taken over control of the SIM bay and was operating it in a real-time mode with us executing it. We tried to run out the rest of the pan camera and get as much mapping camera coverage as we could prior to TEI. My concern at this point was that at the pace we were going, we might not have a chance to even settle down and give TEI the consideration it ought to have. But the Flight Plan all settled down the rev before TEI. We sat there and did essentially nothing but think about TEI, and set up for it in the last rev. I thought that was a super thing to do. I felt really comfortable; I think we ought to always do TEI that way.

YOUNG — I do too. There were a couple of changes to the cue card for TEI that seemed unnecessary to us.

MATTINGLY — I sort of went along with all of the changes we were getting. At the time, we were changing some things because we had to because of a changed time line. It seemed to me that we were also changing other things that didn't have to be changed. A good example was the suggestion that the circuit breakers for pitch and yaw, gimbal number 2, be pulled out during TEI burn in order to protect against transfer from primary to the secondary and then a drop out due to the alternate circuit on the secondary. This condition has existed as long as Apollo has. It's been discussed and rehashed as long as Apollo has been around. It seems to me it's a very poor practice to change those kind of procedures in real time unless some material condition is developed that causes you to alter your standard practice procedures.

YOUNG — I agree 100 percent.

DUKE — The pugs operation. We didn't touch it as per directions from MCC, but before the burn they put the gauging system into the auxiliary position versus the primary. We flew with it in the auxiliary position. You could tell every time that you uncovered one of the point sensors the quantity would jump and the unbalance would ,jump. It would then settle out. At shutdown, we were looking at something like 3 percent on the fuel and 5 percent on the oxidizer; maybe 3.5 to 5 percent. But there was about a percent difference or so in the unbalance meter. We had just started responding to that when we got shutdown. I think it was going toward the decrease position, but I don't recall. But except for that anomaly I think the thing worked as advertised throughout the burn.

YOUNG — I don't think there's any further comment on TEI. The burn.

DUKE — Right on.

YOUNG — Within a couple of milliseconds, it seemed like. The residuals were nothing. Didn't even have to trim.

DUKE — Yes. All you trimmed were X and Z and they were small. We had 1 foot per second.

MATTINGLY — Post-TEI, we did a lot of the photo stuff. It was one of the places where we got interested in what we were looking at. I guess we missed a couple of things. It was one of the problems that used to catch me in training all the time, but the only time it caught me was this one time in flight. When you call up P20 option 5 to get a maneuver to the attitude, you have something like 8 seconds from the time the V18 comes up until you can get an ENTER on it in order to get the rate drive going. This was one of those times where I didn't sit and watch it. And sure enough it required another maneuver. The ground had to call us and tell us we hadn't started P20. I almost missed it the next time because I was still looking out the window. I would suggest that if you ever need a program like that, again, the computer should have some way to call your attention to the fact that there's an activity on your part required. It is not uncommon to have maneuvers to attitude which take 12 to 14 minutes. It's real easy not to be sitting there watching the DSKY for that 8-second window during that time. Future programs should provide the computer with the capability to ring a bell or light another light or do something which would attract your attention at the time another activity is required.

14.0 TRANSEARTH COAST

MATTINGLY — I'm looking through to see if there are any notes. Most of the things that we did on transearth coast we did under direct ground control. We went about our business with stowage, EVA preps, entry preps, and all that kind of stuff. But we let the ground run the SIM bay real-time calls, and I felt that that was probably the best way to do it, rather than have us all sit there and watch the clock. PTC on the way home turned out to be no more difficult than PTC on the way out. With a light vehicle, PTC was another thing that surprised me. We were able to set up PTCs that would end up with the nose describing a 10-degree cone. System Operation. I don't know of any anomalies we had in systems. You got any on your list, John?

YOUNG — Yes, several, as a matter of fact. The cabin fan stopped.

MATTINGLY — The cabin fan. And that went out on the day after TEI. In fact, I guess it was the morning of EVA prep?

YOUNG — Right.

MATTINGLY — All of a sudden, the fan let out a big moan. It sounded like it dropped several steps in frequency and made a big moaning sound. We never turned it on again. We taped up the cabin fan inlet screen which had an awful lot of junk and trash that was being sucked up against it. I cleaned it off once that I remember. Just an awful lot of stuff got up there. We finally ended up cutting out a piece of the map and taping it over the screen to keep stuff from backing out away from the filter. What else do you have on the systems?

YOUNG — I redid this list last night. Stop the tape and let me get organized. What was the first one of those that you read?

MATTINGLY — We talked about the cabin fan filter. One of the things I see on this list is that the gamma ray boom finally stopped just before entry. We tried retracting it and it didn't come in. They said it was sufficiently retracted so that we could jettison the service module without jettisoning the boom first.

YOUNG — The digital event timer.

MATTINGLY — Yes, the digital event timer started to act up during rendezvous. I noticed it at TEI that it lost track of the ignition time. Prior to TEI, it was running intermittently, but coincidentally, it happened to run nicely at TEI. After that, it seemed like it was more often not running than it was running. Each time it would step, it would count. It sounded like it was making a count every second, but it would lose track somewhere in the tens and units digits. It would just come up with the wrong time.

72-H-637

YOUNG — It's clear to me that people who aren't concerned with space flight don't realize how much the crew depends on the digital event timer. We used it for things like timing the 10 minutes the MEED was supposed to be out, which didn't work; timing her down to the reentry reference time, which it didn't work for; and timing up to every burn, which it didn't work for.

MATTINGLY — The clock we're talking about is the one on MPC panel 2. As far as I know, the one down in the LEB continued to work.

DUKE — It worked for the Skylab. When I used it down there for the Skylab contamination, it worked.

MATTINGLY — At the end of the mission, I noticed when you would go to reset on the main panel DET, that you'd hit reset and it would just sit there and continue to cycle. You'd have to hit stop to get it to stop spinning. But it did that only during the latter stages of transearth coast. It didn't do that when it first started acting up. There was no powder or anything like that visible in the window such as was reported by Apollo 15. You mentioned another problem there, John.

YOUNG — Battery compartment.

MATTINGLY — Battery compartment. Yes. We had just put battery B on charge the morning of the EVA.

72-H-571

DUKE — Bat B had just started charging when we got a slight odor like it was burning insulation. I took the battery off charge and looked at the bat compartment and it was up over about 3. So from then on, we started a series of vents. Up until EVA, we initiated a series of little vents to get that pressure back down. We'd vent one and then close the vent, and it would immediately start climbing back up rapidly to about 1.5 and then maybe 0.1 of an hour from there to about 2.7 or 2.8. I personally didn't want to charge any more batteries and fortunately we didn't have to, because the batteries all had good charges on them. So we just left them like that for the reentry and we didn't charge any more. And the explanation we got from MCC was that it was the battery compartment, the type of compartment. Due to the long discharges on LOI and DOI and then subsequent long recharges, the batteries were venting more than normal. The compartment was tight and that's why the pressure was building up. That sounded pretty plausible to us. The only question I guess we still had was why was the thing still venting so long after those charges.

MATTINGLY — The thing I think was significant was that we had a definite odor, an acrid smell that could be attributed to insulation cooking. It smelled to me like things that you smell around batteries. That was down in the LEB. I'd been on the couch and we shifted seats for some reason, and I crawled down and John crawled back on the couch out of the LEB. When I moved down there, I noticed it immediately. But it's the kind of odor that when you're around it for a while, you quickly wipe it out. I think if it built up any slower, you wouldn't even notice it. I think it was distinctly different than the other cockpit odors that we have.

YOUNG — I didn't smell it because as usual my nose was all plugged up and so was my head. I just didn't smell it.

DUKE — But they were convinced that we had good batteries and sure enough, we did. At entry when we brought them on, they took up the load. Boy, they were just great and worked like champs.

YOUNG — Okay. The IMU CDU glitch.

MATTINGLY — Okay, IMU. I guess we got our glitches on that starting the day before entry. That would have been EVA day.

YOUNG — It was EVA day.

MATTINGLY — But it was late in the day. And we had an ISS light and a PGNCS light came on. The first time it came on, I didn't notice the ISS light as it went out. And I think both you and John saw it. All we had was the 3777 alarm remaining on the DSKY. We compared the FDAI with the NOUN 20s, and they seemed to be in agreement. We tried VERB 40 and that didn't make any difference. I think we then had another master alarm with the same sequence of events, where the ISS light came on for just a flicker and then it went out and left us with a PGNCS alarm. We did all kinds of things. The ground gave us some angles to fly through, to try to excite the CDUs. We went through all that and never did come up with anything that could cause the alarm again. I don't think we had another alarm on it until entry morning.

YOUNG — That's right..

MATTINGLY — We must have had a half dozen on entry morning. On one occasion, the ISS light came on and remained on, although everything else continued to operate normally. Then finally, it seemed like one of them went out and John heard a click or clang of some sort in the LEB. We looked around and it looked like the tool E, which had been strapped to one of the handholds on the nav station, might have banged the panel.

YOUNG — It banged the panel just above where - implausible as it seems - it was hanging on the right optics handrail. It banged the panel just above where I think the PSA module is for the computer. At the same time when the ISS light went out, it was underneath it.

MATTINGLY — I didn't see the tool E hit the panel. But when you kicked it later, we did see where that was. That was underneath the verb-noun list.

YOUNG — Yes.

MATTINGLY — But it was right next to that rails of the structure that goes between the verb-noun list and the star codes.

YOUNG — Yes.

MATTINGLY — That's where you'd have to put the light out. At least when you tapped it, the light went out.

YOUNG — Yes.

MATTINGLY — That was the last ISS light we had. Our feeling on that is the fact that practically every problem we had with manned space flight has been due to some kind of contamination floating around in zero gravity. That's probably what it is. Since they are getting it back, they ought to be able to find it.

DUKE — One time the light was on steady though, Ken. I took the tool E and tapped all over the panel, which had no effect on it.

YOUNG — The trouble is, that isn't where any of the gear is located.

DUKE — I know.

YOUNG — It's located in that box behind the close-out bend.

MATTINGLY — In general, we had several bags that didn't spill when you put them on the food port. I don't know whether it was coincidental or what, but it seemed like there were more of these bags that had been thermally sealed right across the entrance valve, more in the cocoa packages than in all the others.

YOUNG — That's right, all put together.

MATTINGLY — Whether that was really the way it was statistically or just...

YOUNG — At least four of them

MATTINGLY — It sure catches your attention when you put that hot water all over yourself.

YOUNG — Not only were the bags sealed across the entry port, but for almost every cocoa bag that we opened, part of the bag itself was sealed so that you couldn't get any water in it. Remember where the bag was just flattened together and welded together? All it meant was that you get more cocoa per ounce of water.

MATTINGLY — I think one of the things that the people that build these bags need to recognize is that you can't make a visual inspection of the bag before you fill it to determine these things. Because a good bag looks exactly the same way under the vacuum package. They all look like they're sealed together. It's only when you put pressure on it, you find that it won't inflate. Then you know that you have a bag that's sealed over.

DUKE — One of them that failed did fail right up there near the filler neck where it looked like it was vacuum sealed. When I put the water in there, the water just squirted out to the side.

SPEAKER — One of the holes?

DUKE — Another one failed around the seam. It was where the seam was crinkled. I put a couple of ounces in it and was starting to mix it up, when I squeezed it and it broke through. We also had a failure of a juice bag on the lunar surface. That was a grapefruit one, I think. Besides one we couldn't fill at all, we had three that once we got water in them, they broke.

SPEAKER — Any other systems?

MATTINGLY — I guess one of the things we need to talk about is the chlorine injector. Some time during the EVA, perhaps it was post-EVA, we lost it. It may still be in the cockpit. I had taken the chlorine injector in the little bag that it comes in and snapped it up on the compartment right over it. I kept it there for quick use. Apparently I knocked it off sometime around the EVA. We never did find it again. So we did not chlorinate the water the last two evenings.

DUKE — And it was delicious.

YOUNG — That's right.

MATTINGLY — It was really good. We ought to confess to that anyhow.

YOUNG — Yes, I want to discuss this other thing that we had. We had an O2 flow high light, right after we finished dumping the water out of the cabin. The ground kept asking us if we had sealed the side hatch. There's just no way you can not seal the side hatch. There's just no way. Because the little plug goes in there and that's the end of it. The problem was that the inflow valve was all covered with dust, and of course that makes the flow go up. But that should have been the first thing that occurred to them. But they asked us at least twice and maybe three times if we had sealed that hole up. I was nervous as a cat about opening it up in the first place. I thought that was unnecessary harassment. (Laughter)

YOUNG — The plug is right there, Deke. And it's as big as the end of my finger. You can't miss it.

MATTINGLY — Okay. I think that's all the systems things.

YOUNG — Yes. The DET, the O2 flow. What about the umbilical cord?

MATTINGLY — Okay. It's really not systems. It's a problem which is not new, but somebody sure needs to remember in space flight that you just shouldn't ask a man to operate in zero-g with a half-inch-diameter umbilical strapped on. One of the beautiful advantages of zero-g is that you can move around, you can hold a position, and you can do things. The amount of effort it takes to resist that torque, which comes in the form of both rotation and translation forces, is a continual nuisance and makes it difficult when you want to get up next to a window and look out of it. You can't just go lay next to the window. You have to get into the position, and then you have to anchor yourself with your feet to your hands. Your hands are the things that you are trying to use. Those are the tools that make your being there worthwhile. It's just a continual nuisance, something that I realize Apollo is going to live with. But it's something that just never ought to be included in the spacecraft design. Communications shouldn't have to go through an umbilical that size. We practice and train every day with a communications umbilical that has a tiny little wire that goes up to your head. You climb in a real spacecraft and you put on something that looks like it could carry half of the United States telephone communications. It's just one of those things that for long-duration flight, I think it's going to become an extremely distracting influence. It's the one that goes to the suit. The one that goes between the crewmen and the bulkhead, and the one that goes between the suit adapter and the comm carrier. I think we've overlooked that comm failure a little bit. Charlie had an intermittent transmitter in his right mike.

DUKE — It was intermittent where it connected to the CWG adapter.

SPEAKER — But within the mike circuit?

DUKE — It was in the mike circuit, yes. Because by jiggling it, you could make it work. I got that out and then the right mike was loose. You could take the head of the boom mike and pull it out and expose the wires in there.

YOUNG — When Charlie buttoned up for the first EVA, we started to go through the comm checks, and he wasn't transmitting. I mean he wasn't getting out on either mike. I figured it was all over right there.

MATTINGLY — One of the things that we did when we went to the comm carrier originally was to put those zippers in there, so you can have spare electronics and spare heads with the cap itself. They are interchangeable. It seems to me that it would be prudent where so much depends on communications in this business to carry a spare set of electronics. I don't think that would be a particularly big package. We've tried the lightweight headset, which is a terrible misnomer. I think Charlie was the only one who ever made it work anywhere near satisfactorily. The big disadvantage to the lightweight headset is that you have to hold the microphone in front of your mouth. The range of it is not as great as what you get on the comm carrier. The comm carrier conveniently keeps the mike position stable. If you want to use your hands to do something and then talk at the same time, you're forced to wear the comm carrier. This just makes for another nuisance operation. I think that I never did use the lightweight headset except just once, to try it and see how it worked. I think the idea of having everything depend on the comm carrier makes that a justifiable item to throw in a spare for, even though we have two independent electronic sets on both sides. Charlie's failure was in the cable.

DUKE — It was in the cable.

MATTINGLY — That may have some common points in it. We do carry a spare CCU cable and we carry a spare suit harness. It seems reasonable to carry the redundancy one step further.

YOUNG — You'd sure be out of luck if the comm carrier crumps out on you. You're not going to do your mission right.

MATTINGLY — Navigation. We didn't have anything on navigation. We didn't do any P23s on the way home. We talked about PTC just being nominal. The Gamma Ray Boom didn't retract any from the last try. The systems people know about what extension I had. We had no indication. The Mass Spec we had jettisoned in lunar orbit. The Mapping Camera's Retraction and Extension times toward the end of the mission were becoming more and more nominal. I think that puzzled everyone. Perhaps the stellar lens shield damage might explain that. We finally bent it to a place where it didn't get in the way. On consumables it looked like everything worked nominally, and as a matter of fact, towards the end of the mission, we went to half-degree dead band or half-degree-per-second maneuver rates because we had so much propellant.

DUKE — It appeared that EECOM was suspecting some stratification in the H2 tanks and occasionally had us bring the fans on during the last 2 days. But we hadn't noticed any trouble in the pressure. I don't know how they figured that out. But from the onboard side, the pressures looked good and everything went well throughout the whole flight. The cryo system, the fuel cells, and the DPS were just superb all the way.

MATTINGLY — Let's say something about the fuel cells. We noticed that in contrast to the simulator where the fuel cell flows would rise steadily, the fuel cell flows from the real spacecraft continually pulsed in both hydrogen and oxygen. All three fuel cells did it. They never steadied out to any nominal value but were continually pulsed. Fuel cell 3 had a hydrogen-oxygen flow unbalance throughout the entire mission. It started out reading about an eighth of an inch. The two scales are different normally; the needles for hydrogen and oxygen are opposite each other. Fuel cell 3 throughout the mission was not that way. It seemed to me that toward the end of the mission, the difference became greater. We checked all the pressures and they were nominal.

DUKE — EECOM was satisfied with it so we didn't worry about it. I didn't even notice it until we were into the mission a few days. Then it caught my attention. One thing, Ken, that we ought to mention, although we were told about it preflight, is the H2 tank I that had a glitchy transducer on the pressure. It would sit there in the green and all of a sudden jump almost off-scale high. It would ride for a little while and jump back down again. It was intermittent, sometimes lasting for a couple of minutes, sometimes lasting just for a few seconds, and sometimes longer than that. No rhyme or reason to it.

MATTINGLY — I'd like to put in another pitch for future spacecraft. One of the things that always gets you in trouble is timing water dumps when you're supposed to dump waste water down to 10 percent or when you're supposed to purge the fuel cells for 2 minutes or when you're supposed to stir the cryo fans for a minute. These kinds of tasks are forever the sort of things that last just long enough that you get bored watching them. You go off to do something else, and the first thing you know, you come back and you find that you've never done it by a significant amount. It seems to me that any operation like that should have some built-in system in the spacecraft to give you a call. We used a kitchen timer. I think whenever we used it, we always made the operation come out properly. There ought to be some way to build in to the spacecraft a call system that would alert you. You could set a scale that would tell you that you're down to 15 percent on the waste water, for instance, and by just ringing a bell when you got there, you wouldn't have to monitor it continuously.

YOUNG — I think the kitchen timer is really a good idea, at least operationally from our standpoint. You always get interrupted in whatever you're doing during one of these 2-minute tasks. Right in the middle of it, somebody wants you to do something else. It's inevitable, you forget it.

MATTINGLY — But that should be something that's built in to the spacecraft, not something that you've got to carry in your PPK.

YOUNG — Some timer like that sure would be a handy thing to have for those tasks because there is going to be a bunch of them.

MATTINGLY — Okay. Star/Earth Horizons. We didn't ever look at those except for a gee whiz. DAP Loads. There is nothing to say about them. IMU alignments. Again, we canceled some of the P52s because the platform was running so smoothly.

YOUNG — The only thing I don't understand about the IMU realignment was the last one we did, just prior to entry with a full five alarm. I was looking out the telescope and I didn't see any stars. And there were sure plenty to pick from. I mean I didn't see any Earth, I didn't see any Moon, or any Sun. I didn't understand that for a while. I don't understand why it flunked the star pick a pair test. The sky was full of stars. Maybe somebody can explain that to us. That was a super platform.

MATTINGLY — The UV Photography. We did it nominally, with the one exception being the last UV photography we did, which comes up about 2-1/2 hours before entry. It required one CEX frame, and we had stowed all the CEX magazines by mistake. So we have the UV photography at the last frame and that was all we put with it. The Skylab Contamination Photography. We did one sequence which they called sequence B which complements the lunar orbit stuff. However, we didn't do the lunar orbit stuff. The Skylab sequence B we did accomplish on the way home. This sequence and the dump photography had the Moon in the field of view of the camera which concerned us at the time. It's very difficult to understand how we're going to get meaningful data out of long exposures when you have a bright object like the Moon sitting right in the field of view. I guess all I'm going to do is comment that the Moon was in the field of view for those photographs. The dump photography procedures worked out pretty well once we got started. We had a lot of trouble with condensation in the cockpit. We noticed condensation under the PGA bag after LOI. We noticed it again after TEI and then all the way home. It seemed like the condensation in the cockpit was building continuously.

YOUNG — Yes, it really covered the top hatch and the side hatch.

MATTINGLY — The side hatch was really dripping by the time we got around to doing the dump photography. All the windows were clouding up during the dump photography. I think we all had to use tissues to wipe the window off between frames. There just wasn't any way to keep from fogging the window. Whether the fogging shows up on the photography or not, I don't have any idea. The procedure for dumping out the hatch, once we finally got it all worked out, seemed to go okay. The first time I put the water to the side nozzle, I turned the heater on some 10 minutes prior to dumping. The first thing it did was spurt for about a second and then freeze.

YOUNG — Yes, I guess that condensation is inevitable because the temperature on the hatch is below the dewpoint. So any water in the air that's in the vicinity of that hatch is going to act like a water separator and just condense it right down the frame. There's no way out of it. The tap hatch, the side hatch, and the windows are better water separators than anything we've got.

SPEAKER — But it didn't do this on the way out?

YOUNG — No, it didn't do it on the way out. But we didn't have these kinds of temperatures on the side hatch. When you'd grab ahold of the side hatch, that son-of-a-gun would be cold. When you'd grab ahold of the top hatch, that rascal would be cold. We didn't have the LM on the way out, and I don't know why it got so cold on the way back on the side hatch. But it sure did.

MATTINGLY — It may have been the attitude we were going to.

DUKE — We were in some strange attitudes, John.

YOUNG — But it's inevitable with the attitude profile that we had.

MATTINGLY — But it also did not clear up when we went in the PTC, which should have stabilized the temperature again. I'll admit that the hatch was cool, but it was not what I'd call cold.

YOUNG — It's cold enough to get the water out.

MATTINGLY — Obviously, it was, unless it condensed there maybe.

YOUNG — We had water on the glycol lines. There is always water on them. It's a fact of nature. There's no way out of that.

MATTINGLY — I think the point that's significant is that apparently our water separation capability is marginal to the extent that once you get an excess of moisture in the cockpit, even going back to PTC is not going to clear it up. Whether that's due to lack of circulation or what, it just isn't going to clear out once you got it in there.

YOUNG — If you're going to pass saturated air next to something that's colder than what it is, it's got to go out on it. It doesn't ever get to the water separator.

MATTINGLY — It was during the dump photography that John noticed the scheme on the dump photography was that you dumped one water bag, one of the contingency water bags full of water, out the side hatch. We would photograph the nozzle and the dump, and then we would run a stereo-set of pictures looking down-Sun in the direction of the dump itself to watch the rate of dissipation. When we turned and looked down-Sun, down along the dump track, we saw the increased luminescence looking against the dark sky. It looked like it was in the area that we dumped the water. We might say a few words about the characteristics of the water coming out of the nozzle. From what we could see coming out of the cockpit looking out of the center hatch window, it appeared that the water coming out and hitting the sunlight looked like you were firing a sparkler out a very narrow cone, much narrower than I anticipated. Charlie, how did that compare with what was coming out the regular dump nozzle?

DUKE — In the waste water dump, this thing would come out to look like a cone about like that and about 10 feet out from the spacecraft, it would start diverging. It was almost like a jet at first, slowly diverging.

MATTINGLY — Charlie is describing what he saw from the LM looking at the waste water dump that we made in lunar orbit.

YOUNG — What kind of pictures did you take of that, Charlie? Was it a 70 or a 16?

DUKE — I don't remember. I think it must be 70, because I looked at all of the 16 last night from the LM and we didn't have it.

MATTINGLY — So did I, and I didn't see it on the 70 so far either.

YOUNG — Okay, EVA.

MATTINGLY — EVA Prep. The whole scene was set by having to go in just prior to launch and put in an extra hour to allow for suit donning on EVA day. Then we started EVA day off by getting up an hour late. Sort of negated all the effort we put into it. But we did make up 45 minutes of that hour.

YOUNG — The reason we were up an hour later is that we did part of the EVA prep the night before.

MATTINGLY — Yes, we tried to do some of the EVA stuff the night before the EVA day. I would conclude that we really didn't get that much done that night.

YOUNG — I think you need a couple of uninterrupted hours the night before. Don't you think so?

MATTINGLY — Yes. We didn't spend a couple of hours on it. We spent about an hour on it. I'm not sure it was that productive, because the problem is that you go to start out on EVA prep and you quickly find yourself putting yourself in a configuration that you don't want, in order to eat dinner that night or eat breakfast the next morning or when you have a midcourse coming up. So you end up sort of wasting the time. We did the EVA prep the first day down to page 3-2, unstowing from A-2. That's one page of the EVA prep. There were several items that we ended up doing the next day. We had to make marks not to forget to come back. So my conclusion was that we could have put 30 minutes more on the second day and then 30 minutes saved on the total time. That's correct. We spent an hour the night before, and it took us an hour instead of 30 minutes to do it because we had to stop and say, can we really afford to do this tonight? A lot of the things, we had to go back and do over. I feel like we really didn't make as much money as I'd hoped we could. The only reason for doing it was we were afraid we were pushing the time for the next morning.

YOUNG — But you figure it was a full hour that we cranked in there, that you had enough time on the day you did it. If you had used that hour, which you didn't obviously.

MATTINGLY — Yes. I think we could have done it all in the same day. I think Jim Ellis did an excellent job of putting it together. Everything worked out fine; when we got to the other end, I had no concern that we'd overlooked anything at all. I felt completely comfortable.

DUKE — I was going to say the same thing. I was just sitting there watching you two guys go at it. It was just real "attaboy" where Ken had it all laid out, and everything seemed to be there and available.

YOUNG — This is something we don't really get to practice, like everybody else. Maybe you do it two or three times during a training cycle. Ken does it a lot for the stowage temp, but Charlie and I don't get in there. We only did it twice on the training cycle. I thought that was really good, considering.

DUKE — I'd like to make a recommendation that even though the Flight Plan says 200 and whatever hours and whatever minutes for the EVA, that you check this thing logically like we did. Slowly go through it, and if it comes out an hour later than that time, then it's going to be an hour later than that time. Because John and I didn't ever get in there to do those kinds of things. I did not feel too comfortable with a hurry-up type procedure. It didn't work that way, but I felt one time that I almost got a panicky feeling that man, we were really going to make this time, but it never did turn out that way and everything got ready. I thought it was great. But due to the fact that we didn't get in there and train except for a couple of times with you, Ken, we really never did the stowage part in the airplane. You don't have a chance in the airplane.

MATTINGLY — There's only one reason for opening the hatch on the time in the Flight Plan, and that's if you made a commitment to a network TV, and you ought to start it early enough so that you can meet the timeline comfortably.

YOUNG — That's not quite true, anymore than it is any other time.

MATTINGLY — We didn't because we didn't have a choice. But I think if you say you're going to do at 220 hours and 10 minutes (and NASA has a real bad attitude about not meeting those kinds of schedules), the onus is on the flight crew to meet it.

SPEAKER — That's not the case, never has been.

MATTINGLY — Okay, on EVA Cabin Depress. There's one thing that we all saw at about the same time. When you open that side hatch, no matter how well you think you got the cockpit cleaned up, there's always a lot of trash in there and it's going to head for that exit. I saw a little screw come by and I tried to catch it, and I batted it out of the way. I thought it was gone. A little later, John says, don't let that screw in there. I made a dive for it as it went through the valve. That worried me for the next hour and a half because I didn't know whether it came out the other side, and I really think there ought to be some kind of a debris screen that you put over that equalization valve on the side of the hatch.

YOUNG — Yes, because we had a lot of pebbles and rocks that were on the suit that we didn't get cleaned off and on the LEVAs and on everything else that we brought back. We were sitting there and there were at least four or five pebbles and rocks that came floating by. That stuff can go through your dump valve and it gets lodged in there so you couldn't close it. That would be a bad thing. I know that's a big valve and a big hole, but I assume that there's something that size that could get through there. Sure would be good to have a screen over it, I think.

MATTINGLY — Seems like you ought to have a screen, or you ought to have a thorough understanding that the inlet is the minimum cross section. Anything that gets started in there is going to keep going. Even then I think a screen is the proper answer, because there's just no way you could get in there and operate on that thing and clean it out. The only thing I could think of to do was if it didn't seal, I figured a rock would crush in the thing, a lot of mechanical advantage. But for something like those little screws that we saw go by, I never would have gotten those things crushed. The only thing I could think of was to open it up, and go ahead and waste some oxygen, and try to build a flow through it. But that was one that I just had not considered at all preflight.

The Hatch Opening. Let me go back and catch one thing on that hatch opening. In order to cover all bets, we did our own thing with the nitrogen, instead of venting the nitrogen bottle after lift-off. I hung onto it until after rendezvous and we got rid of the LM. I then vented the hatch nitrogen bottle into the cockpit. So we had it available then. Just about the hatch opening time for EVA, I vented it again and there had been no subsequent buildup.

YOUNG — I'm really interested to see how the red blood cell mass comes out. I don't know whether it was a factor or not. I don't think the potassium has anything to do with the blood cell mass, but maybe the nitrogen does.

MATTINGLY — When we opened the hatch, we had taken the counterbalance off with the hexnut instead of disconnecting the pip pin as we did on 15. We did this so we'd have the hatch open lock available to stabilize the hatch for the MEED experiment. And it all went just super fine; it was real easy. One of the things that I was a little bit surprised at was when we opened the hatch, we got a little impatient and still had a little more pressure in the cabin than I expected and it opened rather rapidly. I had my hand on it so it didn't go far. There was still a little more residual cabin pressure. When the gauge gets down under a half a psi, the cabin pressure gauge is difficult to read and with parallax, it probably isn't very accurate down there anyhow. The hatch opened with no friction and the same thing was true when it came time to close the hatch. No friction evident at all. It really worked nicely. And movement in the cockpit was no problem. I thought the checkout procedures, again, were just perfectly adequate and gave me a lot of confidence that we had done all the right things in the suit preparation. On the suit integrity checks prior to cabin depress, it took a long time again for the oxygen flow to come down in the suit loop. I finally got it to come down by opening the equalization valve a little bit

72-H-607

and dropping the cabin pressure to build up the suit DELTA-P. We had spent what seemed to me like 2 minutes with the O2 flow going during the suit loop integrity check that you people did. You just have to bump it some way to get it up there. It takes a suit integrity check of about 4.5. That's pretty high pressure, and there's a lot of volume in those suits for that little bleed valve to make up. Moving in and out of the hatch and operation around the spacecraft with the OPS on and all, your helmet and visor are really not much of a problem at all at zero-g. It's so much easier than all the things you do in training, and it's very comforting. Charlie, you want to comment on your maneuvers?

DUKE — In the zero-g airplane and in the simulator, when I got pressurized down, I felt like I was trapped behind the X-struts at the foot of the couch. But in flight, I could just zip right through those and I could bend the suit. It was very good mobility, from my viewpoint. I never felt like I was tangled up in Ken's umbilicals. I never even felt the tether tugging on me and I went out as far as I could. I wasn't about to let go but I still had my feet inside. One thing I wanted to mention, Ken, is the hatch opening. We'd always had this trouble with the temp in valve and the glycol loop. We got it tweaked up before we started depress, and during depress, sure enough, it must have been that water freezing on the glycol lines or something because ECOM started going crazy with these glycol loop temps; we thought sure we were going to have to do some adjustment or something but it stabilized. That would have really been a bear if we had to get down there and fiddle with this 382 panel then. We'd just have had to go with what we had, because it would have been impossible to get down there. That sort of gave us a moment of panic when that thing started acting up. I think it was probably water freezing on the glycol lines. I looked in it and there was a lot of water. All during the EVA, we kept getting ice crystals as big as a penny floating by me.

YOUNG — Right after we opened the hatch, there was a great rush of ice from that region down there. It passed right across me and went right out the hatch. I figured it was coming off the floor but it seemed to be generally coming from that region over there. I'm sure it came from around the glycol lines and around that region where the control panel is.

MATTINGLY — Okay. TV and DAC Installation. That was as simple as you would have expected it to be. We spent very little time adjusting the TVs. John took one look at the monitor and said it was okay the way it was, and we left it alone.

YOUNG — Yes, the ground wanted to change it but they changed their mind on that.

MATTINGLY — I guess this is as good a time as any to talk about the only two EMU comments I have. We had two things before I went out. I found out it was very difficult to move my left wrist. My first thought was that maybe all this stuff about how you lose your strength when you lay around like a marshmallow for 10 days was really true and that I had just gotten super weak. And so I wasn't about to complain about it. Now, I look back on it and I'd like to have someone look at that glove because I think there's something wrong with it. At no time was I able to comfortably move it. Occasionally I could move it with my wrist, but generally, in order to make my left wrist move very much, I had to use my right hand to push the glove over to where I wanted it. And then I could keep it there.

DUKE — It sounds like an Ed Mitchell problem to me, like he had on Apollo 14.

MATTINGLY — Yes. It was a nuisance sort of thing.

YOUNG — And you really had a hard time getting the mapping camera cassette because of it?

MATTINGLY — Yes. The problem is that the glove has two stable points normally, and it's only friction that holds it in the intermediate position. They will stay flopped over towards you or flopped away from you. Every time I tried to put the little wrist tether hook on the mapping cassette, I'd hit that rail around the side of the door, push the glove over to the other side, and I had to keep pulling it out and resetting it.

DUKE — Yes. When you said you were having trouble getting the mapping camera cassette, I couldn't figure what it was because you normally were down there and slapped it on there in the water tank.

MATTINGLY — The other EMU problem had to do with the visors. I went out and the Sun was just bright as all get out. So the first thing I did was to pull down the inner and the gold visors and that was pretty good until I got in the Sun. Then I still wanted to get that bright Sun out of my eyes, so I pulled down the hard covers on the outside, and John had forgotten to tell me that they didn't go back up.

YOUNG — During our last EVA when I got back in, I couldn't get the visor up. I couldn't see what I was doing. I was going in here and there with my eyes closed.

MATTINGLY — Fortunately, they got that little trapdoor in the front of the visor and I could handle that, but I never got the side blind down and I never got the thing pushed back up.

YOUNG — When you are in the lunar module, you butt your head up against this thing, you pull the trapdoor open, then you put your head back down, and the trapdoor comes back down again.

MATTINGLY — And that really worked pretty well. It only got in the way when I wanted to try to close the hatch. And on ingress, every time I tried to close the hatch, I knocked that little visor back down in my face. We went around and around with that one. I guess it's worth checking that the visor goes in both directions before you open the hatch. It seems like a natural failure mode with all the dust and dirt and stuff in there.

YOUNG — Yes. That is what it was, I am sure.

MATTINGLY — And it's just one of those things you ought to be aware of. The Pan Cassette and all that stuff came out super neat. The Mapping Cassette did the same thing. Because of the trouble I was having with my left wrist, I didn't carry it in the mode that I normally would have. I was not going to let go of the cassette. I was just going to slip along the rail with it, holding it in my hand, but it was all I could do to hold on with my left hand. As Charlie says, that sure is black. I probably left fingerprints all over those rails.

YOUNG — When you passed that last cassette in to me, the first thing I did was reach over there and hook that hook onto your tether instead of the cassette.

MATTINGLY — Yes, I was watching that and I had visions of our two tethers hooked together and the cassette floating free.

YOUNG — Yes. I think the message there is just take it slow and easy.

DUKE — I might add here that we had developed a technique, Ken and I, in the airplane where I did not ingress the command module during the film transfers. What I would do was move up on top of the hatch and sit up on the upper part of the hatch with my feet just barely in the hatch, and Ken would come in next to the hatch and he would have plenty of opening even for the large cassette. Between arms, legs, hands, and hooks, there was no way for that thing to float out of there once we got it inside. Something that Apollo 17 ought to look at in their training is to have the LMP sit up on the top of the hatch there around the upper hatch seal. They can hold on to that. You've got plenty of handholds up there, and it worked great.

MATTINGLY — Yes, I thought the transfer was very simple and right. I can say one thing about the visibility when you're out there. You really run into this stark, bright contrast between bright areas where there's direct Sun and complete shadow. When you look around the side of the spacecraft across the shadow line, it just gets flat black. The only light that you're going to get down there is either what's coming from the Moon which is probably very small, or what bounces off of your suit as a reflection. And you can see bubbles all along the thermal coating. There was an awful lot of bubbling around the service module coatings. The radiator's that we looked at all looked nice and clean. I didn't see any surface contaminants on the radiator panels that I saw. Around the RCS was a lot of surface bubbling, particularly on quad B. It was my impression, however, that there was no higher concentration of bubbles at the nozzle exit or in the immediate vicinity of the jet plume than there was scattered around the rest of the thermal panel there. The housing on the side of the quad, the little box that builds up around the nozzles had bubbles on it, just like the skin underneath; so maybe if there is a heating problem, maybe the heating that started the bubbling is either solar heating or it's a heat soak type of thing rather than jet plumes. I looked at the mapping camera first on the way back. The stellar lens shield had apparently been out and had stayed out from the first extension, and it was partially retracted so that the folding lip of the stellar shield was up against the handrail and had been mashed down against it, sort of folded the whole stellar shield as the mapping camera retracted. I had been asked to look at the mapping camera and look at the mechanism for retraction and extension and other features, the cables that went to it. It's very difficult to see, because it's down in a shadowed area. I suppose if I had taken the time to position my body differently, I might have gotten more reflected sunlight down in there, but that's a pretty hard shadow and I just didn't take the time to try to set myself off, although I think if you want to look down in the SIM bay, if you'll point your head so it's toward the Sun, I think that gold visor combined with the white suit make a wonderful reflector. I think you can probably see down inside some of those deep shadows pretty well if you just turn around and face the Sun. I just didn't take the time to do it. In retrospect, now I kind of wish maybe I had.

I thought that we had seen what was wrong with the mapping camera when I found that stellar shield smashed against the rail. I went back and looked at the aft end of the SIM bay. The mass spectrometer was gone, of course, and the shield that goes over the top of it was sticking straight up. I suppose that's the nominal position for post boom jettison.

Gamma Ray. I guess at this time we probably had noticed that it wasn't doing all of the right things because the cover was slightly ajar. It was sticking up at about a 15- to 20-degree angle and it was loose. I could wiggle the outer end of it like plus or minus 0.5 inch. It did not appear that the instrument itself was touching the hatch or the cover door. So somewhere the mechanism was holding it open. The instrument appeared to be down inside and it was on the rails. You could see the tapered end of the rails sticking through the instrument. It appeared to me that it was fully retracted, from what I had seen of the instrument preflight. So it was only the door that hadn't closed. The Alpha/X-Ray Door was closed and it all looked completely normal. I looked at the aft end of the spacecraft, back around where the service module separates from the SLA and that was not as clean a cut as the SIM bay. The SIM bay, where the door came off, was a really clean cut. I looked at all of the surfaces, and I couldn't see one jagged edge sticking out anywhere or any loose metal.

Looking back at the SLA/service module interface cut, there was quite a bit of material to the plus-Y side of what I could see. It has some sharp edges and pieces of metal that were bent out and still hanging on intact. And I'd say they extended out maybe 0.5 inch. Looked like a good area to stay away from. Getting back in, we took the TV pole down. That was no problem. I passed it inside, removed the TV and the DAC, and put the MEED on it. MEED Installation was no problem. When we got the MEED in place in the hatch, we ran into our first difficulty with the MEED. The MEED has a little Velcro strap which has been used to hold the lock in a locked position, just to keep you from knocking it. When you release it on the ground in one-g, this Velcro strap, which is about 6 inches long, will hang down. When you get in zero-g, this Velcro strap has an internal memory, and it goes right up over the Sun site. I was going to cut it off. I got talked out of wasting time on that and found a place where I could look around it and see the Sun site. The Sun site worked beautifully. You could really see the target. You could see where you were on it, what you needed to do to fix it. This MEED position was very difficult to get out to. I had to get completely out of the hatch in order to see it. And that was partly caused by the fact that with the visor down, I couldn't get my head high enough to look up at it.

YOUNG — Did holding your feet help you any?

MATTINGLY — The reason I wanted you to hold my feet was that the only other way I had to get out there was to climb up the pole. In training, what I had done was to plant my feet, one on the hatch and one on the couch strut inside, lean back, and look up at it. But I couldn't look up because the visor was down. The only way I could see the Sun site was to get up so that I could look down at it. And the only way to do that was to walk my way up the pole, and I was afraid that my hanging onto the pole was going to cause the hatch to move back and forth and change the pointing angle. And I don't know how much in angular measure that thing was going to move, so I had Charlie hold my feet down and that let me let go off the pole for a minute. And it turned out that when I let go of the pole, it didn't move any from when I was hanging on to it, so after that, I just went ahead and held onto the pole. And initially, the target showed that we were outside in pitch and we maneuvered in minimum impulse to put it in and got it. We had an error of about 2 degrees in pitch and about 3 degrees in yaw. And we didn't take any more time to dress it up after we did that. Opening the MEED was no problem. I pulled the little ring out and twisted it. Charlie backed up the timing with his wristwatch, and Hank backed it up on the ground. So we had two clocks going on the MEED timing. When it came time to close it, I had been a little concerned about having to climb up the pole in order to close it again, but that turned out to be a pretty simple thing. I closed the experiment and then I went to lock it, and there apparently was just enough extra inflation with the seal, that I couldn't get the lock to lock and once again the problem I was running into was that I was trying to use my right arm to kind of put a half nelson on the box and hold myself to the pole and squeeze the box closed, while I tried to use my left hand to lock the lock and started running into problems with the glove again. So I'm really not sure, everything working the way I had expected, of how many problems we'd have run into. The end result was that I pulled the MEED down without its being locked and passed it inside where John and Charlie locked it with apparently no trouble at all.

DUKE — I thought it was a lot of trouble.

YOUNG — It took four hands.

DUKE — And I had to push it against the couch strut.

YOUNG — If it's a two-hand operation and it takes essentially five arms to do it, that's troublesome.

MATTINGLY — I would guess that we gave it an extra UV exposure of 5 to 10 seconds at the most.

DUKE — Yes, and that wasn't direct.

YOUNG — Indirect. I think that was on the order of 3 seconds.

MATTINGLY — I don't think we disturbed the data at all.

YOUNG — It may have done something to the data, but I don't know how you'd evaluate it.

MATTINGLY — Okay. Comm During EVA. I thought it was super. I could hear you guys and I just didn't have any problems at all with the comm.

YOUNG — In any case, the comm during the EVA was a lot better than it was in the altitude chamber, I thought.

MATTINGLY — Yes, there was a lot less noise.

YOUNG — A lot less noise.

MATTINGLY — Ingress was simple again. Every time I raised the folding visor, I closed the hatch on it and knocked it back down. I was having a hard time looking at the seals. Finally, John and Charlie checked all of the seals to make sure they were clear, then we pressurized it. Once again, I want to mention how nervous you feel about closing the hatch without some visible indicator that you have really gone over center on those locks. One of the technicians showed me how you could check the clearance and at the very last stroke of travel, you could see a place where the dogs come across. But you sit there and you're counting on that thing holding pressure on you, and the only indication you have that it's latched is the fact that it quit stroking and the little bar falls down, but you don't really see all of the dogs go over center.

YOUNG — With the visor down, neither Ken nor Charlie nor I could check that the seal was in fact clear, so it's the kind of thing where you could have the hatch closed and it would be dogged as much as it could, but something might be in the seal and you couldn't check it. This indication that Ken had with the last stroke of travel that shows the dogs over center that the technician showed him is a good way to do it. The only problem was, with the visor down again, he could not make that check and that's a valid check. I think in future spacecraft design, there ought to be positive indication to the crew that the dogs are in fact over center. I guess that's a standard gripe with me and the hatch, and has been ever since I participated in the design of it. I think in future spacecraft design, somebody ought to worry about it.

MATTINGLY — Once we got the cabin pressurized, we didn't have any problem on that. It takes awhile. You just have to be patient while you build up the cabin with the purge flow. It takes about 10 minutes to build it up to where the repressed package brings it to where you have 3-1/2 psi that they'd like to have before you open up the purge valve.

DUKE — I'd like to say one thing. During EVA, I was always very comfortable. In the suit on the surge tank, the activities that I did never felt like I was overheated and we hardly used any oxygen out of the surge tank, the entire hour.

YOUNG — It was still up around 850 or so, I guess. We flew the spacecraft a little in pulse to get the MEED lined up, and I didn't think that was any problem, although you can't handle the hand controller like you normally do because you've got it over in the left hand and it's all in a cattywampas angle to your wrist, but it's still no problem. I don't think flying a spacecraft in that manner for that kind of thing is hard.

MATTINGLY — On the subject of the OPS, another thought passing on that is that we give, going outside, an OPS and a purged umbilical. And we give the people inside two relatively low-pressure oxygen bottles to live on. And the crewman outside has a lot more protection with that OPS than the crewman in the suit loop. Because not only does he have more makeup capability with the purge umbilical just to start with, but he's got a backup system with almost unlimited makeup capability. If there's anybody that has a minimum of backup protection, it's the two crewmen on a suit loop in the spacecraft, where they have nothing but a surge tank and a repress package to work from. Neither of those have a particularly high flow capability. And as long as the umbilical is flown, you'll never get those things pumped back up. It seems to me it's a bit inconsistent in our approach to backup systems.

YOUNG — Yes, but I think everybody's realized that all along, Ken. No question about it. The crewman on the outside is a lot better off than those on the inside, and it makes everybody feel warm.

MATTINGLY — We write rules that say, if you don't have an OPS, you shouldn't go EVA. You've got more to start with than the crewmen in the cabin.

YOUNG — I agree. I think if worse came to worst, and you wanted to get the pan camera and film but for some reason, you lost OPS, I don't see anything more risky about going without CPS, I really don't.

MATTINGLY — Okay. We used the OPS to pump the cabin up. And we used the CPS two more times as a way of bleeding the OPS down. We had something like 1600 psi in the OPS after we finished pumping the cabin up the first time. I think it was 1300 when we finished. That evening when we pumped the cabin up, it had built up again to 1500 to 1600. One of the things that we had overlooked on our post-EVA stowage was that I turned right around and stowed the OPS, only to have to go and unstow it for the presleep checklist to get the cabin pumped up again. That was an unnecessary amount of shuffling of materials around. On the post-EVA, we tried to get started toward an entry stowage, and the technique that we used was that I went ahead and got unsuited, and John and Charlie stayed with their suits on and just laid on the couch. I could move around and we decided not to let everybody unsuit, because that was having six people in there plus all the junk, and the only way you can get in out of these boxes is to have one of the crewmen that's fairly mobile and the other two people passing stuff. I guess we spent about 2-1/2 hours on that. At times there, you couldn't see anything except bags and boxes and junk floating around. I looked up from the LEB once and there was just nothing in sight.

DUKE — I couldn't see you either.

YOUNG — Yes, that was very interesting. We were getting this thing where it's almost like Gemini as far as the stowage is concerned. You can't take one box out without taking the whole mess of other boxes out. It gets pretty tight.

MATTINGLY — If you would take it methodically, it's only a nuisance. It did bother me when I looked around and saw how much stuff we had floating free that if you ever had any reason to have to do something in a hurry, if you had some hardware problem, you were really going to be thrashing trying to get to some of the corners of the spacecraft because there is just stuff everywhere.

YOUNG — I'd like to say that I thought Ken really had the stowage laid out well. It's just by the numbers, and that's really the only way to do it. In the order that you have to do it to stow it, that's the way he had it laid out. Really, the only thing that we did that was probably inconsistent was that we were working the entry stowage list after we finished the EVA, and we were taking it very methodically box by box until we got down to the PGA stowage, at which time we assumed that all three of the suits belonged in the stowage bag. We pushed them in there, only to find out that they didn't want all three suits in the stowage bag. We wanted one under the right-hand couch. I assumed we could have probably left it that way since we got them all in there, except that we didn't have any place to stow the LEVAs. Charlie was sleeping under the right couch, so we ended up taking one suit out and just sort of letting it float down in the LEB, which was in the way of everything. But I guess the thing to do is to remind the crew to sort of have a delta entry stowage after you finish the EVA.

MATTINGLY — We got ahead of ourselves in a couple of cases, such as putting the OPS and burying it underneath a lot of stuff, only to have to pull it out to pressurize the cabin that night and the following night. And we left the TV stowed which we needed the following day.

YOUNG — I guess the way to handle that might be to put asterisks on the entry stowage list so that it sort of indicates that there's a delta stowage you're going to have to do after you finish the EVA or something that you're going to have to restow those items for, reentry maybe. Something like that.

MATTINGLY — I'm not sure that there is a much cleaner way than the way we did it.

DUKE — I thought it was great.

YOUNG — I thought it was great, only it was inconvenient to restow that stuff, that's all.

MATTINGLY — One of the things you have to keep in mind is that three crewmen in three couches don't sleep very successfully. I think that you really need to lose that. Because when you sleep, you kind of sprawl out a little, and I think if you tried to put the third person sleeping in the center couch, that he probably wouldn't sleep very well. That means you have to put one person under the couch, and with all of the rock bags stowed under there and all the other things (and by this time, you've filled up another jettison bag full of trash and you got just enough stuff so that that extra suit has to go to somewhere) so you can leave the poor fellow a place to go to sleep.

YOUNG — We had three items that were our delta stowage after we woke up the last day. The pressure suit we had to restow, the jettison bag which by this time was enormous, and that black fecal bag which we had to move every time we did anything. What was the last one?

DUKE — The goodie bag.

YOUNG — The goodie bag, and it was a lot bigger than we'd anticipated too.

DUKE — It had the Flight Plan in it.

YOUNG — And it had the LM Flight Data File. I'm sure we didn't have a CG exactly.

DUKE — It was all in the LEB, and I tied it all down.

MATTINGLY — Mass Spec. There was nothing to say except we jettisoned it. We did another Light Flash thing and once again Charlie sandbagged it.

YOUNG — Yes, Charlie had me 3 to 1 on the light flashes. I really don't think it was fair.

MATTINGLY — Flight Plan Updates. They settled down a little bit by this time. I don't think we had any particular problems with it, since the ground was running most of the calls in real time. That seemed to work pretty well.

DUKE — It was great, especially during that suit doff on the EVA post.

MATTINGLY — I think you really ought to take advantage of the people on the ground. They've got the time and they've got the manpower to watch clocks and remind you of things and let you go about your business and not be a slave to the clock. I think that frees you to do a lot of things that you'd otherwise have to let go.

YOUNG — Especially since our clock wasn't working.

MATTINGLY — Okay. This takes us right on up to entry. Maneuvering to the Entry Attitude and all those things. We got up on entry morning, and I felt we had a very leisurely, very comfortable time, even with our ISS lights thrown in. And I thought that was the kind of day it ought to be, nothing extra thrown in there. I think that whole day ought to be preserved, just in its present form.

DUKE — John and Ken were in the cockpit making sure we were up to speed on the entry checklist. I was going about the final stowage things, getting the suit tied up under and the sleep restraints and things like that, and I thought that was a good way to do it. Another comment on the EVA. We mentioned our failures, except one I failed to mention was that my watch blew a crystal on EVA-3 and it stopped running at that point. So the flight watch went belly up and I got it brought it back to let them examine it. But the crystal either blew out or broke, and I don't ever remember hitting it. The face of the watch doesn't look like it's scratched. So I think what happened is the crystal just blew out and we got dust in it, and the thing was just not running and that happened on EVA-3. That's it.

15.0 ENTRY MATTINGLY — We brought the batteries on 15 minutes early.

DUKE — El minus 45.

MATTINGLY — This was because of the concern over the battery compartment pressures.

DUKE — Yes.

MATTINGLY — The entry went completely nominal. Everything worked out right down the checklist. We did take a look at the horizon check time. It was dark and we didn't see anything. Nothing unusual in the separation from the service module. It's a big bang, as you'd expect. One of the few things that I think the simulator didn't reproduce faithfully was the dynamics of the command module, after separation. Apparently, the water boiler imparts a torque that causes you to yaw left and pitch up. This was something I hadn't anticipated. I kept finding every time I put in a minimum impulse, putting in some rates to maneuver back to the entry attitude that it was drifting off in the opposite direction. It seems like the water boiler was the most likely candidate for these torques.

YOUNG — That baby is really springy once you get it separated. But the rest of it tells you the water boiler is working.

MATTINGLY — The control system flew exactly like the simulator. The entry was exactly like the simulator, I thought. The thing that caught my attention first on entry was that it was a lot brighter inside the cockpit than I had anticipated. I don't know whether this was the time of day we came in that caused the Sun to be coming into the window. Right after entry interface I became a little bit concerned there was so much bright glare on the 8-ball that I might have a hard time focusing on it, and being able to read the 8-ball. It was really bright. It turned out there never was a problem. It didn't get any brighter than it was right then. It was just like having a big floodlight shine down on there. It's rather hard to watch.

YOUNG — That's really a bright white light. I told you to turn the lights up bright.

MATTINGLY — Be prepared for the ionization. It will be a lot brighter than you expect it to be. I thought you ought to fly the entry on the centrifuge or something so you could have a closed loop system. I thought that not getting a chance to practice that was sort of short changing the CMP. But after flying it I have to retract that. I think the simulation you get in the CMS is just all you could ask for. The thing flew exactly like all our simulations. The EMS was running as well as any entry monitor system I've ever seen in any test or simulation. The EMS profile and the G&N profile were just identical. Think I could have flown either one. The control response was such that it would have been absolutely no problem with flying an EMS entry.

YOUNG — Yes.

MATTINGLY — We built up to right at 7 g's. All our g sources checked. John was calling out the DSKY read-outs and I was cross checking the EMS and the g meter. I was cross checking the two 8-balls. The first g felt pretty heavy.

YOUNG — It's hard to talk at 7 g's.

MATTINGLY — I felt like there was absolutely no problem at the 7 g level. I was concerned about how much peripheral scan might go down. I had a scan pattern bigger than I normally have flying an airplane.

YOUNG — I was looking out your window.

MATTINGLY — The CMS just gives you a real super preparation for that kind of operation.

YOUNG — It's really good training.

MATTINGLY — Sounds - when we pressurized the RCS unit you heard the squibs go and you heard something that didn't sound as much like a gurgle as the plumbing filled up as I had anticipated. There's definitely a sound there that you know it's going in. Firing the engines is a very comfortable little sound.

YOUNG — The propellant usage was nominal. The RCS pressures were coming down very slow. We used hardly any.

MATTINGLY — The spacecraft was good and stable. Picked up the aerodynamics just prior to 400,000 feet. I think that's very faithfully reproduced in the simulator. One thing that surprised me a little bit; I picked up some oscillations that I thought were typical of the transonic region. They were right towards the end of the entry. Turns out that the transonic oscillations were much greater and they are unmistakable. When we put the drogues out, I was really surprised at the oscillations that we got there.

YOUNG — The automatic system did that.

MATTINGLY — The automatic system put them out right on schedule. But we really got some big excursions in attitude. I was watching 30 degrees, plus or minus, on the 8-ball. I don't know if that is typical or not, but it was a great deal more than I anticipated seeing. It wasn't until almost before it was time to release the drogues and go to the mains that we got damped down to a fairly nominal state. When the drogues released and the mains came out there was very little spacecraft motion associated with that.

YOUNG — I had forgotten it was so erratic. You need to be tied down good in that couch if there is any doubt.

DUKE — Looks like those little beauties are going to get ripped right off there.

MATTINGLY — The CMS has very small oscillations when they simulate the drogues coming out. I wouldn't change the CMS to put those in, just commenting to the fact that you're going to see bigger oscillations for real. Main chutes came out and we watched the dereefing in two stages. That was no problem.

YOUNG — Beautiful.

MATTINGLY — I thought it was really uncluttered and very comfortable. One of the things that made as big an impression on me as anything else was that we've got that altimeter that goes from 100 000 feet to sea level on one rev. I anticipated it being off by a great deal as we got down low. That thing must have been right square on. We hit everything. I was calling out the last couple of hundred feet as we came down and that thing couldn't have been off 50 feet from the time we said we were going to hit the water until we did.

YOUNG — That could be the difference between the Atlantic and Pacific sea levels.

MATTINGLY — The comm throughout was crystal clear. We had the recovery guys just with no trouble. We ended it up with a good Navy landing.

YOUNG — Really, it was harder than Apollo 10.

DUKE — I didn't really expect it to be as hard as it was. When I got my eyeballs recaged we were already in stable 2.

YOUNG — There was no way we could have got those parachutes off before we went over.

MATTINGLY — We hit and in one continuous motion we just rolled right over. I was prepared for it to be even harder than it turned out to be. I got a chance to watch us roll over. The stable 2 was no problem. We collected a little water in the rendezvous windows between the outer and inner pane. I think that's probably typical of these birds. It shows up that way in the boilerplate down there in the tank. I think probably that is just the way the bird is put together.

DUKE — I was just going to say it looks exactly like the boilerplate to me. I was watching the ticktock during the thing. The Ionization, the orange glow started before RRT - couple of seconds before RRT. At that point I started the camera, which was a little early, the 16-millimeter that is. I started my watch right at RRT and everything went right on schedule. Visual Sightings - the bus power and everything looked good, so I started looking outside a little bit during the max-G area. What surprised me during the whole thing was the rapidity with which that command module - when it decided to roll, boy, it just took off. You could see the horizon through the ionization sheath, both out window 5 and the rendezvous window 4. There was Mylar on window 5 that was flapping back and forth across the window that was there at touchdown. It's still there. It had come up right at CM/SM sep. I had seen that strip fly by. When we started getting the g's it flopped up over the window, sort of stayed there and wiggled the whole time. Which amazed me. The communications blackout was just about right on time. John started talking to Houston and to ARIA. Comm was just outstanding there. After the mains came out, I saw a helicopter go over us. I guess that was Recovery One. John was talking to them. The only thing that was off nominal was the steam pressure was 32 seconds late. I started my watch on the time that steam pressure pegged. If we had to call the times based on that, we'd have been late on the 50 000 and the other times.

YOUNG — I think that's good enough for a backup system. I think the fact that it would be late is a known thing. They can't really predict it very close.

DUKE — I really think the best backup system you have is that cabin pressure, if you don't have the altimeter. The cabin pressure increasing at about 24 000 for the drogues. Also, the eyeball out the window could help if you know the cloud level. I could see the clouds and everything outside.

YOUNG — I'd be careful of that.

DUKE — The cabin pressure coming through 10 is probably better than the clouds.

MATTINGLY — I got fooled on the clouds at 90 000. I looked out my window; that is the first time I had a chance to look out there. I looked out and there were clouds. The sensation was very vivid that the clouds were bubbling. I thought, my God, we're down to 20 000 feet here and we hadn't done anything yet. It really took a long time for me to convince myself that we were really at 90 000.

YOUNG — I was looking out my window, I had no doubt that we were way up high.

MATTINGLY — You could really fool yourself.

DUKE — It's a fascinating view. We had plenty of backups is what I was just trying to say. With both water boilers going during entry, the glycol loops stayed in good shape.

16.0 LANDING AND RECOVERY

YOUNG — When we opened the hatch, I bet we increased the cabin temperature 10 to 20 degrees. The only thing on stable 2 that was off nominal was that it may have taken us 4 or 5 minutes to upright.. The center bag apparently didn't fully inflate. It's supposed to be the one that inflates first. But the other two bags were certainly inflated. It uprighted just like normal.

DUKE — On the postlanding checklist - you can run through that too rapidly and end up without comm. All of a sudden, it hit me that if I pulled these two breakers, we just lost our comm once we upright it. Right after we got on the mains I started rattling off the checklist. I was really talking to myself because all the breakers were over on my side. There's about seven or eight breakers that you have to pull. I was just rattling them off and getting no acknowledgment. Ken said, "Wait a minute, what are you talking about?" I was really going, too. I apologized for that, you guys. I didn't let you know that it was all on my side.

YOUNG — I thought you were reading things for me to do and I hadn't done the first one yet.

DUKE — I was about five steps ahead of you. One thing I'd like to mention that was just an oversight I guess by the Recovery Division. The hatch opened and Lt. Tashita pitched a bag in and said here's your life vest and NASA's heat sensor. We pulled this thing out of the bag. We broke it out and it was a heat recording drum. It says tape to the instrument panel. Well, we didn't have any tape. Finally, he said there's tape in the bag. We looked in the bag and there's a big roll of tape. It didn't seem feasible to put it on the instrument panel, so I just taped it to my couch. They had caught us by surprise. We had never heard of that.

YOUNG — That caused us about 5 minutes. We'd have been out 5 minutes sooner and in the raft ready to be picked up.

MATTINGLY — It looked like a clock you put on your back.

YOUNG — As it was, from stable 2 splashdown to the deck was 37 minutes, a super outstanding record.

DUKE — Even for an old Air Force guy, I never felt seasick at all the entire 2 days on the carrier. The Navy really treated us great. I had no discomfort at all once we hit the water.

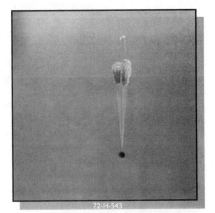

72-H-543

YOUNG — We elected not to use the postlanding vent valve because you don't want a hole in the spacecraft.

DUKE — We did not do that.

MATTINGLY — Those guys out there obviously had really trained heavily. The whole operation was well thought out. But even at that, the frogman that came out to help us get out of the spacecraft really wasn't as familiar with the opening and closing of the hatch as you might anticipate. Maybe he was typical of a frogman approach. He wanted to use brute force to close the hatch. I think the CMP just as a matter of course should help the guy close the hatch. We had apparently knocked the dog partly down when we opened the hatch, before I got the handle back to neutral or something. I may have gotten a partial stroke on it when we opened it. There was just enough of a dog in the way that he couldn't push the hatch closed. I went ahead and recycled the hatch to the open position. Then we closed it. A lot easier than trying to bend all that metal.

72-H-473

YOUNG — He was going to do it. He could have. He wasn't very tall but he sure was strong. Good thing you didn't have your hand between him and that hatch.

MATTINGLY — We had a beautiful calm sea state. If you had any sea state, you really don't want to spend much time with that hatch open. You want to open it and close it as quickly as you can, but methodically; be sure you do it right

DUKE — That was a Navy comment. I didn't think it was so calm.

YOUNG — Except for the waves, Charlie.

DUKE — Pretty calm except for the waves.

72-H-544

YOUNG — Outstanding pickup. That was really good.

17.0 TRAINING   YOUNG — Outstanding training.

MATTINGLY — There aren't enough nice words you can say about Nelson Temple and Gerry Stoner for the way they fixed up the CMS cockpit interior. I asked them when we first went to the Cape if they would keep the cockpit in the flight stowage configuration and put as many things in there as possible. It was our opinion when we started that stowing and learning to handle all the little miscellaneous items and knowing where to get them would be the secret in keeping the orbital time line going. I think that was a valid assessment. We received a lot of good training by going into the simulator and seeing it stowed with cameras, boxes, the flight data file, and all the things you want. We made up a good orbit stowage list which is distinct from the launch stowage list, which most people have in the checklist. These men went out there and put that stuff in so that when we went out to train, we didn't have to spend the first 30 minutes getting stuff out and putting it around the cockpit. I really think they did a good job of making the cockpit look like a spacecraft. I felt completely at home. When it came time in lunar orbit to go after things, it was perfectly natural to reach to the right place to get a film magazine. If you wanted to know where the lens was, it was just a natural thing. We really didn't do that much special training to make it so. I think those men really bent over backwards and we always had a piece of gear that worked. CMS availability - I'll bet we didn't lose 5 hours out of the whole Cape training period because of the machines being down.

YOUNG — I don't think we did.

MATTINGLY — Just super availability. The men that ran the show down there stayed ahead of it. We worked out a schedule and they tried to have some kind of a script available so we didn't waste any time when we got into the machines. In particular, I'd like to say that Bernie Suchocki probably has as good a handle on the software as anyone around the program, and Dave Strunk knows more about flying an Apollo mission and the Apollo hardware than any single individual. Those two men made a very valuable contribution. The visual system, in our case, was up most of the time at least in the rendezvous window and in the optics. Those are really the only two that are worth spending your time maintaining. The rest of the windows were periodically up or down. They did go to the trouble to try and look ahead at the scheduled session to see if those things would be available. They would come in and talk to us about them if we were not going to be ready and find out whether that was going to affect our training. The CMS we used here in Houston was primarily on systems and just general crew proficiency training.

YOUNG — That's all Skylab now, isn't it?

MATTINGLY — I think so. You can't say enough nice things about Roger Burke and the work that he did in building up our software. When I got through with Roger's little course, I came back after Apollo 13 and I was concerned about finding enough interesting things to just stay in that CMS for another 100 hours. With Roger on the console, he always kept us thoroughly occupied and completely interested. The man has an imagination and an understanding of the hardware that is beyond comprehension. He really knows his software.

YOUNG — Yes, he's a great guy.

MATTINGLY — That's all I have to say about the CMS.

YOUNG — Let me say one thing about something you recommended. It became apparent a few weeks prior to launch that we weren't doing enough three-man work. Ken suggested that we get in the simulator and just run a burn or two for half an hour a day and practice three-man teamwork. I recommend that the follow-on crews take time to get in there as much as possible to get team coordination down. Of course, Charlie's and my concern was the lunar surface operation. We got all wrapped around the axle on that. It's a matter of operational judgment as to how much of each you want to do. The fact that we spread the SIM network SIMs out so that we only had to run two a week allowed us more time to work on individual LMS burns. A lot of SIM network SIMs have more to do with exercise of the ground than they do with exercise of the crew. You can run four or five times as many dynamic phases when you're just working together as you get in an average LOI and DOT SIM. I really think that's an important phase. Our IOS men recognize it as being an important phase, too, and I guess we agree with them.

DUKE — In the LMS here in Houston, we concentrated on systems work. Bob Force gave us an excellent review. Maury Minnette and Bob Jones all did a good job for us. It was a good system's review. Operationally, the LMS at the Cape is head and shoulders when it comes to operational training for the mission. At the Cape we just had outstanding luck with the LMS. It was in good shape all the way. The visuals were usually up. On our landing site, I had such good landmarks out my window they developed a scheme where at pitchover we could start out with a view in my window coming straight at me and I could get a quick reference and then they would switch it over to John's window for the last 6000 feet. I thought that worked great and it helped to give me an input into the visibility, as to where we were at pitchover. The landing site looked exactly like the L&A right down to the size of crater of Lone Star, Dot, and End Crater. The fidelity of the LMS was good. Being a fixed-base simulator, you don't get the motion cues that you do in the vehicle. There's no doubt in your mind in the vehicle, when a jet comes on, which jet is firing, and that you're moving. The engine sounds in the LMS during descent was a very good simulation. It was probably a little bit more pronounced than you got in the vehicle, but you needed something, anyway. All in all, I want to just say an outstanding job for Charlie Floyd, Dave Bragdon, Bob Pierce, and everybody that was involved in our training. We felt right at home in old Orion when we got there. I never felt so at ease and so at home in any airplane the first time as I did in that lunar module.

YOUNG — We really did. We really felt as if we were there. It was really comforting. We had a lot of flight time in the old LM. It was just as natural as if you had been doing it all your life. Let me say something that I recommend for training the last 2 or 3 weeks. They really threw the book at us on every dynamic phase that we have. I mean we were really having failures. If you had all those failures put together in a real situation, you would be in really bad shape. I think they should throw the book at you on burns, and every so often they should slip in a burn that is completely nominal. Not a 100 percent of the burns are going to be, in the real world, up to date. As far as the actual dynamic phase of the burn, we have never had a manned spacecraft burn or boost phase that wasn't nominal or wouldn't allow you to perform a nominal mission. I think that one out of three or four burns they should have you spring-loaded to take corrective action and not give you anything during that burn, because I think that's the way it's going to be in the real world. You should never forget how a nominal burn looks. Boy, we were on the verge of forgetting. We were having IMU failures, platform failures, computer failures, and backing it up with the SCS and down moding on that. I think, close to launch, you want to keep the crew relatively calm. I'm not sure with that much pounded into them that they are going to be that way. The visual systems and the L&A was tremendous. There was no doubt in my mind as to the landing site or where we're going at pitchover. There is a shortcoming in the L&A in that at 20 000 feet in the lunar module, I was able to get my nose up to where the COAS is mounted and stick my head by the window. I could see South Ray Crater and the edge of Stone Mountain. So, I had no doubt that we were targeted right into the landing site. In the real world the visibility got better as we came on down. Passing through 10 000 or 11 000 feet, you could start seeing Flag Crater and Spook. In the actual vehicle, you can look 88 degrees down, but you can't look 90 degrees down the X-axis. With that eye position, you can look down 88 degrees vertically and with a 55-degree pitch angle, which is what you get just before you get to high gate. You can see the whole landing site before you get there. Now, that's a good thing if you had any problem with the landing site recognition, which was always a concern too. This was sort of a blah landing site, but you're going to recognize where you are. At least you are going to recognise where you're going if you are really targeted off the nominal spot. You don't get to do that in the L&A. It's something we can't do anything about, but the capability is there if a guy is worried about recognizing the landing site in the real vehicle. The other thing they did for us that I thought helped us is if we had had problems with the landing radar, they built a shadow that gave you an absolute altitude cue. You're looking out of zero phase and here comes a shadow which you can see in the photographs. That is the absolute altitude cue. Man, there's no doubt in your mind how far off the ground you are when you see that shadow. The dust, of course, wasn't as bad in the real world as we made it, in the simulator. In our case we could see rocks and dirt all the way to the ground. That hasn't been the case in all missions. The software in the LMS was just superb. We have software programs and erasable memory programs. I think the LMS is really valuable for checking out things such as having the right stars for the various platform alignments, are there in the right part of the sky, how much attitude maneuvering does it take to get there, and for working out the timeline. We felt as if we had been doing it all of our lives when we finished that training.

MATTINGLY — There's one thing on that whole subject of the CMS training...

YOUNG — We never did a normal DOI burn for the whole mission.

MATTINGLY — We have done one normal DOI burn since we've been together. That was at the Moon.

DUKE — I sure am glad that there is only one allowed.

YOUNG — Over 2 years of training with only one nominal DOI burn and that was in the spacecraft.

MATTINGLY — I think there should be somebody who puts together a training package for you, that lays out what it is you are going to do in the CMS or the LMS, as the case may be, makes up a script and reviews it to see what is appropriate. As it turns out, I had to write down a list of things to do and I had to keep looking at the things. This is fine in emphasizing the mission phases you want to look at, but it also takes away some of the training factor, I think. I had to go out and say what kinds of malfunctions we wanted to see, and where we wanted to see them. I sort of tried to back out of that because that takes away from the ability to go in and face a cold situation.

YOUNG — Close to launch, you're the only man that, really has a feel - or you're one of the men that really has a feel for what you missed.

MATTINGLY — Well, that may be but you know there are only so many hours in a day, and there are only so many things you can do. I felt like the time that I had to spend in laying out a detailed simulator training script was just one more task that I don't think the flight crew should be doing.

YOUNG — That's right. You shouldn't be.

MATTINGLY — I think I probably asked for more from the machines and the people than perhaps we have always done. I think they have done a super job on providing the things we did, but the groups are cut back themselves. We're all in the boat where there is just enough time to do the things you have to do. I really think maybe the backup crew could lay out the script for the prime crew because they know what the prime crew needs to do. Maybe you could do it for each other. I think there needs to be more thought to laying out an organized and a planned way to accomplish the training so that you do all the right things.

YOUNG — The simulator people do have squares you have to fill.

DUKE — Exactly, they have a master plan to work with.

YOUNG — Yes. They have a master plan and there is a lot of training we had to do to fill all the squares.

MATTINGLY — You can say we should do so many TEIs and LOIs, but there needs to be a spread in there so you do them under certain conditions. Certain failures that are only important during certain times should be looked at. If you do it on a random basis, and the missions become more complex, there really is no way you can keep it all in your head as these things change from day to day. You can't afford to go out there and start every burn at ignition minus 20 minutes, because if you want to look at one discrete thing, there are some things you should look at. You should crank up the burn and start it at ignition minus 5 minutes and go do the burn and then drop it right there. When you go to entry, we finally got a good entry point that was EI minus 30 seconds in that 1 or 5 minutes, and that was a good training point. As you go along, you find that there are some reset points that are more useful to you than others. They aren't always predictable when you make up the original training tapes. When the men come up with the reset points from MPAD they're just guessing at what is going to be a good one.

SLAYTON — Well, we're really not, because we've been working on those things for, how many missions, now?

YOUNG — You know how it is, Deke for the last couple of weeks you could do every thing you have to do arid never do the same thing twice.

SLAYTON — Yes, that's right.

YOUNG — You don't really want to do that the last couple of weeks though. You want to home in on your mission phases and tackle them one at a time.

MATTINGLY — I had a feeling that I couldn't really let go of keeping track of what we had done and what we were going to do. I really feel as if that wasn't the kind of thing I should be doing.

YOUNG — We already recommended that the crew look at a couple of things late in the training schedule and they may not even have time.

MATTINGLY — I spent a lot of time flying the Flight Plan.

YOUNG — I don't think we should work more than 8 hours a day the last 2 weeks but we did it.

MATTINGLY — I guess part of my comments about the Flight Plan go along with the simulator. The orbital Flight Plan flown in the CMS with all the activities in it, is to the CMP exactly the same as the EVA surface is to the lunar surface men. The difference is that we have one simulator and two crews to train. I can't practice my orbital activities and then do the nominal mission proficiency kind of training back to back like you. You can fly the EMS in the morning and then you can go out and do an EVA in the afternoon while the backup crew flip flops with you. We have one simulator to do both. You can't do this thing in piecemeal very well because you have to run real time on the real trajectory. You can't practice it in pieces because the things you're looking for are, do I have sufficient time to maneuver from this attitude to the next one? Does this attitude go through gimbal lock? Does it really get you to the right place? You have to fly those things in real time and they're not very glamorous things to do. But, I think to do justice to any Flight Plan and to go out of there professional, saying I know that I can accomplish this mission and I'm going to do these things. There is no way out except to have flown every one of those revs.

YOUNG — Yes, even then we had three or four times where we had to take over to avoid gimbal lock.

MATTINGLY — But, the things we did, we did as new things that were thought up in real time, not to the things that I verified in the Flight Plan. I think our mission is too complex and we've got too much wrapped up in it to go off half-cocked. I don't launch without having flown the Flight, Plan from A to Z to know what's in there, or if I can find any way around it. That's the reason I don't think you should do that kind of stuff inflight. We found lots of things when we went through, and it's just too complex to spit it out and write it down the first time and think you have all those answers. I guess all I can do is point out the problem and also that you really need to maximize your utilization of the CMS even if you get another 1000 hours lying on your back. There's just no way to find that you have an excess of time. You've got to use the thing and you got to optimize the amount of hours you spend in it.

YOUNG — Yes. I think the message there is that you can keep it nominal to the real-time Flight Plan for the sake of optimizing a couple of data points on somebody's gamma ray curve. It's foolishness as far as I'm concerned. It may be necessary, but it sure is foolishness.

YOUNG — I think the LRV nav simulator was a system's assurance test. It sure convinced us that the thing worked. I'm not sure that's necessary in the future because Gene has already done it. They won't have to do it again. It's a nice piece of gear but the LRV nav system, as you know, works like TACAN, when it's working. You can't beat it.

DUKE — It was a good verification of the maps and I think you should have that for the flight. Somebody should go through those traverses drawn on those maps and then navigate in the simulator. When we first started out, the map's starting point was different from the landing point starting point.

YOUNG — CMS, LMS simulations - The comm was never worth a darn. We almost had to yell at each other through the hatch. They should do something about the comm. The other thing is a lot of CMS/LMS simulations are probably unnecessary when it involves the CMS being a passive vehicle all the time, don't you think?

MATTINGLY — Yes.

YOUNG — I think laying around in the passive vehicle is wasting your time.

MATTINGLY — If you have nothing else to do, that would be all right. As I've been trying to emphasize, the CMS is my equivalent to the EVA training, as well as a proficiency trainer. I can fly a lot of rendezvous and a lot of rescue things any time you do a rendezvous, there's always one passive vehicle. The kind of things you should do in the CMS/LMS training are those things which would require coordination between the two vehicles. I don't think you should go out and fly a CSM active rendezvous and have the LMS just sit there as a target. Nor do I think you should do the other way around. You should fly just enough of those together that you get accustomed exchanging data between the two vehicles, so that you know what a nominal rendezvous is going to look like to you, and go through the decision making process. I feel that's another area where the scripts were not adequately prepared. I don't think anyone looked at what was going to be done during the CMS/LMS session to make sure that it was worth both parties participating. I think that is one place someone could look into it and save us a lot of time.

YOUNG — Simulated networks simulations - Invariably, if you added the freebees, you got more than you asked for and then some.

MATTINGLY — One thing that goes along with what you said about flying a few nominals towards the end. I felt like our last launch SIM was not the right foot to start a mission off on, where we did so many things that were way out and then we just kind of backed ourselves in a corner. I would like to have made our last SIM with Houston one of those nominal things where everything works like you're going to see it in flight and you build up your confidence.

YOUNG — Yes, the last launch TLI SIM was pretty near a disaster. It's too close to the end to be doing that kind of work.

DUKE — All in all, it's good training, though.

YOUNG — I really think you should slack off a little during the last of the training. This is in line with stressing the crew preflight. I think you should be limited as to how much of that you should impose on the crew. Missions are tough, but you don't want to bend the guys around the axis just before they get ready to launch. I was bent there after that last launch abort SIM. I don't know how everybody else feels.

YOUNG — DCPS - I thought that was valuable for preliminary launch aborts. But, once we got down to the Cape - I think the simulator down there is certainly excellent to do it, plus they get all three men in the pressure suits and you start realizing the real limitations of doing launch aborts in pressure suits; and, they are serious.

MATTINGLY — I don't think you can overemphasize the fact that you should fly every launch abort, period, suited.

YOUNG — Yes. Sure is miserable though, isn't it.

MATTINGLY — CMPS - I used the CMPS a great deal. I flew every rescue and rendezvous combination known to mankind, I think.

YOUNG — Plus entries.

MATTINGLY — Well, that's a separate operation. I thought that I had flown everything they had. I did this all before we left Houston, so that all the time I spent on rendezvous and rescue review was done as just sort of a paper exercise at the end, and that's the way it should be done. I think the McDonnell-Douglas people running that CMPS have a really super operation. They can show you all the characteristics of the different trajectories and step ahead. I can't think of any way that the CM would ever become an active vehicle in a rendezvous, I would have felt terrible going up, there without knowing that I could effect a rendezvous from any place in the sky. I have complete confidence that I can rendezvous with that hardware with any malfunction that can happen to the command module and still effect a rescue, and stay within my fuel budget doing it.

YOUNG — Except when you start in the low orbit with the TPI from apogee and you rendezvous from below. I didn't think much of that.

MATTINGLY — I still think you could do it.

DUKE — There he goes behind the hill.

MATTINGLY — I think you should skip Gulf egress training.

YOUNG — Yes, I agree.

MATTINGLY — I think if you have ever been out in the Gulf, there's no reason to ever go out in the Gulf again. As much as I cherish my time, I thought to spend a Saturday going out there was terrible.

YOUNG — Do it in the tank on a normal working day, do STABLE I and Stable II in the tank and that's certainly adequate. The Gulf egress training is a waste of time, money, and effort and in particular when you're so dependent on the weather and the sea state. It's not worth the effort.

MATTINGLY — I didn't use the planetarium this time, but I used it on 13 and thought it was valuable. I did go out and schedule some sessions in the CMS to get used to finding my way from one star to another in a starball sense.

YOUNG — Yes. I think the planetarium could be deleted from the crew training, if you know the stars. There are much better star simulators in the LMS and CMS, perfect stars, better than you would ever get in a planetarium.

DUKE — I took about half an hour every couple of weeks to look at stars in the simulators and I think that's quite adequate.

MATTINGLY — I guess that is what we have been talking about and that is one area that really should be looked at. I don't mean it as a function. I don't mean that the guys are not doing their jobs. I think that their job description is changing. As the mission gets more complex, we're adding not only more requirements to the crew, but to the people that are responsible for training the crews. They have to be as sophisticated in laying out the mission training as well as all the hardware and everything else. A lot of times I think we leave the simulator people out of the picture and they're sort of technicians that run the machine and the training, but they don't ever get involved in the flight planning. I think they are short-handed and hard pressed, but somehow we have to include them in the total flight planning.

DUKE — As far as the LMS goes, I can count on my hand how many times we changed our training plans. I never did a thing, it was all their "square filling" and I thought it was good.

YOUNG — I think it was good, too.

MATTINGLY — I think you have an entirely different problem.

DUKE — We do. We don't have our Flight Plan verified. We have a set of procedures and systems to learn and that's it.

MATTINGLY — If you had as many hours in the LMS as I had in the CMS, my CMS hours would have to be proportioned over 290 hours and your LMS would go about 10.

DUKE — Yes. About 10 flying hours, you're right. I felt redundant many times, as if, man, I've done this before. But, to back you up Ken, we do not have the problem with the LMS, since we don't have the Flight Plan to verify. All we had to do was look at new procedures and attitudes, and you can do that in one session.

MATTINGLY — Systems briefings - Man, I didn't have any this time around. The only system I really hit on as a separate thing was the G&N software.

YOUNG — You had it once, but, you never got them logged in your training plan. You were in here on weekends talking to Roger Burke. I count that as training; but, no one else did. Did you log it?

MATTINGLY — Yes, I did.

YOUNG — You didn't log it as a system?

MATTINGLY — Again, that's one of those things, and it depends on the crew's background as to whether it's profitable or not. Some people would find it useful.

YOUNG — The thing that I think we should emphasize here is that it would have been nice to know about such things as SPS and pressure anomalies prior to flight. The way we were working was that we were having a close-in systems' briefings on anomalies. You know when you get close to launch, you're going to take a little different view. We knew about the digital event timer skipping and we knew about things of that nature. But, we sure didn't know about the SPS. I wish we had. Before, Charlie Kline had been there on launch day. That's the kind of thing the crew should be cut in on.

SLAYTON — Don Arabian's little poop sheet didn't have tact on it? Not that one.

DUKE — All in all, except for that one thing, I thought that Arabian's thing plus Dave Bollard's briefings were adequate.

SLAYTON — Yes. That one fell through the crack.

YOUNG — Orbital geology - That's a life subject in itself. I can't say enough good words about El Baz.

MATTINGLY — I can't say enough good things about Dick Laidley. Dick Laidley came in and set up a training plan. His first task was to convince me that when you carry a map and pan camera, there's some reason why you should spend your time looking out the window, and use the man's time at all. I'm really glad that he did because the proof of the pudding came a couple of hours ago when I went over and looked at the photographs again this morning. The things that I saw and recognized with my eyes are not on the photographs. I think if I can record those things that it would all have been worth it. The training was set up, again, solely by Dick Laidley. I think it's a unique thing, as we don't have a system that handles this very well. We

72-H-430

had one man, with the unique capability to take his capability as a pilot, and his ability to fly and recognize what he sees from orbit - that made him especially useful. He gets my vote for being the most organized and most thorough person I worked with in any training capacity anywhere. He'd set up a trip, run through it, make sure that all the kinks were out of it, he'd fly the things, prepare maps and briefings before you went, and he'd provide for critique. He provided for briefings by local geologists on the area and he proofed the whole thing before he ever wasted 1 minute of crew's time on it. And you just can't say enough for his thorough preparation. At the end, El Baz got back in the loop and Dick didn't know as much about the Moon as he did about general geology. El Baz on the other hand, probably knows about all there is to know about the Moon on a human basis. The problem is that is a single-point failure, if there ever was one, in the Apollo Program. El Baz can't afford to sneeze, cough, or anything, because he's the only person that has expressed any interest in that. Putting it all together, it's frightening to me that one person is in that position. It's fortunate he's the super kind of guy that you can afford to have in that kind of job. I just don't think we should be running a program as expensive as this with only one person carrying the ball for such a large portion of the activity. In this case he got wrapped around the axle on Apollo 15 until very late in the game. By the time he gets through reducing the data on 16 and gets through with all that stuff, he'll probably be late getting back into the Apollo 17 picture. I don't know what you can do about it, except to recognize that fact. Gary knows how many hours we spent on this stuff, but I'm sure we spent maybe about 50 percent more than is recorded.

YOUNG — But, Ken, let me tell you something. I think you're the kind of a guy that's really interested in that, and I'm not so sure that the 17 guys are going to be that way.

MATTINGLY — Okay, I'll buy that. I'm not sure. That may be a personal thing.

YOUNG — He sure is a good man. There's no doubt about that.

MATTINGLY — Landmark identification training - We spent a lot of time on that at the end. That's one of the things that you'd like to start working on early, but you can't until the ground track is firm and then you have to wait until you get the maps out and all that sort of thing. It almost by definition comes up in the last month, when everything else is starting to fill. It just seems like it all hits you at one time. I think that the time we spent on it though was really well worthwhile. I was able to operate by looking out the window and seeing things, and knowing where I was without referring to the map. And, I think if you're going to make any useful visual contribution you're going to have to have that kind of operation. Your time on target is so short that if you waste your time cross-checking your position against a map and looking back and forth, you're not going to get around to doing anything except verifying that the pictures on the map are on the right place. That probably goes along with how much emphasis you want to place on visual recognition. If you don't, I'm not sure that you really need to do those things. They kind of go hand in hand.

SIM bay training - We didn't do any formal SIM bay training, to speak of. And, again, that's because of the background we have, since we were in on the hardware development, on the first SIM package. It was sort of like we started our SIM bay training 2 years ago, and if you picked up the program at a different step, why you'd have to spend some time on it.

YOUNG — I sure thought we were well trained. I don't know how we could have gotten any more, unless we just trained for another month. We were as close to 100-percent trained as you can get. Maybe we let some things go, but, by golly we were in good shape.

DUKE — Lunar surface training one-sixth g and KC135 - We had about two or three sessions on that and it was good. The Rover seatbelt fit was excellent, and it turned out to be perfect, and then right at the end we had another couple of sessions looking at pallets and things like that and I thought that was good.

72-H-439

YOUNG — I thought it was valuable. I got the idea that the suit would really be very mobile in one-sixth gravity and sure enough it was. I had the idea that there was going to be many times when I was going to be forced to get down close to the ground and pick up something that dropped and sure enough the next minute I dropped the flag staff and a few things like that. I had to pick up rocks and the lunar portable magnetometer reel which was always in the middle of the dirt. Sure enough, based on one-sixth g and the work we did in the KC135, it was no problem. And the one-g walkthroughs, and we had a minimum of those, we had one on the ALSEP and one on the MESA. I think you should do a minimum on the unsuited work, as little as you can to get the idea of what you're doing, and then plunge into the suited work, for the EVA training.

DUKE — At most, two ALSEP exercises was enough to learn the procedures, and I think that after that you were wasting your time as far as effectivity goes, because it's all that suit that encumbers you. You don't want to learn enough about it so that you break the training gear. We had EVA training. After EVA training, exercise, and 13 or 14 ALSEP deployments since last June; and, gosh knows how many EVA-2 and -3, and it's a backbreaker, and you cuss all the time. But, it really paid off. We felt right at home. Everything worked.

YOUNG — We got 350 hours in those pressure suits, according to my records, since the day we started this business. I think it paid off. I think that learning to work in a pressure suit is the most important thing you do in the lunar surface operation, because you've got to learn the limitations of the equipment. And, no doubt there were some serious ones with the pressure suits. You can't move your fingers very well. One thing that was different in one-sixth g was that I ended up on the first EVA leaving a hole in my wrist with a wrist ring. I had some wristlets that I had taken up there with me, but unfortunately, they were in my lower pocket and they were all full of dust. So,

I borrowed one of Charlie's and used that on the other two EVAs, but it just kept getting worse. I just tried to ignore that, but I should have worn a wristlet. I didn't anticipate and because of the fact that you had to bring the pressure suit up like that (gesture) to do V and you're doing some motion that I hadn't anticipated max in the one-sixth gravity, with the controller. I guess that's just a fact of life. I should have worn wristlets.

DUKE — The geology field trips were outstanding. The monthly trips that we did from the time we started on the crew was just right. In 2 or 3 days, you would come up to speed. I thought they were excellent.

YOUNG — Also, it helps you to get a teamwork pattern and I think that's real important. You are not very effective unless you're working as a team up there. I thought Charlie and I did a lot of teamwork when we were working together in our own peculiar ways. And, I thought it worked out real well. It's really important. Otherwise you're just going to be spinning your wheels on the Moon and that is not where they want you to spin them.

DUKE — Once you got there, there really wasn't going to be the geology to see that you see on the field trips to the San Gabriels or the Sierras. But the Moon has its peculiar geology and there really are things to look for. It's mostly in descriptive terra on a broad and a narrow scale. By broad, I mean describing mountains, such as Stone Mountain in general describing rocks in detail and estimating. Getting your eyes tuned to estimations of slopes and percentages and sizes and things and the only way you get that is on a field trip.

YOUNG — I know how important that is and it makes the geology guys feel good.

DUKE — I thought we did a better job up there than we would have without it.

YOUNG — That Moon is really looking at a geology field trip through 6 feet of dirt and it's kind of tough. The LRV training was outstanding, once we got that seat cushion put in there where we were elevated up to the same position that you were on the Moon. I had the feeling when I was up a little higher than I was on the seat, but not much, just a little bit.

DUKE — I felt I was maybe riding a little higher on the Moon than I was in the trainer with the yellow seat cushion. The qual unit we had for a week to look at off-nominal deployments was very useful. We had a nominal deployment, it's good to see those things.

YOUNG — As you know, we had to fix the walking hinges, the wheels, and the pins. It's good it wasn't more serious than that.

DUKE — But the saddle came right off.

YOUNG — Yes. The saddle came right off. Of course, we were pretty flat too, when we landed.

DUKE — The CSD Chamber - We did one too many on that.

YOUNG — Yes.

DUKE — Two would be adequate.

YOUNG — And, I hassled with those guys, trying to get this thing brought around to our way of thinking. I think that's just too much work for the good you get out of it. It's unfortunate. What we're ending up doing is spending all day long to run an altitude chamber test during which we'd actually be running equipment for 40 minutes, with the prebreathing; the pressure drops; the added test that they were running (which didn't have anything to do with what we were doing); the way the ingress and egress; the chamber; and all. We would like to start right out with the prep and post in there, and going to the EVA and working the gear. With the uncomfortable and unfortunate suspension of that rig - the backpack in the chamber - compared to the amount of good you get out of it, you're really spinning your wheels for a whole day. Since the equipment does work so good, I'm not sure that's worth it.

DUKE — I think there's some good in the two on the PLSS; not in the LM cabin part, but in the other part of the chamber, where you have the PLSSs on and you're exercising the PLSS modes.

YOUNG — I think it is important to know how to do the PLSS water recharge. You could do that over there in the lab where the guys are showing you exactly what's going on. You want to make sure you get your water serviced right or you are not going to do much EVA. But they can certainly do that adequately in the lab, both here and on the flight unit down at the Cape, later on if you're concerned about it, which we certainly were. When we first started, my concern was that we never knew if we had the water serviced properly. So, we made a special effort to make sure we did. I don't think you need to do that in the altitude chamber.

DUKE — I don't think you need to do the Buddy SLSS in the altitude chamber, either.

YOUNG — No, I don't either.

DUKE — And that took I hour to get that thing on and off because of those connectors they had in there. It's just not worth the effort.

YOUNG — Contingency EV training, EVT training.

DUKE — We did one in the WIF which I thought was a lot of fun.

YOUNG — But one time when we all got in the command module there, we had suits and bodies and hoses and I had a feeling that if something had happened, nobody could have gotten in there with us. It's kind of crowded that's something. I was lying on the bottom of the pile when they closed the hatch. I had the feeling that if anything happened, it was a bad scene.

DUKE — Closed the hatch and lost comm or whatever we did that day.

MATTINGLY — It's not clear how much value you'd get out that training, even if you had to do a real EVT. It's so much harder to do things in the water tank than it is in real world. But, it does kind of help to keep you from being overly ambitious in your ideas about what it is you might be able to do.

YOUNG — I do think you need preparation for EVT, but you could do it in one g.

MATTINGLY — That's a one-g thing.

YOUNG — Yes, because you want to make sure you got the thing on, and hooked up right.

MATTINGLY — I'm not sure that if a guy has never looked at the EV transfer pad or something, he shouldn't do that once. But I'm not sure there's any reason to do it twice. All the things we did there I can see them coming up. If you ever had to do one I could see you coming up wanting to know why we didn't do one. Like if you go the outside route with the OPS, you can't talk a fellow through the things and I think you really need that visual image stuck back in your head somewhere that says "Oh, yeah, I remember now, I gotta go back in this corner" because there's no way you can get to the fellow to tell him what he should do.

YOUNG — Well, there's some things that we hung up on in the WIF, crawling across the vehicle with all that line on you. It might have been valuable to have done that if you are going to have those lines hooked onto you, to make sure that you remember how bad it was. You may not get hung up on them in zero g, but, I don't see any reason why you wouldn't, you know. Twenty minutes of OPS - You know that's a little last ditch thing. You should do that once.

MATTINGLY — I think it's worth an afternoon of your time. You should do it earlier in your training, perhaps, than we did, By the time we got around to doing it, it was getting down to the point where we were pressing to do other things that were more important.

YOUNG — EVA prep and post was outstanding.

DUKE — We had about three times on each prep and post and really felt at home by the time we left there.

YOUNG — We were able to easily regroup toward sleeping first. Of course, the procedures called up were just right down the wire, and they were good. But, the fact that we had run our prep and post so many times just sort of made them seem natural.

MATTINGLY — Yes. I'd like to throw in a comment on the same subject, Jim Ellis set up that prep and post training for the inflight EVA. Again, I think it went well, primarily because I had done it many times. It's a pain in the neck to do because it's not an interesting thing, stowage.

YOUNG — Especially in one g.

MATTINGLY — But it really paid off.

YOUNG — It's painful.

MATTINGLY — Yeah. I'd like to comment on that chamber thing, too. We ran the umbilical in there and I think it is important that you run the umbilical system in the CSD chamber before you run the altitude chamber at the Cape. And exercise the same system such that you have some idea of what a nominal system performance is like and because there's such limited instrumentation that no one except the guys that's using the gear is really likely to know what's going on. And I thought doing that once was a very worthwhile thing.

YOUNG — Mockups and stowage training.

DUKE — Great big plus for Jerry Stoner and Joe Dougherty. Those guys with their EVA, MESA pallets, and Rovers. We never had any problems.

YOUNG — We sure didn't when we finally got all the gear together. As usual they are still on Delta right down to the last time we did it.

DUKE — I didn't see a thing on the lunar surface that looked different than the training gear.

YOUNG — Yeah, that in itself is a miracle. Photography and camera training was adequate.

DUKE — I naturally had zero confidence in the camera gear going up there, but it worked like a champ. Those guys had given us little hints that would help on the 16, especially, and we checked every mag and it just worked great.

MATTINGLY — I think Dick Thompson really did a lot to help, to put all that stuff together. Before, I had never had anyone sit down and go through the malfunction with the camera and take pictures and critique what you did with them and all those things. I think having Dick available to look at all that camera stuff was a big help. He took the time to review each photo we took, and what we could have done better about it, particularly the ones we took in geology training and then sat down and make sure we talked about all of the malfunctions. We went through all the procedures. I thought that was a super thing

DUKE — The training Hasselblads were getting a little bad towards the end. It was frustrating. The camera briefings were all good. We had a good deal about how it should operate. The training gear was getting a little rusty by the time we finished with it. And, it left some frustrating moments, and I guess that's what led to my lack of confidence. But the flight gear was really good.

YOUNG — You know, where we really let it fall down in the crack was on those bags.

DUKE — Lunar surface experiment's training. We had briefings from all the PIs and had them witness the deployment of their experiments and at the end everybody was happy. I understood how the equipment was supposed to perform and I thought that was good.

YOUNG — If I had to do it over again, I'd put all those cables up in the air so they would not get in the way of the crewmen. I sure didn't think of it. I know that everybody said you had to step over the cables very carefully and I thought I was doing it, but I sure wasn't.

Lunar landing, LLTV, LLTVS, LMS - Up until the mission, I would have thought that the LLTV might have been pushing it too far. But, after I flew the thing all the way to the ground, I spent the last 200 feet with my head out of the window. LLTV is mandatory as far as I am concerned. It's really a nice feeling to have flown a vehicle that responded just like the regular lunar module did. I had mixed feelings about it because I thought it was a pretty risky way to train, but, I really think it's essential. It really helps you get prepared for that last 200 or 300 feet, and the LMS, of course, supplemented it with the L&A. It sure is a good gear. The planning of training and trying program - I thought it was excellent.

MATTINGLY — There is nobody that looks after all the command module experiments. There's nobody that lays them out. There's nobody that puts them together in any kind of a reviewed fashion to see what it is we are trying to do, except the flight crew. There is no one to point out who we should talk to. There's no one to set it up and yet you have to do everything through the backdoor. For instance, take an experiment checklist that's 1-1/2-inch thick of experiments. There's nobody that's screens how those things are done, how they should be done, what it is you're trying to do, how you should take them, when they should be done. We've pulled every bit of that stuff out like pulling teeth.

YOUNG — But you made that book up, didn't you?

MATTINGLY — I am one of the few people that knows what's in that book.

YOUNG — I was continually surprised.

MATTINGLY — I should be given that book with a set of procedures to learn. Instead, I'm one of the few people that knows what is in there and why it's there.

YOUNG — That's the only way to make it work, Ken.

MATTINGLY — But, I don't think the flight crew has the time to do that stuff. I feel like I put so much more in this thing than you should do.

YOUNG — I know you did.

MATTINGLY — I started on this thing 1-1/2 years and that's the only reason I got there, because I sat down and I talked to everyone of those PIs and every one that had an idea. I had to call them in and I had to look them up and find out what they were trying to do, try to put it in the context of where our spacecraft was going to be and what we could do and how we do it.

YOUNG — What you're saying is that there's probably no single point of contact from S&AD working on it.

MATTINGLY — The things we got weren't useful. They weren't usable. The requirements would come in and they'd be ding-dong. You look at it and you would say, "Well, this is not going to do anything because I know what the guy wants to do." All they're doing is reading from a mission requirements document, which doesn't do a doggone thing for you. The whole area of orbital activities has totally been lost. No one is looking after it, except the flight crew. Perhaps the proper approach is to drop it. But I would never recommend that another crew spend the time that I spent on it, and if I were doing it over again, I would not spend the time on it.

SLAYTON — The problem is that they all assume that all you have to do is turn them on and off at the proper time and that's it.

MATTINGLY — The SIM bay takes practically no training and no effort. We did that. We covered that part of it with minimal effort. It's all that other stuff that takes so much of your time and there isn't anybody to do it.

SPEAKER — Well, what happens to the ALFMED, Skylab contamination, all those?

MATTINGLY — That dim light photography takes such an inordinate amount of your time, learning to do that stuff. There's nobody to screen some stupid procedure. Somewhere I've got a copy of the original procedures we got. They are humanly impossible to do. You're handed this stuff, and told, "Here, go do it." Well, I'm the poor cat that's responsible. I have to sign off the yellow sheet and say I'll fly this airplane and I'll make it work and there is no way. No one's looked at it and the final crowning insult is that you go back and you have to argue with some ding-a-ling who doesn't know a thing about it, about whether this is a reasonable thing to do or not. S&AD and every bit of this has been a filter in the middle that has been a time delay. They should bring a set of experiments that have been thought out and practiced. There's no excuse for anyone ever bringing me an experiment that they haven't sat down and done themselves, and I don't think we got a single one that was brought to us that way.

SLAYTON — Sure, you're right.

MATTINGLY — I will say that the guys on the ALFMED put more effort into their experiment and did more work and did more to make our familiarity with it useful than anyone else. But, they are the only ones that volunteered to come forward and work on it. The rest of the stuff we had to go pull out tooth and nail. And in many cases, I think we had a lot better handle on what could be done and should be done. They come in with something that takes a lot of your time. I probably spent 15 hours in the CMS alone on low-light level photography, learning the techniques because you're sitting here changing shutter speeds, changing the aperture. You're sitting there in the dark doing it because we have inadequate gear to properly do the thing, You've got to practice how to do all that stuff in the dark. Some of the stuff I finally had to end up having to write and talk into a voice tape, which I could then turn on and play back to me. There's another thing I had to build. You know operating on the surface that way, if they want you to do something like that on the surface, some surface guy goes up and works out all the things, makes it up and let's you try it. You kind of work out the rough edges. In this world, the command module experiments which I think should be done, that's the reason I did it, because I think we can contribute a lot of good stuff. But I just don't think the flightcrew should have to be the experiment interface and all that stuff. I think he should be handed stuff to learn and to do.

SLAYTON — You're right. But, as long as you're willing to do it and get it done, it isn't going to get done any other way. The only way you can get it done the way you say it should be done is to throw it right back in their lap, and say, "Give me a package, or forget it."

MATTINGLY — You may be right.

YOUNG — If it hadn't been somebody as conscientious and methodical as you are, the stuff wouldn't have gotten done, Ken.

MATTINGLY — Well, there is no question in my mind that we wouldn't have done it.

SLAYTON — I wish you had bitched about it before the flight instead of after. We probably wouldn't have flown it.

MATTINGLY — I'd do it because I think we've got an orbital platform up there that should be used. I think we can learn things. I wish I had had the authority to do the experiments that were worth doing and not the ones that aren't. We threw off a lot of good experiments.
YOUNG — You never get that opportunity.

MATTINGLY — No, you never do. You have to get the ding-a-ling ones, too. I guess I wouldn't mind if I thought someone had ever put these into the priorities and weighed one against another for the time involved. That never got done, never.

SLAYTON — Nobody knows what time is involved until you get into the nitty gritty like you did with that one. That's the time you should blow the whistle on it.

MATTINGLY — But, there's no guy to go back to. There's no one that is responsible for all the experiments. You only talk to the guy that is responsible for this experiment. I want to throw this one off; this is a stupid thing to do.

SLAYTON — There are people that are responsible for all the experiments. There're three of them. One is Tory Calio, one is Jim McDivitt, and one is D. K. Between the three of us, we're responsible for all of them.

MATTINGLY — I shouldn't bring my nitty-gritty problems to you guys.

SLAYTON — That's what we get paid for.

MATTINGLY — There should be other people that I can interface with.

SLAYTON — There are people that are responsible, if they are doing their job. You should be raising hell about it, which you're doing now, but I think it's kind of after the fact. Maybe it will help Apollo 17. But, I think we could have helped 16 if you would have screamed about it earlier.

YOUNG — We have a quiet period of time there going to the Moon. I thought we were going to sit around and take it easy, but Ken's over there making up his tape so he can do his dim-light photography. And he had to do it. It took about 1 hour to do that going to the Moon. And, of course that's the only way he could have run the dim light in lunar orbit, near as I can figure. Because I don't know how you work in the dark with a clock.

MATTINGLY — Let me add one other constructive thing to this, it's not just complaining. If you're going to do this kind of stuff, once again, there needs to be someone who takes all these things out and finds out where they go into the Flight Plan and says now, here's an area that you need to practice. This sequence is important and you should go do that. And lay out a little training package for you. That's something else where there was no such thing. All this stuff just adds up to a lot of extra time you've got to spend.

YOUNG — In specific instances, if the mission had been nominal, it was a week before flight that Ken had to delete some stuff from rev 72 and 73 because they were physically impossible to cram into the time to do them. That probably is too late in to be doing that. But when you slip the launch a month you're going to run into that problem because the stars move and the Sun comes up at a different time, or something.

SLAYTON — You slip a month but you don't gain a month. It takes about 6 weeks to recover.

MATTINGLY — And, we couldn't have picked a worse time to slip that month.

18.0 COMMAND MODULE SYSTEMS OPERATIONS

YOUNG — Well, you would have been worse if it had been later.

SLAYTON — You've got a point there. My only point is that if you weren't so conscientious, we wouldn't have that problem. Some other guys would say to hell with it.

18.1 GUIDANCE AND NAVIGATION - PREVIOUSLY DISCUSSED.

YOUNG — I knew you were having a problem, but I didn't know how to attack it either. Maybe if you hadn't been worried about that you would have been worried about something else that you didn't need to worry about.

MATTINGLY — That could well be possible.

18.2 STABILIZATION AND CONTROL SYSTEM

YOUNG — Charlie was always worried about everything.

MATTINGLY — Let me make a remark on thrust vector alignment. It is my impression on the circ burn and on the plane change that there was more attitude excursion than I anticipated. Maybe I was just supersensitive to excursions.

18.3 SERVICE PROPULSION SYSTEM - PREVIOUSLY COVERED

YOUNG — Maybe that's what the short burn logic does for us.

DUKE — One of the pitch or one of the yaws, we had a little tough time hearing it on checkout.

18.4 REACTION CONTROL SYSTEM - PREVIOUSLY COVERED

YOUNG — RCS?

MATTINGLY — Back in the command module when you checkout the minus pitch, the farthest away engines are really hard to hear.

18.5 ELECTRICAL POWER SYSTEM

DUKE — We commented on that fuel cell on transearth coast - the cycling the flows back and forth, which was something the simulator does not do.

YOUNG — DC monitor group, AC monitor group, AC inverters, main bus ties - all worked as advertised.

DUKE — Nonessential bus switch; never got out of MAIN A.

YOUNG — G&N power switch, cryogenic system, cabin lighting and controls, split bus operations, gimbal motor transients.

DUKE — Before lift-off they're better if monitored on the fuel cells. After lift-off they show up better on the batteries.

18.6 ENVIRONMENTAL CONTROL SYSTEM

MATTINGLY — The cabin ran at 5 psi when we started, which is a little lower than some of the others have been running. At the end of the mission, it was running down at 4.8 on the cabin pressure gauge. Which says that both regulators were low or maybe the gauge was reading low.

YOUNG — I'm not sure there wasn't a bias in the gauge, because when we pumped the cabin up to 5.7, it looked to me like it was reading ...

MATTINGLY — We never popped the relief valve. If it were biased, it was reading lower than it should be. Once I got it up to 5.9 on the gauge and didn't pop the relief valve, so it doesn't sound like a linear shift. It's kind of interesting that the cabin runs that low since you got two regulators and both of them are set higher than that.

YOUNG — Cabin atmosphere. Amazing how much methane that cabin can handle, although at times, we overpowered it.

DUKE — Only momentarily.

YOUNG — Call 2 hours momentarily? I guess it was momentarily.

DUKE — I thought the water system, once we got the separator on there, was excellent. The hot water was really nice and hot and the food tasted real good with that hot water in there, stayed warm for 15 minutes or so. The cold water was real cold, especially when you got up in the morning to take a drink.

MATTINGLY — Or, if you went ahead and ran it out, it got colder as you drank it. I understand that.

YOUNG — I think the only thing we noticed on the suit circuit is that it takes a long time for the suit volume to build up on the suit pressure test. You got to wait on it for the suit flow to come back to nominal value, In fact, on the EVA checkout, you ran the cabin pressure down in order to get the suit flow down.

MATTINGLY — The purge flow is under the cabin so that was one of the things that you can get a box with. That purge flow will go up faster than that little regulator can pump the suit up. So you're falling behind - you never catch up.

YOUNG — That may be a problem that the Apollo 15 guys had when they couldn't pass their suit integrity check.

MATTINGLY — It could have been. It fits the sequence.

YOUNG — That's what it was on their sequence. Ken thought of it right away. He dropped the cabin pressure, and the suit flow came right down. Waste management system, urine and fecal disposal problems. Let's have at it.

MATTINGLY — Why not - it's a spacecraft problem as much as a medical problem. Take them separately because they are somewhat distinct problems. Fecal disposal. We used the black Skylab bag first. And we had it completely filled. I guess I really can't say that I ever smelled any odors unless you went and put your head right next to the bag; I vented it using the waste management hose and the waste management vent. Everytime we did a urine dump, I'd put a couple of minutes of vent on the thing to try and suck the gases out of the bag.

YOUNG — We had long periods of time where we weren't allowed to vent the bag.

MATTINGLY — That's right. We would go for 12 or 24 hours at times without getting to do that. The bag always had the appearance to me that it was puffed up, except for the times when I would actually go and vent it. Unless I was right next to the bag - my head within 2 feet of it, I didn't smell the odor from it. But if you got within 2 feet of, I think it did.

YOUNG — Charlie slept within 2 feet of it the whole night long.

DUKE — By my feet.

YOUNG — On an operational mission you shouldn't have to keep that kind of stuff in the cockpit with you unless you can vent it to vacuum. You ought to be able to vent it to vacuum whenever you want to, to get the smell out of the cockpit. The rest of the time you ought to jettison it. You ought to jettison it at the most convenient time that comes up. I just don't think that guys ought to be wandering around with a bag full of feces in the cockpit.

MATTINGLY — Our concern was that with cabin depressurization, that the bag would blow up.

YOUNG — Boy, would that have been a mess!

MATTINGLY — I vented the bag to make sure that the big bag didn't burst. That had nothing to do with the little bags. As far as I know, none of them burst. I didn't open the bag to find out either.

DUKE — Fortunately, you can't really get an airtight seal on those fecal bags.

YOUNG — No you can't, no way.

DUKE — That probably saved us. They had 5 psi when we started; I'm sure they went down. We filled up that black bag.

MATTINGLY — Then we put some more in the side compartment - three more.

DUKE — Three. I did two entry morning and you did one the night before.

MATTINGLY — I guess the rationale for using the supplementary bag first was a holdover from the desire to be able to throw it away, which we weren't allowed to do for other reasons, but I really think that's what you should do.

YOUNG — You should have been in the LM when we got rid of it.

MATTINGLY — I really think so. And you ought to plan to get rid of the next batch when you do the EVA. I just don't think you ought to carry that stuff around, if you can avoid it. I think it's a health problem if you ever get some of that stuff loose in there.

YOUNG — I do too. I don't know what data is going to come out of this, I bet none. For an operational mission you shouldn't do it. Skylab maybe a different problem, but even in that, it's going to give them trouble.

DUKE — We had three types of bags.

MATTINGLY — The two bags are really more similar than I thought they would be.

DUKE — The first time I had to go was rigid after waking up on the first day, after the first sleep period. Ken broke out one of those Skylab bags, and I tried that the first time. I thought it worked pretty good. I took the stickum off the little ellipsoid opening, and used the stickum and the Velcro strap around the legs. I didn't feel like it was tight enough with just the Velcro strap, so I undid it and pulled the stickum off. Once you performed the task, the clean up was still as horrendous as ever. I later found out that it really didn't help that any. It's a learning curve on this just like anything else. Shutting the bag off was probably a little bit more secure feeling with that Velcro across my legs than with the other bag. I felt the positioning of the bag was easier due to the hard opening than with the Gemini bag.

YOUNG — I made some study of this problem in depth, starting back in the Gemini Program. I still don't see any use for that finger in the bag.

DUKE — That was one thing I was going to add. You want to get that finger out of there.

YOUNG — Get the finger out of there to keep the feces from hanging up, which it does every time the finger's in the way. All that's going to do is give you a bigger cleanup problem than you already got.

MATTINGLY — I tried doing it the way they suggested - pulling the finger thing out first and then use it afterwards. All that does is smear. Absolutely no advantage to it. It looks to me like you could simplify the bag and remove one more potential weak spot in it by just deleting that whole thing.

DUKE — Our technique was to abandon the LEB to whoever had to go, get naked, and go. That was about a 30 to 45 minute task.

YOUNG — You got to take off all your clothes because we don't have two-piece underwear, which I think the Skylab guys are smart enough to have done.

MATTINGLY — In comparing the Skylab proposed bag with the standard bag, it seemed to me that the finger thing was better than the one we have now only because it was easier to pull out. If you deleted the whole thing, both of them would be better bags. I thought the belt on the Skylab bag was a luxury that we don't need. I didn't think it did anything for you. You've got to have some sticky back around the seal. You've got to have that. As long as you've done that, then putting that Velcro on there was just window dressing, I thought.

YOUNG — What it does - it gives you a much bigger package to dispose.

MATTINGLY — And it has the disadvantage that the hard part that goes in your crotch there gives you something that's very hard to wrap up into a small package when you go to stuff it in an overbag to stow it later. Given a choice, I'd rather not have all that hard stuff in there to have to stuff into the bag later on. Just work a little more carefully and try to get a seal on the plastic bag to begin with. I thought it also made it easier to seal. Concerning the hard stuff, I was never able to get a good seal because of the rigidity of that plastic or whatever that stuff is. When you fold it up and put the two surfaces together, along the axis of the fold, there was always an air passage, You could never seal that completely just because of the material properties. The thin plastic bag didn't have that thing. I felt like you could get pretty close to a gas-tight seal with the original bag.

YOUNG — You don't have the time or the ability in zero gravity to mix the dye in that thing. There's no way you can get those things all together.

MATTINGLY — I understand in their desire to make that bag so that it doesn't inadvertently break the germicide, or whatever that little blue package is. For me, one of the most difficult things was to put that thing in there and break that little bag inside. I was always afraid that instead of breaking that little bag, I was going to break the outer bag. You really have to push on that thing. Like Charlie said, you've got to just take the heel of your hand and just really smash it. I always had visions of the big outer bag coming open when I did that. I really had a hard time mustering the courage. There ought to be a better way of doing that. Seems like you ought to put it in one of these packages where the outer coating does like these pills do where it sort of eats itself away like pills do in your stomach.

YOUNG — It would be more convenient, too, if it was packaged down in the bottom of the bag where it's supposed to be. Every time you put it down in the bottom of the bag the first thing it does is float back out.

MATTINGLY — Instead of something you add afterwards - it ought to be something that's in there with a coating that will come off when it's exposed to something in the fecal matter. It would just take care of itself. You shouldn't have to add a pill or a germicide or anything like that to the bag when you get through with it. The other thing that appears to me that no one had considered is the large amount of tissue that; is required clearing yourself up. These fecal bags are big enough to hold a volus [?], but somebody forgot about all the tissues you're going to stuff in there. You're not very efficient with them because you just take one swipe, and, if you got a loose stool, it takes lots and lots of swipes. When you get all that in there, then you go to put it in that overwrap bag. We had a couple of them that were really a tight fit. You either need to provide a different disposal or you just have to allow for it. There is no reason to make those bags real tiny. I'd like to reiterate my comment about the URA. At least the way ours operated, it needed more vacuum to pull the fluid in - to keep from collecting a bubble around the rim. Your best technique was to get close to the honeycomb. That would minimize the amount of splatter you would get most of the things inside. When I got through, and put the cap on it, there was always a ring of fluid that would - surface tension would pull back up around the side and get on the cap, and when you opened it up, out came a bubble. I don't think that this is unusual. This has been commented on before. You have to be prepared that when you open the cap, you've got to have a tissue handy right then and there to get hold of it and do something with the thing. I felt like it was a much cleaner technique all the way around to use the bag with the roll-on cuff. Given a choice of all those devices we had, even if I could use it at any time, that the bag with the roll-on cuff was a much cleaner thing to do. You need lots of cuffs and lots of spare valves.

DUKE — Yes, we only had one spare valve.

MATTINGLY — You ought to have more than one spare valve.

YOUNG — You're absolutely right. On an operational mission, you don't want to wake the other guys up if you have to make a head run at night. You got the bag with you and you can use it right there. If you're going down into the LEB at night you're going to have three guys up watching you go. There's no way you can get down in the LEB without waking up everybody.

DUKE — Well, with the SIM bay, once you jettison that door, you're on the bag anyway.

YOUNG — The actual performance of the system was nominal. We didn't have any blockages, although one time we changed out a filter, a short filter.

MATTINGLY — That was because we were getting ready to dump. We didn't see anything come over the side. We changed the filter, but it turned out we must have had the bag empty anyhow. It's one of those big white urine bags with a Beta cover around them. Those things are difficult to tell when they're emptying because the overwrap bag is semirigid. You can't see the interior bag to know when it's all out. You kind of have to watch outside and see when it quits dumping.

YOUNG — I thought the real-time test that we performed on telling whether or not the doctors could tell how much urine was dumping was really begging the issue. It indicated that they had little operational knowledge of what the heck goes on when you dump. The stuff comes out in spurts. It doesn't come out all at once. The line clogs up, and then maybe it breaks loose or something, but we saw it coming out sometimes in spurts. The urine is always dumped that way. The waste water was a continual flow, but urine seems to dump in groups, in a sort of a random manner. We were supposed to tell them when we started and when we ended. That meant you'd have to have somebody down there who could tell when the white bag was empty. And as Ken says, there's nobody in the world smart enough to know when that thing was empty. That took however long it takes a bag to dump, somebody on it full time. That's got no place in an operational mission. I don't know how you would tell even if you had somebody on it the full time. The procedure is to wait until 5 minutes after you see the dumping stop out your left-hand window before you shut the thing off. It's a subjective thing because even after you finish, and you know the bag is pretty empty, that thing is still over there dumping stuff over the side. Where it gets it from, I don't know. From a line, I guess.

MATTINGLY — After we quit dumping, that thing would still continue to vent on the order of 10 minutes, a minimum.

YOUNG — I just don't believe that for an operational mission that the guys should be saddled with measuring that kind of thing, testing it, and looking at it and carrying it around. That's a mistake.

MATTINGLY — I think it's an important thing. Concerning what you do with the other trash that you have onboard the spacecraft, we use the jettison bags for a trash can. We talked about that being larger than you'd like to handle. The spacecraft launches with an awful lot of paper and bags, particularly the food bags and all that stuff. Number 1, you need to minimize the number of bags, disposable things that you take in a spacecraft because you can't open the door and throw it outside. Number 2 is that you have to make plans to handle all that. We jettisoned, during the course of the mission, all three jettison bags, which were filled just about to their total capacity. That's an awful lot of trash. We didn't eat all the food. If we had eaten all the rest of the food, we would have had even more trash in there. I don't think we have really paid enough attention to this. I don't know what you can do for Apollo except to be aware of it. Future spacecraft must provide a better way to handle that kind of trash. You need an intermediate-size bag that's easy to use with a snapping lid. Either that or someone could make a snapping lid to go into the jettison bags so it would be easier to handle.

YOUNG — Goodness gracious, don't let them start on that. No money in the world could build that bag.

MATTINGLY — That might be.

YOUNG — It would be an advantage to have a jett bag that you pulled open, like that, with maybe a bungee in it - so that you put the stuff in, then close the bag and not have everything come back out.

MATTINGLY — You need some self-closing thing. CO2 absorbers - we had one that stuck very tight. Previous to that we had one that was a tight pull, but it wasn't really stuck. The one that did stick - we must have worked on that thing 5 or 10 minutes, pushing it in, pulling it out until finally we were able to break it free. When we got through, we set it side by side with the one that was clean, and there was hardly any visible difference.

YOUNG — I didn't think Ken was gonna get it out. I really didn't. I felt we were stuck with a CO2 cartridge that wouldn't come out. It was that tough. He'd slide it back in and slide it back out, slide it back in, making maybe a quarter of an inch at a time. The cartridge's stowed as ordered in the Flight Plan A6 or A4 - A6, I think.

MATTINGLY — I don't remember. Not A6.

YOUNG — The second one inboard from the port side.

MATTINGLY — I really didn't know that there is a mission rule on that when one of those is stuck. It appears to me a prudent rule is, make sure the other chamber doesn't stick, because if you ever get that thing stuck halfway in, you can't close the door. There's no way to open the door on the other side and you're out of business. I really think you need to be ginger about it, and all the assurances in the world that you can't break that little string are hard to swallow.

YOUNG — You're really torquing it, man.

18.7
TELECOMMUNICATIONS

DUKE — Everything was fine. From my side, I thought the comm was great.

YOUNG — The VOX was good.

MATTINGLY — The VOX was excellent.

YOUNG — We didn't use the USB emergency keying.

MATTINGLY — Some time during the mission, the Sony tape recorder would not record and then later on it did. I think it was because you couldn't depress the red button, and I have no idea why or what happened, but I think probably something floating around in it caused the problem.

DUKE — There was a failure that we had that we haven't discussed so far and that was the failure that we fixed by taking the NORMAL switch and going to OFF and then back to NORMAL. I don't know what we fixed there.

MATTINGLY — I remember we did that a couple of times.

YOUNG — Whatever it was, the ground said they'd fix it. They were having trouble with the up-link or the down-link or something.

DUKE — Oh, they got themselves out of sync with a series of commands, and I think it was an up data link. We cycled the up TLM switch a couple of times and that fixed it. By going into COMMAND RESET, it didn't fix it, but going down to the NORMAL position and back to OFF fixed it. What that was, I don't know exactly, but they admitted a problem or two.

MATTINGLY — One of the things that's been a perennial problem with Apollo - the warning tone in the C&W warning system is not loud enough for some people, and I happened to be one of those people that likes a nice loud warning tone that's unmistakable. I think that our spacecraft was probably a little quieter than some of the others have been that I've heard, and I prefer to have a tone booster on as a way of warning myself. Several times I saw the warning light and only then was I aware of the fact that the tone was on. That happened preflight, and it happened again in flight. I guess the rest of you found that the warning tone was completely adequate.

YOUNG — It scared me to death.

MATTINGLY — I just didn't hear it.

YOUNG — Especially when Charlie purged the fuel cells.

18.8 MECHANICAL

MATTINGLY — I have no mechanical comments. Everything worked like it was supposed to.

YOUNG — I thought the mechanical worked well. The couch removal was no problem. We had trouble locking the Y-strut for entry.

MATTINGLY — Yes, we did, the Y-Y-strut on my side. We had to open it in order to get into 382, and I had a considerable amount of trouble getting the thing back in. When you got it in, I'm not sure what you did to get it to lock - how you got enough room to twist it.

YOUNG — I pressed the button real hard while you were twisting it.

MATTINGLY — Yes, I was really beginning to wonder if maybe we were going to make your entry without that.

YOUNG — Yes.

MATTINGLY — I was getting ready to find something to wad in there so it wouldn't rattle so much.

19.0 LUNAR MODULE SYSTEMS OPERATIONS

19.1 PGNS

YOUNG — PGNS, Inertial - beautiful, Optical - tremendous, Rendezvous Radar - outstanding, Landing Radar - the same way. Computer Subsystem - great; G&N Controls and Displays - beautiful. Procedural Data was the only place where we got hung up procedurally. After Charlie put in the P27, we didn't have a CSM vector and the thing was trying to find out where it was. I guess it would be a good idea to remember to remind the guys if they have to do that again, to put in a VERB 66. It'll save them some wondering maybe.

19.2 AGS

DUKE — The AGS was practically nominal, Modes of the Operation. Initialization and Calibration were outstanding. It took the auto updates and the radar and I think it had an acceptable solution for midcourses and the TPI. We never let it run the engine so I can't comment on that. We never had any of the problems that were associated with Apollo 14 and 15 of the AGS warning lights, it was just perfect. Burn programs we did not do. The controls and displays were quite adequate. In fact, when we used this pointing, the 507, during P20 lockup on the rendezvous, the AGS needles were centered right on so it knew exactly where the old command module was.

19.3 PROPULSION

YOUNG — Descent system was nominal and we have commented before on the 2 percent difference between the fuel and oxidizer, before touchdown. Ascent, beautiful.

19.4 REACTION CONTROL SYSTEM

YOUNG — Except for the problem with the Reg. failures, the performance was as like a champ.

19.5 ELECTRICAL POWER SYSTEM

DUKE — Everything was as advertised. The batteries were all up from gyros to Battery 6. The DC Monitor was good, the AC Monitor was good.

YOUNG — I think you took battery 3 off the line there, before descent, remember that:

DUKE — No. That was before descent and I did that to let bat 6 take a little more of the load and heat up a little bit more before we started descent.

YOUNG — Oh that's the reason.

DUKE — The battery management scheme that was passed up left me uneasy. I prefer, before I took both batteries off the one bus even though they were tied together, to keep at least one battery on the bus. That's just a feeling I have, I don't really think it's necessary. But everything just worked exactly right. All the pyros went off, we could hear most of them when we were not suited. We did not hear the SH tank go at DPS start, but it was working, everything went great. The LM lighting I thought was good. We used the utility lights maybe once or twice in the dark, on the back side passages.

YOUNG — I think the lighting on the LM was great. Sometimes you get in the lunar module simulator and you can't see any of the switches and gauges. We never had that problem in the lunar module.

19.6 ENVIRONMENTAL CONTROL SYSTEM

DUKE — Okay, one thing for the follow-on crew, it's really not a gripe, is that at cabin repress, you ought to warn the CMP before you activate that valve because it really gets your attention, especially when you're going back to close on the thing.

YOUNG — Yes, that close-out is like somebody fired off a shotgun.

MATTINGLY — Sounds exactly like a pistol shot.

DUKE — The LCG cooling is outstanding. We used that whenever we got hot, and in 30 seconds you would be okay.

YOUNG — I'm not sure that that maybe a factor in keeping your pump running, because any time you can cool without sweating in the system where you are part electrolyte, you're in pretty good shape. Whereas running air cooling the whole time, you're going to be cool by sweat.

DUKE — The glycol pump probably ran a little louder than I'd expected, and it leaked on activation. I guess that's probably because the suit loop wasn't up. But once you get the suit fan going the air noises drown out the glycol pump at that time. One thing I would like to comment on, in the Suit Circuit, is the cabin gas return valve. When that valve was in AUTO, our suit gas converter to push cabin, cabin gas returned to AUTO, it felt like we were getting a chattering in the suit loop and the flow was pulsing out of the hoses. It just seemed to me that there was something that was opening and closing intermittently in the loop. We went around changing things and figured out what it was. It was the cabin gas return. When we put it to OPEN, the chattering stopped and the pulsing stopped, and the suit loop sounded nominal to me, like it did in the chamber. Later on, for all our sleep configurations on the surface we had the cabin gas return OPEN. Before EVA-3 I wanted to check that valve and I went back to AUTO on it. It had apparently fixed itself because it did not give us that chattering and pulsing again. So apparently that flapper valve, in AUTO, was hanging up. But it then fixed itself. By the end of the mission we could have stayed in AUTO and it would have worked okay. In order not to perturb the procedures, we decided to let her go in OPEN.

DUKE — The only thing we already mentioned, is the steerable antenna, would not steer in YAW. It never did come out of the stowed position. We tried it many times in the activation, and during orbital work and on the surface right after we landed, we tried it. Before the EVA I tried it and before lift-off I tried it, and it never worked. So we never did try it after we got in orbit again. I don't think it would have ever worked. We ran the tape recorder partially, but we had it open most of the orbital time and during the descent, due to the power conservation. We operated on down voice/backup most of the time, with biomed off to improve the circuit margins which gave us a hot mike. It was annoying at times, but we usually remembered it.

YOUNG — I think the ground told us whenever we were on hot mike. We had one case there where we had a stuck mike button on my audio panel. The button on my umbilical was stuck and that gave me a problem or two, but cycling the button fixed it. On the VHF, either Ken had a garbled VHF-B transmitter, or we had garbled VHF-B. It wasn't unreadable, but it was not as clear as VHF-A. I recommend that you go back and check his transmitter to see if that was it. If not that then it was probably our VHF-B receiver. They had done some work on our VHF a couple of weeks before lift-off. They changed out the whole thing, due to some high "g" relays, or something like that. I don't remember whether it was VHF-B or not, but it was certainly a garbled reception on VHF-B. I thought omni's were outstanding. We had clear ground comm the uplink was clear all the time.

YOUNG — Okay, we have already discussed the deployment, and the problems we had associated with it, and the setup, mounting and dismounting. A couple of times when you got on, Charlie, your backpack hung up on that rubber rim on the seat so you had to get out and back in again.

DUKE — I could feel it. Some times I overcame that by just bending forward and putting my feet on the floorboard. I had not wanted that floorboard when we had first started, but it turned out that that floorboard was a good deal. I'm glad I had it, because it gave me a good leverage to push that backpack back in against the seat. These LRV operations are all covered in EVA-1 and LRV setup, earlier in the debriefing.

YOUNG — Yes, I thought it was a great deal more controllable than I thought it would be. We'll have to go back and analyze the data and see what kind of slopes it was climbing, cause you really didn't have a feeling for them but it was getting up there. Crew Restrictions, Limitations, and Capabilities, Hand Controller Operations. We didn't experience this lateral PIO that we ran into with the trainer at maximum speeds. It just didn't seem to be there. The reason, I guess, was that the suit was rigid and having a seat belt that cinched me down real good probably prevented any feedback into the hand controller and the normal operation of the speed also didn't seem to be a problem. The general tendency was to leave it at Vmax or very close to that and to take what you got. The next best reference that I had for speed control was to look at the speedometer. I really didn't have a feel for the difference between 7 kilometers and 10 kilometers without looking at the speedometer. There probably wasn't that much difference the crew moving within suits - I don't think we moved in the suits at all, to amount to any thing.

DUKE — Very comfortable riding in the Rover

YOUNG — Beautiful suspension system. If it hadn't been, we probably would have walked a long ways.

DUKE — Really outstanding! You could have 3 wheels off the ground at once and the thing would just recover smartly and it was just beautiful.

YOUNG — On at least three different occasions, we bounced up in the air and came down on a rock, which we were passing over, and it didn't seem to affect the operation in the slightest. The LRV Systems Operations - Nav System - we covered that, and it worked perfectly except for that failure between station 8 and 9. Power batteries were excellent although the ground kept having us turn off Bat-2, because they said it wasn't getting enough cooling.

DUKE — We did get an over temp on that.

YOUNG — We got an over temp caution and warning at station 11. Steering and traction drive was excellent, except for initial startup and, at the time where we accidentally went to PMW 1.1

don't know what the problem was on the rear steering not working, when we initially powered everything up. Voice Communications and Antenna Management, TV and TCU - Did we talk about the antenna measurement, Charlie?

DUKE — Yes, how we had to GCA each other in on the thing occasionally. The Earth was almost overhead, about 80 degrees at elevation. They said it was going to be difficult to point and it was. You had to have the other guy GCA you in because you couldn't see the Automatic Gain Control while you were moving the antenna. You had to really arch your back and then you could barely see the Earth out of the top of your visor. Sometimes you were lucky and it came right into view, but the crescent that we had would just fill up about 10 percent of that center circle. We always had good signal strength. It was always greater than 3-1/2 on the signal strength. Everything went on just as advertised on the TU/TCU. Easier than I expected.

YOUNG — Electrical Mechanical Connections, you didn't have any trouble with those?

DUKE — We already mentioned earlier the power cable from the LCRU on the Rover.

YOUNG — Dust Generated by the Wheels: We've got plenty of photography of what happens when you lose a rear fender. And the Grand Prix says what's going on the rest of the time. There's plenty of dust on the radiators just from opening the battery covers. It's very difficult for me to reach across there to close the battery covers. You would think that the next time anybody designs a vehicle, they'd put the openingvand closing mechanism on the outboard side, instead of on the inboard side of the radiator. I was always afraid that I was going to end up falling right in the middle of the batteries.

21.0 EMU SYSTEMS

YOUNG — PGA Fit and Operations, I think we've covered that a number of times. Biomedical Instrumentation we've covered.

DUKE — I had one thing I haven't mentioned, is that I got some - I got lesions from those sensors that I was sensitive to. It looks like it's from the tape and the doctors have seen it, so they are well aware of it.

YOUNG — Liquid Cool Garment - a good piece of gear. Helmet. We discussed the problem associated with taking it on and off when it's full of orange juice. LEVA Operation. It was okay, except I didn't mention previously that I got it stuck on the last EVA and couldn't get it off for ingress - undoubtedly due to the dust. Gloves. The only thing I can say is that if we're ever going to have a manned space operation where a guy is going to do useful work during long periods of time in space, they ought to develop a glove where he can use his fingers. That's all I can say, we aren't there yet. UCTA Operation.

DUKE — I used the transfer from the suit to the can and the tank in the LM and it works. At 2-1/2 psi we could effectively empty the UCTA in just a minute or so.

YOUNG — I never used mine, because I was always afraid that if I did I would have a very bad pressure suit odor, and I'm sure I would have.

DUKE — In fact, right before lift-off that's what happened to me. I got a very warm, left leg.

YOUNG — And for the rest of the time Charlie's pressure suit smelled like an old head. EMU Maintenance Kit.

DUKE — Not enough.

YOUNG — We said we needed some more tube in there for the surface operation and the dust. Drink Bag. I make recommendations to keep it from leaking. The Anti-Fog. Worked as advertised. They packed the anti-fog differently for the two EMU maintenance kits. The ones that we had were packed so that it was flat. The ones that Ken had in the command and service module were packed apparently at 15 psi, or something, because his EMU maintenance kit was three times the size of ours. When we got it out, it was fully 4 inches in depth, and it should be about an inch in depth. And it was because the anti-fog pads had all pressurized. I think both of the EMU maintenance kits ought to be packed the same way. You don't want to break that seal, because if you do the anti-fog dries out and it's no good. PLSS/PGA Operations, we talked about that. Pressurization and Ventilation we've discussed. Liquid Cooling and Circulation we've discussed. Communications, we discussed the problem of stowing and unstowing the antenna.

DUKE — The only time we forgot that was after EVA-2. We felt like the ground would want us to really get back in, and we just flat forgot to put the antennas down. Then I broke one - John's broke off about 2 inches from the top. It was a little jagged, but I taped it up, and we used it the rest of the time. The Comm was just as good on EVA-3 as it was otherwise.

YOUNG — Connectors and Controls were adequate. Whether or not you can reach them is a function of where that old backpack is on your back on the PLSS. That would vary from donning to donning and vary throughout the surface

DUKE — The two that you had to reach, like the ice water and the coolant, were always there.

YOUNG — Right.

DUKE — The ones you couldn't reach were the primary water and the oxygen.

YOUNG — We only needed to reach them in an emergency.

DUKE — The RCU got dusty due to our dust fender problem. I had a tough time reading mine even though it had the plastic over it. They really did a good job putting that plastic over the top of that thing or we never would have been able to read it.

YOUNG — They told us to give a PLSS check, and I couldn't read my RCU numbers because I made a mistake reaching up with my finger and tried to wipe the dust off. Apparently, the dust acts like an abrasive as it just completely clobbered the RCU and I couldn't read what percent oxygen I had from then on. I think they ought to do something about that.

DUKE — I dusted mine all the time. It got dusty.

YOUNG — I couldn't read mine. OPS as advertised. Foot Restraints?

DUKE — I would like to see the Velcro taken off the flight floor, because it sure got dusty.

YOUNG — Sure did.

DUKE — Made it terrible to clean. If you took that Velcro off you could take a damp cloth and swab the floor.

YOUNG — Get all that dust and mud and throw it in a jettison bag.

DUKE — But with the Velcro there, you couldn't do that.

YOUNG — For the short time you'll be in there in zero-g, you could use the tie-downs.

DUKE — That's right.

YOUNG — You're going to be in there suited anyway.

DUKE — In fact you want to float free in zero gravity at least I did. I felt more comfortable floating. And the hoses really restrain you to some degree. The Velcro didn't work. I tried that and I just came right up off the floor.

YOUNG — I guess I agree, Charlie. I don't think you lose anything by getting rid of the Velcro, but you would sure get rid of a lot of dirt.

MATTINGLY — We had the event timer problem which we already discussed, but the one on the MDC-2 cromped on us about halfway through the mission. And that's really a nuisance - you can work around it but it's awfully difficult to work some problems in our DET instead of on a countdown basis. Trying to set your watches is a difficult thing since it works in 30-minute increments when the sweep second hand goes around. So you really need those things. And for future spacecraft, there's got to be a better way to set those clocks. You should never have to spend all the time trying to sync a clock to the computer for instance. You go to make a burn; it can take you a minute or two by the time you reset the DET to count up when you ought to really is just push some kind of a sync button and have the thing start counting. And that's just a terrible nuisance. We're always joking about the fact that there isn't enough time to set the clock, but it's more than just a joke. On future birds, that's really got to be an included item in the design. Crew Compartment Configuration - I think all you can say is that the J-configuration with all the stowage in there, really is a full house by the time you tie down the rock bags on top of the stowage compartments. It gets to be very frustrating when you find that there's no way you can stow so that you don't have to go back into those compartments. You just have to get used to the idea that you have to untie the rock bags and then go on down in there. When the PGA bag is full, you just cannot get to the connectors where the rock bags tie down to the top of A-1 and A-7. The PGA bag gets so full that you can't even get your fingers in there much less the probe that ties them down. One of the bags I tied to the top of the PGA and the other one, I just had to leave it hanging free. The back of the PGA bag is up towards the side hatch which represents a volume of quite a few cubic feet. The best we could tell, this was totally unused. Could be used for more deviated volume or whatever but it's back in an area where it's difficult to get to. You can't put large items down through the couch to get into it but it's one of the few unused volumes in the spacecraft. We used the tunnel a great deal for our stowage place and that's very convenient until it comes time to do a LM oriented maneuver when you have to go into the LM or something. Now you've got to push all that stuff out of the way to find some other place for it. This is particularly true when you go to don suits. The Mirrors I don't think we used for anything except shaving.

DUKE — I did to put my sensors on.

MATTINGLY — Did you?

DUKE — It worked pretty good. I like those mirrors.

MATTINGLY — No sweat there. I think one thing that would help is either a portable mirror or a mirror you could take with you down into the area when you have to defecate. I think there's some application for a mirror for that.

YOUNG — It will help you in cleaning yourself off. You know where to rub. It wouldn't be so hit and miss.

MATTINGLY — Well, at least you wouldn't miss something.

YOUNG — That's the difference between using half a carton of tissue and using the whole pile. You're rubbing blind. I think it's a serious problem. We really thought when we started the mission, we were going to be out of tissues inside the phase.

MATTINGLY — You really go through those things in a hurry. IV Clothing and Related Equipment - I thought the jacket and trousers worked real well. That pair of long johns, again you ought to use two pieces and I guess Skylab has gone to that we don't have to say anymore about it. That's a much better way to operate.

YOUNG — I guess everybody has their own temperature associated problems but none of the three of us were at any single time wearing the same amount of gear. I wore the top part of it and the lower part of the skivies with the boots on to keep my feet warm. And Charlie, he was running around in his underwear and Ken had the whole business on and I think that worked pretty good. I think with that much clothes you're able to set your own temperature.

MATTINGLY — Yes. I think in the spacecraft cabin that's probably the way you have to do it. You just can't go around controlling the cabin temperature at will.

YOUNG — You don't have any control, near as I can figure, except you could control the spacecraft cabin to an extent by either turning up the floods or by putting the window covers in.

MATTINGLY — So we've already talked about the Couches and the Mechanical business - the problem of the "Y" strut. Maybe a procedural problem for all I know, but it was something that caught us by surprise. Restraints were certainly adequate. Inflight Tool Set, we only used a couple of tools out of it. It was nice to know they were there. That tool E is certainly an invaluable thing and you really would hate to lose that. We have another one in the tool kit. I guess I worried the whole flight that we were going to lose that thing, particularly on the day that the LM left. I was afraid John would have it in his pocket. We had a shakedown before he - before he could get out of the cockpit. Same thing goes for pencils by the way. You can really run yourself out of pencils up there and then you're in trouble - Probably a spare pencil ought to get stuffed away somewhere. I wouldn't carry anymore, but I'd take some of the ones you have onboard and stuff them away for use after you've lost all the ones you started with. We lost a couple out the hatch during EVA. Data Collection, I'd like to say some words about data collection in general. Data collection on an operational mission ought to be something that takes the minimal amount of crew concentration. I think you ought to put the minimum number of entries into the Flight Plan. And, the ground can watch the DSKY or the ground can copy things before it goes on the DSE. I think that's the way it should be done. One of the problems we had with our tape recorder is that you're never sure when you've got the tape recorder to put the data on. You can be pretty sure if you're on a J mission if you're on the front side of the Moon that they're dumping the tape and you have to go on the radio. Translunar and transearth coast I never knew in building future hardware if we're going to operate this way, it seems like a darn good thing would be, instead of the tape motion light, at least you ought to have the complement, one that says whether it's recording or rewinding. I think you need access to a taped voice channel all the time. Because it turned out that I wanted to make a lot of comments on things I saw, things I didn't want to forget. Taking the time to write them down, you don't get your job done. And you just really ought to be able to just say the words and then that'll jog your memory when you get home. Data collection in the Flight Plan just isn't the way to go and I finally resorted to calling it out to the ground. Hopefully it can be sorted out but it's gonna take the air-to-ground playback plus some flight plans to - sort out where all the different frames went. Thermal Control of Spacecraft. Far as I can tell, we didn't have any. Took whatever we had. And, in lunar orbit, that represented a rather large swing of excursions between hot and cold. That's no surprise to anybody. Camera Equipment, I thought that the camera equipment operated beautifully. I still have some very strong comments about the idea of using a camera like the Hasselblad and having to use a ring sight on the side of it. I have several pictures where I had the target in the ring sight and unfortunately the lens was looking at part of the window frame. And I think through the lens is the only way to prevent that. You have to be careful when you use the 250 because it makes the camera so long. You really need to emphasize making any optical equipment as short as you can when you want to get up next to the window with your head. Just anything like that and the same thing is true of the binoculars. They turned out to be small enough that they were very useful. We felt like we couldn't get enough uniform illumination inside to get good photography. If you can get the Sun shining in the window, that's fine. But that says anytime you want to take a picture of something, like demonstrating how a meal preparation goes, you have to try and time it when the Sun is coming through the window. And it just turns out that on an operational mission you just never have the luxuries to do those kind of things. You really could use a portable light in the cockpit if you want to do photography or anything like that. SIM Bay Equipment, I don't think there's any comment at all on that.

YOUNG — We used to have a portable light.

MATTINGLY — I don't remember ever having one. I remember asking for a...

YOUNG — Well, we talked about using it and carrying it down in the LEB with us and everybody said well, you can use a flashlight so they did away with it.

MATTINGLY — The flashlights don't give you enough illumination to take photography.

SLAYTON — We probably ought to have a higher priority on it.

MATTINGLY — I think the Skylab is going to be seeing a lot of times when they'd like to take internal pictures. I don't know how they illuminate Skylab.

22.2 LM

YOUNG — Crew compartment configuration.

DUKE — That we just talked about, the Velcro. I thought the launch stowage was excellent. We had a couple of changes that we decided to do to give us a minimum time for getting ready on the lunar surface. That was to put the footstool harnesses on for lift-off and things like that. Minor things that just saved us a few minutes, and all that went great. We used the Restraint Systems for landing and you really were anchored. We had a pair of pliers and a pair of dikes that never were out of stowage.

YOUNG — The emergency APS start system.

DUKE — We never touched those either. And we had a emergency tool B, I guess it was to open the command module hatch that was never touched. The Camera Equipment all worked great, just great.

YOUNG — One thing that we just sort of arbitrarily imposed on Grumman was to put a strap in every compartment behind the equipment that's in there so you can pull on a strap and get the equipwent out. I would say if you send one engineer in there and judiciously look at the compartments where the things are in there tight, those would be the ones you could leave the straps in. I bet there's 3 to 4 pounds of straps in that vehicle which serve no useful purpose if they're looking for weight savings. There's an awful lot of compartments in there and every one has a strap in there that pulls something out and you don't need them. There must be a strap that's as long as this table, that's 10 feet long that pulls the LCG bags out the LCG compartment. You don't need it.

23.0 FLIGHT DATA FILE

23.1 CSM

MATTINGLY — The Launch and Entry Checklist were super, you can't say anything about those things. They work fine. They've been proven over many flights. I sure wouldn't change any of that now. Cue Cards; I had no complaints about the cue cards, but I have some complaints about the system that gets them done. I didn't really care what the cue cards said, as long as they were basically correct. I always wanted to look up and see the same words. You don't use it as an item that you go through and check each item like you read it out like you would a checklist. But rather, it's a reminder, and your eyes need to be able to get used to looking in the same location for the same cue. But our system somehow gets wrapped around the axle of correcting spelling, correcting typographical errors, introducing typographical errors and reprinting them. We reprinted a whole new set of cards because they changed a Velcro patch on the back of one. Then you got to go through and proofread the whole thing all over again. Finally, a week before the flight we got a whole new set of cue cards, and they didn't fit the spacecraft. They're like
big things. They sure are a terrible nuisance when you ought to be able to put that stuff to bed 3 months before flight. I feel rather personal about those cue cards. Those are mine, and anyone else that wants to deal with the checklist can. But I don't want them to put it in words that mean more to them on the cue card. I want it in my words.

DUKE — You actually had somebody change the words on your cue card.

MATTINGLY — Yep.

DUKE — That's amazing.

MATTINGLY — They improved it. And I can't argue that it wasn't in fact a clear statement. But you know that's not the purpose of those things.

YOUNG — It's just a reminder for the individual crewmen of the things that get him through the operation.

MATTINGLY — On the other hand, I don't mean to be bad mouthing the people that are doing the job because I think they really have gone way out. This guy Wes Jones really puts his heart and soul into doing that stuff. One of the problems there is that Wes Jones lives in Houston and the crew lives at the Cape. Any time you want to deal with something, you go through a rather lengthy chain of command and channels to communicate with each other. That's kind of difficult. Although I think everybody in the loop recognizes it and works real hard at trying to help. Star Charts were fine. I thought they did a good job on those things. GNC Checklist was good. Systems Checklist was good. The data was all fine. The Malfunction Procedures, we didn't do much with those things this time. I had no problems with them. Time Line Items was a real problem. Let's first talk about the mechanical aspects of the Flight Plan. We ended up with two volumes. I asked them to print it on heavy paper so we wouldn't tear it up while we were handling it. And I think that was a good thing to do. But it also ends up that these two volumes are so large that you can't put a clip around it. When you get in the cockpit, the thing is forever floating closed. You really ought to be able to put one of those little clips on it, even if it means breaking it down into three volumes instead of two or something, so that it stays a more manageable size.

YOUNG — Also, from a standpoint of being able to do simultaneous operations of logging, like medical data and food data, there ought to be separate books.

MATTINGLY — I think we probably could have helped ourselves in real time if we'd have gone through and just pulled out those kind of things like the menus and the medical logs and taped them together somewhere. I didn't think about it at the time. The Flight Plan is always in demand. Another thing about the Flight Plan is the timeliness of it and the thoroughness that you use in preparing it. It turns out that we had a lot of problems. I really never had the feeling right up to the launch that I had reviewed the Flight Plan to the level of detail that I think the Flight Plan should be understood. The primary cause was the fact that we changed our launch month by 1 month at a time when we had a Preliminary Flight Plan and it was being reviewed and screened. Then we changed and we had to wait while the Flight Plan changed. Then when it finally did come, we had only one version of the Flight Plan and that was final. That was 6 weeks before launch. But it was really a preliminary, and we were changing it as rapidly as we could. Again, it was through no one's carelessness or anything else, it was just that you can't do everything at once. And these guys were really doing all they could to catch things that go through and review, make sure they hadn't left something out, correct the numbers in it, but we never got finished. We were still catching those things launch week. On 13, I launched with a Flight Plan that I knew, and I think that's the way you ought to fly. I think I'm paid to know that Flight Plan, and I never was able to get to that level of proficiency with this Flight Plan.

YOUNG — Some of the real time changes were corrections, in the Flight Plan. I knew that.

MATTINGLY — I don't think so. I don't know how you handle that problem, except that the people that decide when to slip launches should be aware of what they're doing. You're never going to do something that's unsafe, because of this problem. We're going to do the big things - LOIs, TEIs, and landings, and that sort of stuff. It's a much more difficult problem than the landing, because you can change the calendar day of landing, but the EVAs remain the same. The only thing you have to worry about is Sun angle and the approach azimuths. But then, we come into what really drives the Flight Plan and why we had to rejuggle it so much. And it's all the inertially-related items in here. I think dim-light or lowlight-level photography and photography of celestial features is the thing that really is the tail wagging the dog. Because, those things change as the Moon moves around in the sky. You might have a sequence of say a gegenschein sequence on a rev 42 and that would be followed by a near-terminator photographic strip of the Moon at the next sunrise and so forth. Then you come up and the first thing you have to do is move the low-light-level photography to a different rev because of the inertial angles were different and the Moon occults on that rev. So you have to put that photograph on a rev 60 now. So, you take whatever was on rev 60 and it moves back, and the first thing you know you've got a completely out-of-control situation, where everything is juggling and you start all over again to fit this Flight Plan together.

YOUNG — Yes. I still think if Tom and those guys had done all that work and got the Flight Plan in shape, then these kinds of changes, boy, are ...

MATTINGLY — These problems were caused before flight, and I give those guys a great big "atta boy," for ever settling it down.

YOUNG — Okay. But they got it settled down, and then in flight we did something else.

MATTINGLY — And then we turned right around and threw it all away.

YOUNG — Yes, and I think that's unforgivable.

MATTINGLY — And I think that's unforgivable.

YOUNG — Particularly after Ken's run through all the many hours of checking the stuff out. We found situations where the spacecraft is going to fly through gimbal lock and not be at the attitude at the time the pictures are suppose to be taken. In real time, nobody's running through all these changes that are coming up on the simulator somewhere and making sure that they have sufficient time to go.

MATTINGLY — Think you should fly this kind of a Flight Plan in one or two ways. Either you should not put the density of data collection in it that we have, which may be a valid thing to consider, or, when you miss something, you've got to work on the principle that I've missed something and that's it and I'm going to stay right on the preplanned time line and do what I know I can do, instead of trying to think originally. Because that's where you can really get yourself in a bind. The other problem with the Flight Plan, and it relates to the experiments in the relationship to the Flight Plan, is the flight planners live in a force in Houston and the flight crew, again, lives out at the Cape. The communications link just isn't adequate, and I don't know what you do about it, except I think the flight planners ought to live where the flight crew does.

SLAYTON — They do. They are down here full time, and there's no other function to perform.

MATTINGLY — But the guys that are writing the Flight Plan are seven strong and they live here. There's only one guy who runs back and forth.

SLAYTON — Whether you got seven down there or one wouldn't make any difference. That's the only one you'd ever see anyway. Because that's the way the system's set up. We don't want a whole gang down there.

MATTINGLY — I guess I never knew what was ever going on. I kept getting a new set of changes, and I felt like I was always out of the loop. It ended up that I wrote a lot of crew notes and changes on my own that, at first glance, I think it looks pretty unprofessional. But I think it's the only way because I put in little notes like, "If you don't start this maneuver by here, you won't get there in time and you'd better get the map out for the next thing at this point, because you're not going to get another chance." Or, where we find a sequence where it just wasn't getting performed properly if you did it this way and we had to go through and change it based on trying it in the simulator. By the time you go fly these things and look at all the things and you run around and you ask questions, there never was a guy at the Cape that had a good enough handle on it that you could just drop the question in his lap. Generally I was the only guy that really understood the problem, and I had to go back and find the guys in Houston that I could talk to, because they were the only ones that could fix it.

SLAYTON — That ain't right; something is wrong there.

MATTINGLY — You're right, boss.

SLAYTON — Let me find out what the problem is. You wouldn't have had that problem either if you would have said something, because here again, that was something we could have fixed in 10 minutes flat.

MATTINGLY — I'm not sure it can be fixed.

SLAYTON — Sure it can be.

MATTINGLY — Tommy and his guys just bent over backwards to really do a super job on this thing, and I feel like they really did. I think they were working from behind and we all were. But, somewhere in there the communication link was driving us up a wall.

SLAYTON — Something wrong; shouldn't have been.

MATTINGLY — We just didn't stay in time together. I would end up reviewing three times, I read a Flight Plan cover to cover only to find out I had just read the wrong Flight Plan.

SLAYTON — But, who was changing them?

MATTINGLY — It's in a constant flux. And I don't think there's any way to stop that.

SLAYTON — There sure is. As late as we got into this new launchment, I don't know that there would.

MATTINGLY — But somewhere there's gotta be a freeze, even if it's wrong. It seems to me that a month before launch, that Flight Plan ought to be frozen.

SLAYTON — It was frozen in another Crew Procedures meeting after that.

MATTINGLY — There sure were a lot of changes.

SLAYTON — That's what I'm trying to find out. Who's generating them? Where are they coming from?

MATTINGLY — Well, I'll tell you, maybe I don't even understand the whole problem. A lot of the things were things we did after we saw what we had. We had to change it to make it work, and we didn't see those things. I don't think I could have reviewed the Flight Plan any sooner. I don't think it was a matter that we didn't get hot on the job. But as a result of flying them in the simulator and flying a page, we've learned that there were things that really wasn't the way to do it. And the guys back here were trying to do that, but they don't have unlimited access to the machine time either, and we just never really got ourselves in sync. And I was really concerned that I didn't have a good enough handle on what was here to make sure that I could get that thing executed. As it turned out maybe some of my fears were uncommon. We did make it, but I thought it was a real struggle and I thought I was putting more into it than you really ought to have to. The Experiments Checklist falls into the same category, because it's subject to the same problems. Like poor old Bob Nute and the guys that put that thing together were really stuck in the middle, because they were given ding-a-ling procedures without the authority to go in and change them. They were told, here's what you should accomplish and I'd get the thing. They'd write it down for me; I'd get it and I'd go try to fly it in the simulator, and it wouldn't work and I'd come back and say, "Why are we doing this?" "Why don't we do something else?" And then they'd run back and see if they could chase it down to this communication link, with me at the Cape, talking to these guys on the telephone in Houston. Then they'd have to go run around and, boy, we just got the Experiments Checklist - what - 2 weeks before flight. But when we got it the last time it was rough. I asked them not to print it until it had been proofread and flown. There was no sense in that, we were just going to have to print it over. So, it's not their fault that the thing came out late. Because I asked them not to do it until we could print something that was right. There ought to be some guy who works out all the details on our Experiments Checklist. Then, if we run something in the simulator and find that this didn't work very well, then I'd try to give it to one of the guys down there to take care of. But by the time I explained all the experiment, went and sat down and take anybody that didn't know what it was about, and explained it to them, hell, I could have saved a lot of time and finally I gave up and I just make the call myself because it took less time than trying to explain it to a third party, who would then go chase it down. And again the guy that does that ought to be living in close proximity to you. And it just got very cumbersome and I don't know if I really have an answer.

I want to give Maurice Walters and Ron Weitenhagen the biggest "atta boy" of the mission. When they got down and got into the loop, they kept that Flight Plan up to date the last couple of weeks there on a minute-by-minute basis. And I finally got some confidence that I knew what the real Flight Plan looked like. We were verifying one Flight Plan in the simulator although they knew it wasn't quite the one we would fly with. But they really jumped on that problem and that problem disappeared in the last couple of weeks. They really were a help. Rescue Book, that was a good job. Gus Wallace put that together in good shape. Trained me in the CMPs. I'm perfectly happy, I think that's a good job.

23.2 LM

DUKE — Contingency Checklist - We used it for the docked deactivation stage. The appropriate pages were clear even though some things could change. It was a good checklist.

YOUNG — The Lunar surface checklist though was excellent.

DUKE — Outstanding.

YOUNG — Cuff checklist.

DUKE — Bob and Roger did a great job on that one. It was only a couple of weeks before flight that we finally got the flight ones printed up. We did get a chance to review it leisurely and use it, and it was excellent.

YOUNG — The Cue Cards and Star Charts, I think were excellent. We did need the star charts.

DUKE — Early in the program we went back to the Apollo 14 cue cards because nobody could understand what Dave had on his - at least I personally couldn't.

YOUNG — They're good, but he had a lot of stuff on there where we'd have trouble getting at.

DUKE — They were too busy, I guess is what it was. But anyway, after that iteration I don't think anything changed on those cue cards.

YOUNG — Systems Activation Checklist, great.

DUKE — That was super.

YOUNG — Subsystem items - G&N dictionary was good. Systems data was good. Charlie was in there and found out right away what the double reg. failure on the RCS was.

DUKE — Malfunction procedures were never opened. I thought they were adequate to cover anything. We had a good feel for those.

YOUNG — Time Line Book, excellent. Rendezvous Charts, excellent.

23.3 CHARTS AND MAPS

MATTINGLY — I'm not sure which one of these they're referring to but there are four lunar orbit charts, the A, B, C, D. They were very good, couldn't ask for anything more. One of the things the guys had a hard time with is: they wanted to hold up printing the ground track on the charts, and holding up putting the charts together, until they had all of the visual targets and all of the photo targets to put on them. As a result we got them late. We finally decided to take what data was available and print it. Any further annotation, we'd pen and ink it. I think that was a good thing to do. I only wish we would have done it earlier in the game and gotten the charts out in time to look at them a little earlier. That's an area you ought to just think about - just print the ground track, just a basic groundtrack, with nothing else on the map. Get it out early and then do the other stuff on an annotation basis. It's hard for the flight planners to get a handle on things like visual and photo targets. You'd like a set of those things 6 months before launch but the way our program feeds back on past experience I doubt that you're ever going to get the right ones more than a month or two in advance. It depends on how much study you do and how much time you have to put into it. I think we need to build in that kind of flexibility.

YOUNG — Sun Compass. We knew how to use it but we didn't have to. You didn't need it. There were so many familiar landmarks around there, we never had any doubt about where we were. Even if the nav hadn't worked, I don't think we would have needed the navigation system at our site. Lunar Orbit Charts.

MATTINGLY — I think that's what I read just talked about.

YOUNG — LM Landing Site Monitor Chart. We didn't use that. I think you only use that in case you don't know where you`re landing. I don't think we had that problem. Ascent Monitor Chart. We didn't need or use.

DUKE — We didn't even break it out.

YOUNG — Lunar Surface Maps.

DUKE — I'd like to make a comment about the Lunar Surface Maps. When your photography is only good to 20 meters, I think you're wasting your time carrying the photos of the traverse.

YOUNG — I agree.

DUKE — Really all you need and all you can use, due to the Rover, is, the topo chart with the headings and distances on the back of it, and one chart that you wedge up there on that camera.

YOUNG — The rest of them you can't get at.

DUKE — The rest of them you can't get out while you're driving and once you stop you don't have time to look at them. We had some tremendous craters that showed up on the topo maps that now that I've just looked at them again in detail, did not show up on the photographs. The photographs were not necessary on the lunar surface map when you have the resolution that we had in our photos.

YOUNG — I don't think the Apollo 17 guys will have this problem, except the problem of accessibility of the map when you're on the Moon is going to be still with them. No sense taking a whole bunch of maps out there if you can only get at one. Look at one while you're enroute. If you're enroute to a place, all you need to know is what your bearing and heading and distance is to go.

MATTINGLY — CSM Lunar Landmark Maps were fine.

YOUNG — They're adequate. Simulated Obliques in CSM Lunar Landmark Maps.

MATTINGLY — I think we ought to save the Government some money and quit making those.

YOUNG — You didn't need those, did you?

MATTINGLY — No.

YOUNG — Simulated obliques are where they take the orbital pictures, or something like the orbital pictures, and rotate them. It must cost a heck of a lot of money to do that. The Contingency Chart.

MATTINGLY — That's just a big scale chart. I think you're probably obligated always to carry that. That's a good one.

YOUNG — EVA Traverse Maps.

DUKE — I just talked about those.

YOUNG — The LM Lunar Surface Maps are maps of the whole landing site area in case you get off nominal and have to redo your EVA on the order of a couple of miles.

DUKE — If you land long.

YOUNG — In which case you're going to be sitting in there drawing on them, I expect. You probably ought to have those. They're really contingency lunar surface maps is what they are.

MATTINGLY — Visual Study Data Package. That's one that Ken Peterson put together this time. He did a super job on it. I thought it was very useful, but I'm not sure that everybody wants to study it to the same level. I'd suggest that before he puts that much effort into it, he better check and make sure. I didn't know he was going to do it. I'm glad he did. But someone else might find that he can save himself a lot of trouble. He ought to check to make sure that the CMP would like to look at it.

23.4 GENERAL FLIGHT PLANNING (FDF)

YOUNG — Level of activity and recommended changes.

MATTINGLY — There is a lot of question in everybody's mind, including mine, whether we had scheduled too much during lunar orbit. A lot of that depends on the individual. After trying it during the solo phase I had no problem whatsoever staying with it. The fact that we could even handle all the real time updates in addition to the kind of things we were doing says that it was a reasonable work load. It is really a drag to live for days in the lunar orbit on a minute-by-minute basis without a break. I don't think I would recommend somebody do that unless they really have a personal desire to want to do that kind of thing. I think you have to take the activities in lunar orbit when three guys are onboard and shoot for about a third of what you think you ought to do. I find that it is just slow operating with three men in there. That kind of took me by surprise. It's just one of those things. When you go to reach for a film magazine I'd either have to explain to John where it was and I always was faced with the choice of trying to explain where to go and what to do.

YOUNG — Or telling me to get out of the way.

MATTINGLY — Or telling him to get out of the way. Which way is going to be faster, I never came to any good conclusions on what was the right thing to do.

YOUNG — Part of the problem might have been that we could have learned where the stowage was in better detail and worked with you more, except I don't think we had the time to do that.

MATTINGLY — No, I think we did all the right things. I don't think we misused our time at any point. I think that it's just one of those things that the whole system has to be aware of - that when there's three guys in the spacecraft, no one is very efficient.

YOUNG — It's kind of like a big Gemini especially when you get all the rocks in there and you get all the suits in there you're just crowded.

MATTINGLY — You got to pull your horns in a little bit and be a little more relaxed about it.

YOUNG — O.K. Level of Details Provided in Onboard Documentation.

MATTINGLY — I thought it was just right.

23.5 PREFLIGHT SUPPORT

YOUNG — Preflight Support, I guess we talked about that.

SLAYTON — Poor, you said.

YOUNG — Up to a point until about the last month. In the LM we didn't have any problems.

DUKE — We had good support.

YOUNG — We didn't make any changes. I mean, we weren't making any changes; all we were doing was going down and landing.

DUKE — My statement is still correct. We had good support. The training exercises were ready. We never had any delay for suits, the Bendix people were always there with the air.

YOUNG — We're talking about the flight plans.

DUKE — Anyway we didn't have any problem with the flight plan or the time line.
YOUNG — It's good to get a good handle on those things early. That's one thing that I would recommend if there is any way, and I don't know if there is or not, is to get all those sightings down and come up with an early time line on the lunar surface station stops. I don't know if you can do this or not, let the guys get into working station stops early in their training. It was the last month or so before we started getting really into serious working station stops. Until we had done it two or three times, I didn't have a lot of confidence how I would handle it. It worked out okay. It gave us something to do for the last month. You know how we kicked that around and how we eliminated all those stations. We cut back and cut back, and we still ended up cutting out things real time. We added a bunch of things. It didn't seem to bother us any, if you can get an early hack on what your stations are going to be, and what the activities are going to be. The tendency, of course, is to overlook that surface operation by about 150 percent. I think we cut them back 60 percent, which was a long, hard struggle because they didn't want to give up a thing. I don't know

that I blame them. You want the most you can get out of a mission. That may be one of the reasons we came out doing pretty good.

DUKE — You single handedly did that John. If it wasn't for you we would have had all that extra stuff.

YOUNG — What you need is a couple of beers and then sit down with those guys and get reasonable.

24.0 VISUAL SIGHTINGS

YOUNG — I think we discussed almost all of the visual sightings that were unique during the flight.

MATTINGLY — There's one or two sights we didn't mention.

YOUNG — Mention them.

MATTINGLY — One of them was ,just the "gee whiz" type. On launch morning right out the window was the Moon. That sucker was rightcentered in there.

YOUNG — What's the other one?

MATTINGLY — The other one was that little flash I saw. Maybe somebody saw it on one of those seismometers, on one of the early revs. It's on the DSE because I remember commenting out loud about it. I was sitting there waiting for a solar corona. I had all the cockpit lights out sitting there in the dark. I could see a nice horizon against the corona. There was this bright flash, and it was below the horizon. I looked out the window every dark pass after that looking for another one. I never saw any more. I don't know whether I saw a meteorite hit, or what, but it was a bright flash. It was brighter than any star or planet that was above the horizon. It was definitely below the horizon. Maybe somebody in one of their seismic things might have seen something about that time. I don't know what good it would do us if they did, but it was there.

YOUNG — You saw your first light flash.

25.0 PREMISSION PLANNING

MATTINGLY — There is one thing to say about Mission Rules. Why do we publish a Mission Rules document one week before launch?

YOUNG — That makes you nervous, huh?

MATTINGLY — I couldn't believe that I had a change to the Mission Rules on my desk dated a week before flight. All those mission technique books that come out of MPAD come out with dates within two weeks of launch, and it's absolutely impossible to read them.

YOUNG — I was looking through my desk and I saw a bunch of questions posed by the Flight Operations Division, dated the 14th of April, like - what we're going to do on battery management for the ascent and descent.

MATTINGLY — When the CPCB says, okay, it's time to quit making changes, that's fine. But it doesn't stop people from making changes. There were a lot of changes that I saw. A good example was on how to check the batteries on the sequential thing if you have to cut the Apex cover disconnect. We'd run into this on the simulator. They had been hashing it over and finally they came up with what they would do. So they wrote it down and proposed it. Gary got hold of it and said no, it's too late. We can't submit changes this late. But in fact, had the situation occurred, that's what we would have done in real time. I felt a little uncomfortable knowing that there were some things which the flight control world had researched, and agreed that that's what they would really do if the situation came up. We had stopped that flow of information. I don't know how you get around it, except to be aware that that kind of a problem can happen. You need to maintain a verbal communication on those sort of things.

SLAYTON — That's exactly what we're suppose to have here.

MATTINGLY — I think everyone has to be aware that it happens that way. I'd get some little tidbit of data that didn't come through the system. It was obtained only because you just happened to be talking to them.

SLAYTON — We had better try to close that loop up. Hopefully most of it isn't all that important.

MATTINGLY — I hope that's the kind of stuff it is.

YOUNG — Spacecraft Changes - I think that Dave Bollard kept us pretty well informed. I think we were well advised as to what those were. Procedures Changes - we didn't have any to amount to anything. We didn't change any in the LM that I know of.

26.0 MISSION CONTROL

YOUNG — I think they were pushing the PDI GO-NO-GO a little closer than I would like to have had it to the NO-G0. I don't know what you do about that.

DUKE — Go baby , go .

MATTINGLY — Don't check the secondary gimbals.

YOUNG — Communications - we must have had terrible communications with the ground without the steerable antenna. The LM and I never had any trouble hearing them. Communications - I thought were real good both from the command and service module to the ground, and back and from the LM to the command and service module, as well as to and from the LM to the ground. Ken said he could hear us all the way on the VHF. On the LM, he could hear us til T-2. T-2. Which would have been real useful to know in case we would have had to use that mode. Thank goodness we didn't.

MATTINGLY — Just as clear as you could be.

YOUNG — I think there should be some use of the CMC DSKY to obtain antenna pointing angles. When a pointing angle changes, I think there should be some limitations as to the number that they pass up to you. If it only changes a couple of degrees, I don't really think there's any sense in taking up crew time copying down what it is.

27.0 HUMAN FACTORS

YOUNG — Preflight - Preflight Health Stabilization Control Program. Healthy rascals. It worked, didn't it? Nobody got sick.

MATTINGLY — It probably is a good idea to hold down the number of people you have contact with the last few weeks but for reasons other than health stabilization.

YOUNG — The healthiest thing is that they are well rested, relaxed, and in a happy frame of mind. I'm just not sure that you get that by being shut up for 3 weeks. Sure you can restrain them all you want to, but I really think it's important for a guy to leave the office occasionally. I know everybody lives, breathes, and eats the space program, and nobody any more than I do. That month or so before launch where things were coming thick and fast, people are always running in and saying, did you hear we can't service the RCS, or we are unable to fix this and we are going to have to take it apart and put it back together again. The crew needs to be in a relaxed frame of mind. They're better mentally suited to approach the program if they get to see their wives, sweethearts, or whatever at night. I think that is very important. I don't know how to put a number on it. Charlie and I were very fortunate in that our wives were there. I think it is important to a guy. I don't know how it is going to be on Skylab, but I'm sure that's going to cause them some problems there. Maybe necessary, but they are going to be there.

MATTINGLY — All the stabilization program does is it takes care of the clinical thing, but it doesn't put the guys in an operational mode.

YOUNG — It doesn't keep you in a frame of mind that you want to be in.

MATTINGLY — You really need to have another outlet.

YOUNG — Medical care - I thought there were a couple of occasions where the medical people unnecessarily mentally torqued the pilots all out of proportion to the amount of physical good they were doing. My personal opinion of that is, unless the crewman obviously has a medical problem in which case he might or might not go see the flight surgeon, I don't believe he should be bothered with that kind of mental stress close to launch. I think that is a real problem. If they are concerned about the morale of the guys that are going to fly the mission, how well he feels and how well they are able to do that, we ought to see if we can handle it in a little better manner than it was handled on the Apollo 16 mission. That's all I'm going to say about it.

MATTINGLY — It seems to me that there is a basic conflict of interest in taking care of the crew - trying to make him best suited to go fly his mission and doing research. You put the flight surgeon in the middle. He's trying to be a personal physician to help you. He's also trying to get research data. I think that's an incompatible set of circumstances for anyone to be put in. I don't think any one person is capable of handling both jobs. At some point you have to decide which is your first choice. I think there ought to be a Flight Surgeon, who is unrelated to the research end of this business, just exactly that. He's a Flight Surgeon that we could go and talk to, and if you got something that bothers you, go see him and tell him about it. I personally feel very reluctant to do so. Before flight I didn't have any problems, but if I had, I felt like I really didn't dare go tell anybody that I had a headache, or that I had something that was bothering me. Unless I was sure it was something I couldn't handle, because that opens all kinds of extra problems that you'd just as soon not add to your preflight workload. I'd sure like to see an independent Flight Surgeon, who is just interested in taking care of me, and keeping me, and keeping me healthy. I also felt like many of the things we did preflight were your battles around eating special foods, were again for medical research. You have enough things to do the week before the mission. If you want to grab a hand full of cookies for lunch, I think you ought to be able to do that. Unless it's going to really materially affect the operation of this mission. I think that when you're flying an operational mission you ought to be allowed to do those kind of things. You ought to be able to go on a low residue diet preflight so that you can minimize the number of bowel movements you have on the first couple of days while you're getting acclimatized to your new job. All of these things, which may have some medical return, just have to be weighed against the total package of how good can I do my job. I looked at it as preflight harassment. I know that the people didn't mean it to be that way.

YOUNG — Time for Exercise, Rest and Sleep. I think we had an adequate amount. We made a point of trying to limit it to 8 hours a day the last couple of weeks. I know we didn't do it, but we were trying to. We made time for exercise, and I think that's really important. I sure recommend that a month prior to launch the crews train this way. I feel like some of the preflight and postflight medical type things are not necessary for an operational mission. The retinal photography, for example, that was the most painful. experience that I ever had in my whole life. I don't think anybody ought to do that to a guy's eyeballs, unless somebody can show that it's essential that you do that. They wouldn't run that procedure on the spacecraft controls. Yet they thought nothing of running it on a flight crew. The vestibular function thing, where you're standing on rails, is clearly a learning curve. I got better at it all the time. Postflight, I'll admit that my ankles were weak, and my ear was plugged. I couldn't stand up very good, but that didn't seem to make any difference to them. They still wanted me to run the vestibular test. It is a very tiring test. I don't think that the data that you get off of something like that is going to prove anything. I really don't. We already have experience that shows that guys recover from spaceflight. I don't think we ought to fill unnecessary squares. The physical examination part of the preflight medicals, in every instance, amounted to about a half an hour of the examination over a half day's time. I don't know what we're proving, to put the crew through that stuff close in. I thought the whole business was a very unpleasant experience. For example, we got five X-rays to tell how a guy's heart is doing preflight and five more postflight. I think that's going a little far on the X-ray business. I'm not qualified to say whether it is or not, but I just think that's too many X-rays for a crewman.

MATTINGLY — I'd like to say one thing about it though. If you have to do all the things that we did, I think Chuck Lapinta and the guys that put it together really bent over backwards to make the maximum use of the time we put in.

YOUNG — Well, I do too.

MATTINGLY — I never ended up wasting any of my time, as far as sitting around.

YOUNG — And when we got close into the flight like, we had to do it on weekends. I don't think we ought to have to do that.

DUKE — It happened that minus 15 fell on a weekend.

YOUNG — Well. Make it F minus 16 or F minus 12. Okay, I guess we've talked about eating habits and the amount of food consumption from F minus 5 to F minus zero. I don't think we ought to have to worry about that.

SLAYTON — A comment on those meals?

DUKE — I'd rather have Lou's cooking, but it was edible.

YOUNG — It was edible? I still don't think the crew ought to be constrained to eat that kind of stuff on an operational mission close into launch.

MATTINGLY — One time I didn't get anything to eat for launch. It was a big deal to go and arrange to eat the special and I wasn't where they thought I was going to be, instead of grabbing a candy bar and things out of the machine - you're not supposed to do that. I just don't think that's the way to operate.

YOUNG — It's not low residue food; that's for sure. If it is, there's something that we don't understand about that diet.

DUKE — To me, I think the most important thing here is that that kind of diet changes your eating habits in food intake. That's probably going to materially affect something downstream. And in Ken's case, it did.

72-H-655

27.2 FLIGHT

YOUNG — Flight, Appetite and Food Preference. Boy, there sure was a lot of food on there.

DUKE — I think we did a good job or tried to do a good job of eating - keeping well fed. You were hungry - at least I was hungry when the meals came around, but after eating a couple of the packages I felt full and, if I had to really complete a whole meal, I would feel stuffed because there was just a lot of food.

72-H-475

YOUNG — Yes, I don't think our appetite in flight changed any. I think we were all concerned about keeping hydrated; we kept drinking every chance we got. One of the reasons you get dehydrated, I think, is that on a spacecraft with that low humidity when you sweat a lot you end up waking up in the middle of the night with your mouth being dry and you want to get a drink of water. I think a conscious effort is needed to keep hydrated and I think we were able to do that. I think it's important for that 7-hour lunar surface operation that you be able to get some liquid inside of you because you do sweat a lot during that thing. I

72-H-467

think maybe one of the factors that we were better hydrated is the fact that, in our prep and post donning, we used the LCG water cooling instead of air cooling so we had long periods of time where we didn't sweat very much while doing a lot of work in our pressure suit. We would have sweated had we used air cooling alone during the prep and post. Also, that would have gotten us behind a power curve and caused us a little heat stowage there just the same as it does on Earth. I guess that I'm not convinced that we don't have too many citrus drinks on board. At least once every 3 days I have gas pains that Z couldn't believe. And I don't know what caused them. I'm sure it had something to do with the food and it was a source of discomfort to me. I never knew when they were going to come back and I never knew when they were going to go. I didn't say anything about them because I don't think it's any big deal, but it's got something to do with the food. And that sort of kills your taste for eating the stuff. And I can attest for the fact that my fellow crewmembers had gas pains or something like it, too.

MATTINGLY — I don't think I had as much trouble as you.

DUKE — Me either.

MATTINGLY — But it sure wasn't my mode of operation.

72-H-468

YOUNG — No, it sure wasn't normal.

DUKE — Well, we swallowed a lot of air in the drink bags and in the food bags.

YOUNG — Do you think it was due to the food and all those citrus drinks we drank. I ,just totaled up my citrus drinks - 27 citrus drinks in eleven days at 7 ounces a cup; that's a lot of citrus. That's about like 27 times as much as I drink in any other normal 11-day period.

DUKE — See, there again is a personal preference, John. I would rather drink the orange juice than drink the coffee. And if I had to drink 27 gallons or whatever you logged up of coffee, I'd be vomiting.

YOUNG — I'm with you - I'm with you.

72-H-469

DUKE — So they ought to get something else. Maybe we ought to have a Coca-Cola dispenser or something to get you a little variety.

MATTINGLY — Let me ask you a question - that we talked about in the flight. John has done this very same mission before without the same response. That tells me that you've got to look for something different between Apollo 10 and Apollo 16.

YOUNG — We sure didn't have it on Apollo 10.

MATTINGLY — And it's not gas in the water because we had that sort of stuff. So, we ought to be looking at something different.

72-H-471

YOUNG — We had at least three to four times the gas than on Apollo 10. Our water bags were half full of gas.

MATTINGLY — And I thought our gas separator was doing a super job.

YOUNG — It was.

MATTINGLY — I thought we were in good shape. So I think you've got to focus your attention on things that are different between what we did then and now. The most likely thing is the food somewhere. I thought the food was good. Rita did an astounding job.

YOUNG — Yes, Rita said it would be the dried fruits - well we had dried fruits on Apollo 10 and that didn't bother us.

MATTINGLY — I didn't eat very many dried fruits because I thought it might aggravate the situation. I left those out.

YOUNG — Well, I ate them sometimes and sometimes I didn't.

MATTINGLY — The size of the food portions, I thought the portions were fine. You didn't have to start on something if you didn't want it. The main thing is that some of those things, like the tuna salad spreads and all those things that they packed in larger cans; the problem that you got with those things is once you started on one of those you got to finish it some way and the most convenient way is to eat it. But I ate a lot of things I didn't really want just because it seemed to me like a better way of discarding it than it was to try to figure out how to keep the thing from sitting in that trash can and smelling things up. That's particularly true of the tin can kind of things. A food bag you can put a germicide tablet in and put it away with partial meals in it. But, those darn tin cans you're committed to finish it when you open it, I think. You don't have a good way - like you take the peaches you've got to eat all of the syrup and everything that goes with it because otherwise it`s going to be flopping around. I don't think we have an adequate way to handle the residue of the tin cans.

YOUNG — Right, not on an Apollo flight. I guess they've got a thing they throw it all down on the Skylab.

SLAYTON — Yes, they've got a tank.

MATTINGLY — Boy, it had never better clog up.

YOUNG — I'll tell you one thing. I don't know if the tank is going to be big enough for two 28's and a 56, or whatever.

MATTINGLY — I also found out that - I don't know if it's true - but I was told that those cans are pressurized at 14 psi.

YOUNG — No wonder they squirted out.

MATTINGLY — No wonder they squirted out when you opened them.

YOUNG — We've got pictures of opening the can right up to the tunnel, beautiful.

MATTINGLY — I thought maybe it was because we were running our cabin at 4.8 instead of 5 and maybe it was that. And she said, oh no, those were 14 psi. No wonder - I wouldn't have opened the first one if I had known that.

YOUNG — Well, when you open a can of peaches the juice leaves the can and crawls up your arm. Charlie said it was a learning curve until he tried it.

DUKE — We opened two of them - didn't have very much luck with either of them. You know, Ken, the individual food portions were good. The total meal though, I thought was too big.

MATTINGLY — Yes, I agree with you.

YOUNG — We only had two meals a day in the lunar module and we ate everything except for a couple of wet packs. We ate everything in there and were hungry too. I think it was probably because...

MATTINGLY — I can't get hungry without exercise and there is no way you can get enough exercise in the spacecraft to be very hungry.

DUKE — I thought the wet packs were the best.

MATTINGLY — What do you mean the wet packs - you mean?
DUKE — Turkey and gravy and things like that.

MATTINGLY — You really need some way to heat those things. Nothing is as unappetizing to me as a hot dog that was cooked 2 weeks ago and was covered with cold...

DUKE — I didn't say the hot dog. I didn't like those.

MATTINGLY — The hamburger was in the same boat. There was also something else I got out that was the same thing. It was all cold gravy that had been laying around in the skillet for 2 weeks and I thought it would have been really good if I could have heated it. But in its present form, I only ate it because it was the only solid thing around. But, I felt like it was really unappetizing.

YOUNG — Food Preparation and Consumption, program's deviation from program menu and eat periods. We had several of those. There were a couple of times when we actually missed a whole meal because we were so busy, a couple of times in lunar orbit and, of course, Charlie and I missed one on descent because we were 6 hours late. It's just unavoidable; you can't stop when you're in the middle of something. Operationally you're up against it and it didn't seem to bother us any. We just took the next eat period and went on with it. We got them all logged. Food Preparation and Consumption, Programs with Rehydration (mixing; gas). We never had enough time to allow the things to sit around. I mean, it says on the package 15, 20 minutes. We constituted them and ate them right on the spot. And we never had time and I don't think we were getting the kind of mixing that we should have had because I ate a lot of food that didn't have any hydration in it at all. But you can't stop to worry about that.

MATTINGLY — There is something wrong when you have a 30-minute eat period and the first bag you come to is "wait 20 minutes."

YOUNG — Yes, you can't do it.

MATTINGLY — You're sort of behind.

YOUNG — Food Temperature - I think the hot water really made a lot of difference in the taste.

MATTINGLY — That's good, it really is.

YOUNG — Effect of Water Flavor and Gas Content of Food. I don't think that has any effect on it. Use of the spoon BOWL package - that was okay until such times you got your soup too soupy.

MATTINGLY — There were two items I have - I don't know what it is but they have an extra amount of surface tension or something. One was that lobster bisque and the other was that tomato soup and they were both very thin and they would crawl right up the side of the package and onto anything they could find. There is a learning curve to that. I found out that your natural reaction - when you open the bag and it starts to squeeze out - your natural reaction is to close it and that's exactly the wrong thing to do because it gives it extra surface to climb on. And as soon as it starts to squeeze out if you pull it apart it'll just climb up to the rim and stop and then you can eat it out with a spoon and it all works fine. But, your first inclination is exactly wrong.

YOUNG — Okay, on those really soupy ones, I cut a hole in the side of the package and sucked it out the side instead of trying it with the spoon because that's a manual dexterity test. Sure you can balance stuff on your spoon and it doesn't get off but that kind of stuff is too time consuming to suit me. Using the spoons as long as you've got something you can put on there is okay. Opening of cans.

DUKE — I liked the spoon idea.

YOUNG — I liked it too, Charlie, except when the thing was too soupy. I saw you over there with that stuff crawling up your spoon, didn't you notice that?

DUKE — At times.

YOUNG — It worked okay. The point is it just slowed you down. Opening the cans - don't, especially the Skylab cans packed at 14 psi.

DUKE — Well, tuna salad is okay.

YOUNG — Tuna salad?

DUKE — Stuff like that. But if it's got any juice in it, don't open it.

MATTINGLY — Well, let me say one thing about the cans. You've got a real disposal problem. You got pieces of metal with sharp edges on it floating around the cockpit and no adequate way to dispose of it.

YOUNG — Yes, you can't mash them; you can't flatten them.

MATTINGLY — You just got to be careful of that stuff.

YOUNG — Consumption From Cans - well, things like pudding and things that didn't leak all over the place weren't any problem.

DUKE — I had vanilla pudding that I thought was good. Butterscotch was good.

YOUNG — Yes, that's good stuff. It's off the shelf from the ground.

MATTINGLY — So were the spreads and things like that.

YOUNG — But the liquids were pretty tough.

MATTINGLY — And this discussion does not include the Skylab food cans. That's a separate subject.

YOUNG — Food Bar Usage During EVA periods - Charlie didn't get around to that.

DUKE — I was going to but after that orbital shock with the juice bag and the food bar I decided to forego the food bar. All I needed was some beverage. And that's true. You really don't need the food bar. And it gets messy when it gets soggy.

YOUNG — Function of the Germicide Tablet Pouch.

DUKE — They function okay.

YOUNG — Extent of Use of Germicidal Tablets - we put them in everything except maybe I didn't put them in mine the last 24 hours.

MATTINGLY — The germicide tablet is hard to put in one of those tin cans. You have to go wrap the tin can in something now to keep the tablet in there.

YOUNG — Undesirable Odors. I don't think we noticed any from food. Quantity of Food Eaten on the Lunar Surface - almost all of it. It's in the log. Quantity of Food Discarded on the Lunar Surface Prior to Lift-Off.

DUKE — We didn't discard anything.

YOUNG — Skylab Fecal Container.

MATTINGLY — We already discussed that.

YOUNG — We talked about that. Water, chlorine taste and odor - I didn't notice any chlorine taste.

MATTINGLY — Not a bit, and I drank some water right after chlorinating it too.

YOUNG — You did?

MATTINGLY — Yes. I wanted to see if it was going to have a taste.

YOUNG — Iodine Taste and Odor - I don't think I noticed any. Physical Discomfort - I don't think we had much gas in the water and neither did you. I already remarked that, when we packed a gas got out of it. I estimated the top 20 percent of the bag would be gas which you could vent right out. Gas/Water Separator - we already commented on those. Intensity of Thirst During Mission.

MATTINGLY — I felt like having a drink about the same frequency as I do on the ground.

DUKE — One thing I did that I never do at home is I'd wake up and I'd want a drink. We had our water bags with us. I usually had one water bag throughout the night - 7 ounces.

YOUNG — Same here, and I think that's due to the air cooling system and the dry humidity. It's like sleeping in the desert. You know when you wake up in the middle of the night in the desert you're kind of dry.

MATTINGLY — Let's finish that Skylab food. We didn't really ever talk about that.

YOUNG — Okay. Why don't you talk about that.

MATTINGLY — The drink bags were the little accordion things that unfold and they looked like they were pretty neat little items. They folded up in small packages when you started. And I think we all planned to use those as our water container. We'd have a measured amount and something to use. The little valve on that thing is really nice because you can pull it open with your teeth, drink from it, and when you're through, you can push it closed. It had one drawback in that all three of them leaked. They all leaked around the valve. We had the same failure mode on all three of those bags. Other than that, I thought the bags worked pretty good. If you could fix that valve, I think this is an improvement over the rollup bags. I always had trouble with the rollup drink bags. I never could stop once I started taking a drink. I could never pinch it off enough to keep it from seeping out once you opened up the crew port. This bag looked like it would do that, if it didn't leak. What did we have in those cans? Chicken and rice or something, some such thing in that. I thought that was a giant step backwards.

DUKE — Yes, I don't understand why you'd want to bother putting chicken and rice in a can. Why not just put it in the feed bag like you do in the Apollo Program.

SLAYTON — They have everything in a can on Skylab.

YOUNG — Do they have a trash masher on there? They don't!

MATTINGLY — I don't understand carrying a tin can just for the sake of carrying a tin can.

YOUNG — It allows them to prepackage longer in advance. That's what it does for them.

MATTINGLY — How?

YOUNG — Why don't they take it out of the tin cans before they fly it.

SLAYTON — You can put the supplies on board 6 months ahead of launch.

MATTINGLY — The thing that surprised me is that the bags - and I understand it wasn't just ours - but inside of that tin can is a plastic bag just like the one we have right now. It's not clear to me what the tin can did for us except to provide another waste management problem.

YOUNG — Yes, I didn't know what to do with it.

MATTINGLY — It's hard to handle, it was hard to throw away. It's a hazard if you get ...

YOUNG — Maybe this really isn't apropos. Maybe the Skylab guys know how to handle the tin cans.

MATTINGLY — Okay. Maybe they do, but these were problems that I had with it. Fredo cut himself nice and clean just like a razor on one on the ground.

YOUNG — Yeah. They're really sharp.

MATTINGLY — Those are really sharp edges. And I think ...

YOUNG — There is no methyolate in the medical kit, either.

MATTINGLY — And when you get through, what you have is a plastic bag just like we have now in a much less useful shape, because there's no zipper top. You can't open it out. Now to use your spoon to get food out, first you take this plastic out of the tin can and you put it on the foodport, hydrate it, just like we do our present bag, and then you squeeze it and squish it. When you are ready to eat it, there is no good way to handle this thing. You have to take your scissors and cut the top off of this thing. If you have any fluid in this bag, surface tension is going to pull that stuff right up around the top and every time you stick your spoon in there, you're shaking stuff off it. It just seemed to me like it was a step backwards from the spoon bowls that we had before. Maybe I don't understand the problem, but that's the way I saw it.

DUKE — The soup bags, the plastic bags, that are normally in the cans - you just don't cut the plastic off the fill valve and fill it. They have a little stopper in there and you have to take a sharp edge and peel it, like you do an orange, to get the cap off. Then you work your fingernail under there to get the stopper out so you can put it on the fill port. It is a heck of a lot easier to have the old style where you cut the plastic off and stick it on than this new improved bag. Another thing on these drink bags, the accordion bags are great but it has a stopper in the fill plug and when you pull the stopper out - I ended up opening the fill port and all of that powder came floating right out the valve.

YOUNG — Yes. I think that's a real problem.

MATTINGLY — Yes.

YOUNG — How do you avoid doing that,

MATTINGLY — You pull the valve out to close it.

DUKE — No, you push it in to close it, but there's a stopper in it, so you can't get it on the fill port. When you pull that stopper out, what happened to me is that the valve opened and the stopper came out and I had enough pressure on it so all the orange juice powder or whatever it is just went pshew! - just like a talcum powder spray in the cockpit and we filled up with orange juice powder.

YOUNG — That orange juice was really out to get you this time. Charlie says, "here, let me show you what it's doing" and he comes over to me and goes pshew!

MATTINGLY — There was another problem with the Skylab food bag in the can. After you got through with all these things, now you go to put your germicide tablet in there and there's no way to seal the bag up when you get through, Because you had to cut into the bag to get into it.

DUKE — And it leaves a big hole in the top.

YOUNG — Yes. It left a big hole in there and now you still don't know where to put the germicide.

MATTINGLY — You put it in there but it comes out all over.

YOUNG — Maybe they're not...

DUKE — I really think those guys are going to have a lot of trouble.

YOUNG — I do too. Is there a vacuum on the other side of that air lock?

MATTINGLY — The whole thing looks like it was organized by General Jubilation T. Cornpone, the advance to the rear.

DUKE — That meal was a little disappointing to us because it really gave ...

MATTINGLY — I think we photographed a lot of it. I'm not sure we took enough footage of it to show what we did. We had hoped to photograph a normal meal and that one got cancelled with the ISS flight or something.

YOUNG — We didn't have any chlorine taste in the water. We didn't have any iodine taste, I didn't notice any.

SPEAKER — Was there iodine in the LM water?

YOUNG — It's good stuff if there is.

DUKE — It didn't taste like it.

MATTINGLY — You guys commented on how good the LM water tasted.

YOUNG — Good and cold. Physical discomfort - I didn't notice any gas in the water - to amount to anything other than what we've mentioned. Gas/water separator - worked like a champ. We talked about the intensity of thirst. Work, rest, sleep. Difficulty in going to sleep - Charlie, did you have any?

DUKE — The first night I felt like I catnapped all night long. After that I was okay. I think that was just getting acclimatized.

YOUNG — I slept like a log.

DUKE — We reported all this sleep stuff.

YOUNG — Yes, we reported it all.

DUKE — The flight doc's already got that.

MATTINGLY — I don't think I slept much the whole mission.

YOUNG — You just weren't that tired. He was only 5 beats down on his post flight heart energy test. You never do sleep much.

MATTINGLY — I get about 6 hours but generally it's a solid 6 hours. I think, without physical exertion, my body just didn't believe it was time to go to bed. I just sat there and lay wide awake and I tried everything.

DUKE — I tell you, a couple of the sleep periods I was tired. I was looking forward to them when we got to them, I did go right to sleep. It was refreshing to have that to look forward to.

YOUNG — Disturbances. Charlie has commented on the disturbance - couple of master alarms on the lunar surface and a break lock.

DUKE — You've also commented that, any time somebody else moves around with a light on, you're going to wake the other two guys up.

YOUNG — Exercise. I think we did our exercise periods.

MATTINGLY — We skipped one translunar and skipped one transearth. That's because we had operational things that we were doing. We were probably getting exercise anyway. Stowing the suits is more exercise than you get from the Exer-genie.

YOUNG — That's a pretty interesting exercise.

MATTINGLY — We've already talked about the exercise.

YOUNG — We weren't sore.

MATTINGLY — I got sore on one of those periods.

DUKE — Oh, did you?

MATTINGLY — Yes.

YOUNG — In Flight Oral Hygiene, Mouth Discomfort -

DUKE — None.

YOUNG — Brushing Frequency, adequate.

DUKE — I tried about twice a day but I never did average that.

MATTINGLY — The tooth paste is also packed in 14 psi.

YOUNG — You open the top and out it comes.

MATTINGLY — The first time I opened it we could have brushed the teeth on an elephant.

DUKE — Out it comes - and out it comes - and out it comes.

YOUNG — Toothbrush Adequacy; it's adequate.

DUKE — We didn't use the dental floss,

YOUNG — Sunglasses or Other Eye Protective Devices.

MATTINGLY — I used the sunglasses.

DUKE — Yes, that's mandatory.

YOUNG — Yes, I think they're mandatory. The first couple of revs in Moon orbit my eyes hurt and I got a headache from looking out the window; I know that is what it was.

MATTINGLY — My eyes were very very tired the first 2 days. I could really feel them being sore. That was the only time I felt like I was ready for the sleep period, not because of sleep but I just wanted to turn my eyes off for a while.

YOUNG — You're gonna have to look in the vicinity of the Sun, and right at the subsolar point on the Moon. You really need something to keep your eyeballs from burning out, because it sure is bright. Visibility of Instruments and Controls Inside the Spacecraft with Sunglasses On.

DUKE — You correct that by taking your sunglasses off.

YOUNG — Unusual or Unexpected Visual Phenomena or Problems Experienced.

DUKE — That's all covered under the ALFMED.

YOUNG — We didn't have any rapid accelerations or decelerations that would cause _____ _____ to _____ count for the splashdown

DUKE — The next one up we've already commented on the Distance Judgment versus Aerial Perspective During EVA. Looked to me just like a desert scene. You think that the mountains are closer than they really are.

YOUNG — Sure.

DUKE — Stone Mountain looked a lot closer than it was.

YOUNG — That isn't an unusual thing.

DUKE — No, but it happens up there just like it does on Earth, I might comment on the outer visor. When I first got out during the beginning of the EVA I wanted my visor down even in the shade. But after we'd been out in the sunlight during the whole EVA and came back into the shadow, on the closeout, I wanted my visor up because I didn't feel like I could see well enough.

MATTINGLY — After the second day I finally got to where I could leave my sunglasses off when I was looking outside and that is a much more desirable mode to operate in if you can. But somehow you've got to get your eyeballs used to all that intensity.

YOUNG — The Medical Kit, Helmet/Visor Reflections - Yes, there are helmet/visor reflections but I thought it was at least a thousand percent better than it was on our training visors. Didn't you, Charlie?

DUKE — Yes, even though I had a scratchy one by EVA-3.

YOUNG — Medical Kit - We pulled the strap off the medical kit in the CSM because all the biomedical sensors were on it. There were a couple times there when I didn't think we were going to get it out to change biomedical sensors. It's packed full of medication. I can't believe that anybody would ever use all that medication.

DUKE — Plenty of stuff there.

YOUNG — There sure is.

MATTINGLY — I guess it would be better to package all the biomedical sensors and stuff all in one place instead of throughout the medical kit, just one package. The thing I thought was missing that belongs in there is some soap. I can't believe that medical kits that are designed to keep you healthy - give you injections to take for everything from heart attacks to gas warfare - and there's not one little bar of soap that you can clean up with.

SLAYTON — I though we had soap in there.

MATTINGLY — No, I remembered it right after lift-off that that was one of the things the Apollo 15 guys said you ought to have and I had written it on my list of things to do and that's the one that got away from me. But I really don't think my PPK is the place to carry a bar of soap. If the medical world wants to keep me healthy, they ought to give me a bar of soap. You really need it.

SLAYTON — I'm not sure it ought to be in the medical kit, though.

MATTINGLY — Okay, I shouldn't comment on where it ought to be but the thing that torques me is that here are things to cure me once I get sick and there's nothing here to keep me from getting sick.

YOUNG — Yes, there's no way you can get clean either. You ought to be able to wash your hands, for example, when you get through going to the bathroom because you may be right in the middle of an eat period but - you put water on your hands and wipe them off and I never felt that the hygiene was the best in the whole world.

SLAYTON — How about the towels, that didn't show up here any place?

YOUNG — Well, we had towels. What you do is wet the towel with water and wipe your face off, but that really doesn't get...

DUKE — And they were too small, not bigger than a dollar - smaller than a dollar.

MATTINGLY — I thought they were a waste of time.

DUKE — The towels were good - the big ones, you know, with the red, white and blue stripes - were great.

YOUNG — Wet wipes, according to something I read, you're only supposed to use one of those a day because of the poison in them. Did you ever read that? (Laughter) I was hoping for the best, Charlie.

MATTINGLY — The whole business of hygiene - all I figure we did was disprove all the theories about the importance of personal hygiene because we had absolutely none.

YOUNG — Yes, pretty dirty I thought. Even though - at least once or twice a day we'd take a towel, wet it down and wipe ourselves up as best we could. Housekeeping.

DUKE — Learning curve.

MATTINGLY — Yes, that's very important.

YOUNG — Shaving.

DUKE — The Wilkerson one we started with - I had a half of a shave with it and there was no way to get the blade clean and it just went belly up on me.

MATTINGLY — I tried the windup and that worked great until you missed a day. If you miss a day, you've had it because that thing feels like its pulling the whiskers out instead of shaving them off.

YOUNG — The Wilkerson worked okay if you'd taken that cream and made a lather out of it.

DUKE — Well, you looked pretty bloody, John, the time you used it.

YOUNG — I really did. I used it.

MATTINGLY — You wouldn't have sold any blades, John.

YOUNG — I really didn't get too good, did I? Pretty bad. The day before launch, I used the Wilkerson. It was about 4 or 5 days growth maybe. Really bad.

MATTINGLY — I got a data point on the mechanical guy when you got back, it'd been about 3 days since we've shaved. I was shaving and I missed one day and I went back to get it the next day and once it got a head start, that mechanical guy just couldn't hack it from there. He just gave up. We even tried cleaning it.

DUKE — Yes, I did clean it out.

MATTINGLY — We cleaned that all out. You really need a plain old everyday razor. Somehow we ought to be able to find a way to let you have a razor that you can open up like any other safety razor and clean off. That's the big problem you get that thing all crudded up and that's it, There must be some way to do that without producing a free floating hazard.

YOUNG — Dust, Density and Effects on Visual and Respiratory Systems.

DUKE — It was dusty when we got back in orbit, we've already commented on all that stuff.

YOUNG — I don't think it was any problem.

DUKE — I got one piece in my eye - a little something when I was over in the LM that gave me problems, but that was okay; it cleared right up, watered it out.

YOUNG — Radiation Dosimeter. Everybody wore their PRDs, except me, who forgot and left it in his pressure suit and sealed it up. But I wore it up until time we got back to suit stowage.

MATTINGLY — Yes, I threw my personal dosimeter away when I changed skivvies. We just never thought about it.

YOUNG — Radiation Survey Meter. We didn't use it. Personal Hygiene. Adequacy of Wipes, Size and Numbers.

DUKE — The one in the food kit is almost worthless.

YOUNG — Yes, they assume you don't get any food on you. The Adequacy of Tissue Size and Numbers. We certainly had enough, and I sure thought we wouldn't, but I'm sure there's some left over.

MATTINGLY — There were only two boxes left, and we were into both of them.

DUKE — Yes.

MATTINGLY — We didn't have any extra.

DUKE — I don't think there was any extra, quite frankly.

MATTINGLY — All you had to do is have a few more trips to the head, we'd have been out.

YOUNG — Yes, I think we would have been.

DUKE — I think like John. Those tissue boxes should have a snap on the bottom of them, so you can snap 'em to the LEB and pull tissues out. We were constantly looking for tissue dispensers.

YOUNG — At very embarassing times.

MATTINGLY — Yes, it's a two-handed operation to get a tissue out, when you ought to just reach up and pull it out. Cause they got a piece of Velcro there which will hold it, if you just want to set the box there. But it's on a side that if you open it, the Velcro isn't available to set it on.

YOUNG — And invariably...

MATTINGLY — You can't yank it.

YOUNG — When you need one, is when your urine system is leaking all over you, or your food bag is broken, and it's running all over you, and you got to hold it close with one hand, and wipe it up with the other.

MATTINGLY — And it takes two hands to get a tissue. So if you didn't get one before you need it, you're in trouble. I really think that's one that they ought to put a snap on the bottom of the bag.

YOUNG — Potable Water Used for Personal Hygiene. We'd wet the towels and wipe up. But that sure is not the way to get clean. All that does is smear the dirt around.

**28.0
MISCELLANEOUS**

YOUNG — PAO requirements - no problem.

MATTINGLY — Actually, I thought that went pretty well. I really would recommend that that interview day here in Houston to be continued as 2 days. I thought that was a lot easier work load than trying to cram it all in one.

YOUNG — Yes, I thought the PAO thing was real good and well handled.

MATTINGLY — I guess one other thing on that recovery thing is that I thought that extra day on the ship, before we came home, was a good deal. If I would have been given a choice preflight, I would have chosen to come home. But I thought having a day to just relax and unwind with no real requirements, get your physical done, and spend your full time on that and just get it over with, sit down and make up the list we made up. I really thought that was a good plan.

SLAYTON — Would you have wanted a tape recorder and a debriefing guide and jumped right into that. We talked about that once and didn't do it.

MATTINGLY — I really think that making the list the way we did was probably the right thing to do. And I just had the feeling that I had been working for so long everyday, every minute of the day, just to sit and do absolutely nothing was the neatest thing in the world.

YOUNG — Yes.

MATTINGLY — I really like that.

DUKE — We had a tape recorder, Deke. A little portable one, but we decided to write on a paper. It took us a couple hours to get it done.

YOUNG — Well, we spent about 3 hours in the afternoon on the ship, and we spent about an hour coming back on the airplane. Maybe 1-1/2 or 2 hours coming back on the airplane. Got 'em all jotted down. We noted a lot of items that we just happened to think of. I'm sure glad we had this list.

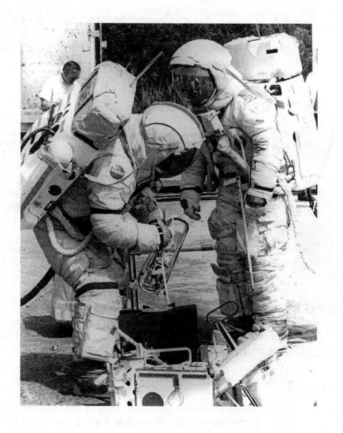

Look for these other important books about space flight at your local retailer!

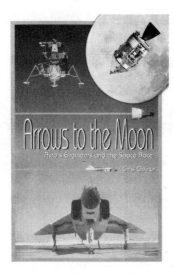
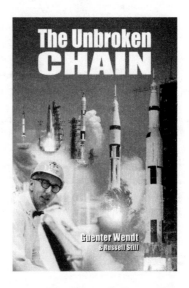
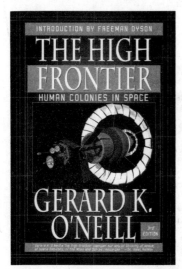

~EASY TO USE!~

CD-ROM

The attached CD-ROM requires no installation.

It is designed to leave no footprint on your computer's hard-drive and requires no special drivers to run.

All of the files are programmed as a web page and only require a web browser to view*. All files are HTML, JPG, MPG and Quicktime MOV.

The disc includes:

The entire Television broadcast from the Apollo 16 landing site. ~Over eleven hours of video!~

Every still image taken by the crew - **Over 2,500 pictures.**

An **Exclusive** video interview with Commander John Young.

Eighteen Quicktime* Panoramas taken at Descartes which allow the user to scroll around and view the landing site in 360°.

* Quicktime is a free program which can be acquired from http://www.apple.com